Second Edition

Endorsed for
**Pearson Edexcel
Qualifications**

Pearson Edexcel GCSE (9–1)

Mathematics

Foundation

Student Book

Series editors: Dr Naomi Norman and Katherine Pate

1

Published by Pearson Education Limited, 80 Strand, London, WC2R 0RL.

www.pearsonschoolsandfecolleges.co.uk

Text © Pearson Education Limited 2020
Project managed and edited by Just Content Ltd
Typeset by PDQ Digital Media Solutions Ltd
Original illustrations © Pearson Education Limited 2020
Cover photo/illustration by © David S. Rose/Shutterstock, © Julias/Shutterstock, © Attitude/Shutterstock, © Abstractor/Shutterstock, © Ozz Design/Shutterstock, © Lazartivan/Getty Images

The rights of Chris Baston, Ian Bettison, Ian Boote, Tony Cushen, Tara Doyle, Kath Hipkiss, Catherine Murphy, Su Nicholson, Naomi Norman, Diane Oliver, Katherine Pate, Jenny Roach, Carol Roberts, Peter Sherran and Robert Ward-Penny to be identified as authors of this work have been asserted by them in accordance with the Copyright, Designs and Patents Act 1988.

First published 2020

24
10

British Library Cataloguing in Publication Data

A catalogue record for this book is available from the British Library.

ISBN 978 1 292 34614 4

Printed and bound in Great Britain by Bell and Bain Ltd, Glasgow.

A note from the publisher

In order to ensure that this resource offers high-quality support for the associated Pearson qualification, it has been through a review process by the awarding body. This process confirms that this resource fully covers the teaching and learning content of the specification or part of a specification at which it is aimed. It also confirms that it demonstrates an appropriate balance between the development of subject skills, knowledge and understanding, in addition to preparation for assessment.

Endorsement does not cover any guidance on assessment activities or processes (e.g. practice questions or advice on how to answer assessment questions) included in the resource nor does it prescribe any particular approach to the teaching or delivery of a related course.

While the publishers have made every attempt to ensure that advice on the qualification and its assessment is accurate, the official specification and associated assessment guidance materials are the only authoritative source of information and should always be referred to for definitive guidance.

Pearson examiners have not contributed to any sections in this resource relevant to examination papers for which they have responsibility.

Examiners will not use endorsed resources as a source of material for any assessment set by Pearson. Endorsement of a resource does not mean that the resource is required to achieve this Pearson qualification, nor does it mean that it is the only suitable material available to support the qualification, and any resource lists produced by the awarding body shall include this and other appropriate resources.

Pearson has robust editorial processes, including answer and fact checks, to ensure the accuracy of the content in this publication, and every effort is made to ensure this publication is free of errors. We are, however, only human, and occasionally errors do occur. Pearson is not liable for any misunderstandings that arise as a result of errors in this publication, but it is our priority to ensure that the content is accurate. If you spot an error, please do contact us at resourcescorrections@pearson.com so we can make sure it is corrected.

Contents

iv

Pearson Edexcel GCSE (9–1)
Mathematics
Second Edition

Pearson Edexcel GCSE (9–1) Mathematics Second Edition is built around a unique pedagogy that has been created by leading educational researchers and teachers in the UK. This edition has been updated to reflect six sets of live GCSE (9–1) papers, as well as feedback from thousands of teachers and students and a 2-year study into the effectiveness of the course.

The new series features a full range of print and digital resources designed to work seamlessly together so that schools can create the course that works best for their students and teachers.

*Active*Learn service

The *Active*Learn service brings together the full range of planning, teaching, learning and assessment resources.

What's in *Active*Learn for GCSE (9–1) Mathematics?

- ✔ **Front-of-class Student Books** with accompanying PowerPoints, worksheets, videos, animations and homework activities
- ✔ **254 editable and printable homework worksheets**, linked to each Master lesson
- ✔ **Online, auto-marked homework activities** with integrated videos and worked examples
- ✔ **76 assessments and online markbooks**, including end-of-unit, end-of-term, end-of-year and baseline tests
- ✔ **Interactive Scheme of Work** brings everything together, connecting your personalised scheme of work, teaching resources and assessments
- ✔ **Individual student access to videos, homework and online textbooks**

Student Books

The Student Books use a mastery approach based around a well-paced and well-sequenced curriculum. They are designed to develop mathematical fluency, while building confidence in problem-solving and reasoning.

The unique unit structure enables every student to acquire a deep and solid understanding of the subject, leaving them well-prepared for their GCSE exams, and future education or employment.

Together with the accompanying online prior knowledge sections, the Student Books cover the entire **Pearson Edexcel GCSE (9–1) Mathematics course**.

The new four-book model means that the Second Edition Student Books now contain even more meaningful practice, while still being a manageable size for use in and outside the classroom.

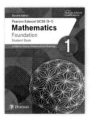
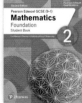

Foundation tier

Higher tier

Pearson Edexcel GCSE (9–1)
Mathematics Second Edition
Foundation Student Book

Building confidence

Pearson's unique unit structure has been shown to build confidence. The **front-of-class** versions of the Student Books include lots of extra features and resources for use on a whiteboard.

Master

Learn fundamental knowledge and skills over a series of lessons.

*Active*Learn **homework**

Links to online homework worksheets and exercises for every lesson.

Students can make sure they are ready for each unit by downloading the relevant **Prior knowledge check**. This can be accessed using the QR code in the Contents or via *Active*Learn.

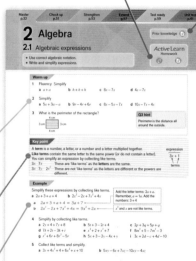

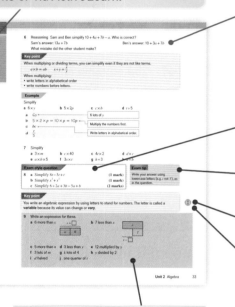

Warm up

Accessible questions designed to develop mathematical fluency.

Key points

Explains key concepts and definitions.

Worked example

Step-by-step worked examples focus on the key concepts.

Crossover content between Foundation and Higher tiers is indicated in side bars.

Click on any question to view it full-size, and then click 'Show' to reveal the answer.

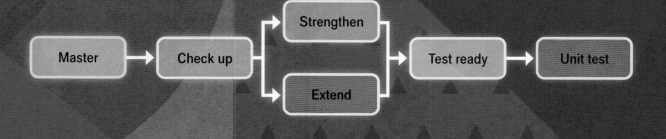

Master → Check up → Strengthen / Extend → Test ready → Unit test

Problem-solving and **Reasoning** questions are clearly labelled. **Future skills** questions help prepare for life after GCSE. **Reflect** questions encourage reflection on mathematical thinking and understanding.

Exam-style questions

are included throughout the books to help students prepare for GCSE exams.

Exam tips point out common errors and help with good exam technique.

Teaching and learning materials can be downloaded from the blue hotspots.

Helpful videos walk you step-by-step through answers to similar questions.

Check up

After the Master lessons, a Check up test helps students decide whether to move on to the Strengthen or Extend section.

Strengthen

Students can choose the topics they need more practice on. There are lots of hints and supporting questions to help.

Extend

Applies and develops maths from the unit in different situations.

Test ready

The **Summary of key points** is used to identify areas that need more practice and **Sample student answers** familiarise students with good exam technique.

Unit test

The exam-style Unit test helps check progress.

Mixed exercises

These sections bring topics together to help practise applying different techniques to a range of questions types, which is required in GCSE exams.

Interactive Scheme of work

The Interactive Scheme of Work makes reordering the course easy. You can view your plan for your year, term or lesson, and access all the related teaching, learning and assessment materials.

*Active*Learn Progress & Assess

The Progress & Assess service is part of the full *Active*Learn service, or can be bought as a separate subscription. It includes assessments that have been designed to ensure all students have the opportunity to show what they have learned through:

- a 2-tier assessment model
- separate calculator and non-calculator sections
- online markbooks for tracking and reporting
- mapping to indicative 9–1 grades.

Assessment Builder

Create your own classroom assessments from the bank of GCSE (9–1) Mathematics assessment questions by selecting questions on the skills and topics you have covered. Map the results of your custom assessments to indicative 9–1 grades using the custom online markbooks. Assessment Builder is available to purchase as an add-on to the *Active*Learn service or Progress & Assess subscriptions.

Purposeful Practice Books

A new kind of practice book based on cutting-edge approaches to help students make the most of practice.

With more than 4500 questions, our Pearson Edexcel GCSE (9–1) Mathematics Purposeful Practice Books are designed to be used alongside the Student Books and online resources. They:

- use minimal variation to build in small steps, consolidating knowledge and boosting confidence
- focus on strengthening problem-solving skills and strategies
- feature targeted exam practice with questions modified from real GCSE (9–1) papers, and exam guidance from examiner reports and grade indicators informed by Results**Plus**.

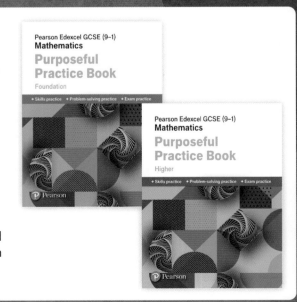

1 Number

1.1 Calculations

Prior knowledge

- Apply systematic listing strategies.
- Use priority of operations with positive and negative numbers.
- Simplify calculations by cancelling.
- Use inverse operations.

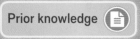

Warm up

1 **Fluency** Which of these calculations have

a a power   **b** the same answer as 3×2?

A 2×3 **B** 3^2 **C** 3×-2

D -2×-3 **E** $2 - 3$ **F** $2 - -3$

2 Write the missing number facts.

a $-8 \times 4 = -32$ so $-32 \div -8 = \square$ and $-32 \div 4 = \square$

b $-7 \times -6 = 42$ so $42 \div -7 = \square$ and $42 \div -6 = \square$

3 Simplify **a** $\frac{12}{2} = 12 \div 2 = \square$ **b** $\frac{24}{4} =$ **c** $\frac{45}{5} =$

4 Anthony has a red pair of shorts and a blue pair of shorts. He also has a green T-shirt, a white T-shirt and a yellow T-shirt.

a List all the combinations of shorts and T-shirts that Anthony can wear.

b How many different combinations are there?

> **Q4 hint**
>
> RS: red shorts, GT: green T-shirt
>
> RS
> /|\
> GT

5 A holiday club has 4 different art activities in the morning (A1, A2, A3 and A4) and 3 different sports activities in the afternoon (S1, S2 and S3). Each day, students can choose one activity in the morning and one activity in the afternoon.

a List all the combinations of art and sport activities there are each day.

b **Reflect** Write a sentence explaining how you made sure you listed all the combinations without missing any.

c How many different combinations are there?

| Holiday club activities ||
Morning	**Afternoon**
A1 Drawing	S1 Football
A2 Painting	S2 Golf
A3 Scrapbooking	S3 Table tennis
A4 Cardmaking	

6 **Problem-solving** These squares and circles represent a calculation.

$\square \bigcirc \square \bigcirc \square$

In each square, there is the number 2, 3 or 7. In each circle, there is the operation $\times$ or $-$. Each number and operation can be used only once per calculation. List all the possible different calculations.

7 Use the priority of operations to work out the answer to each calculation you listed in **Q6**.

8 **Problem-solving** Here are six calculations.

A $12 - 3 \times 2$ **B** $12 \times 3 - 2$ **C** $(12 - 3) \times 2$

D $3 + \frac{12}{2}$ **E** $12 \div 3 + 2$ **F** $2 \times 3^2 - 12$

Use the priority of operations to decide which of these calculations have the same answer.

9 Write these calculations so that their answers are in order of size. Start with the smallest.

A $2 \times 3 - 10 \div 5$ **B** $5 - 2^2 \times 3$ **C** $10 - 5 \times 3 + 2$ **D** $3 + 10 \div (3 - 5)$

10 Work out   **a** $(9 - 1)^2 + 2$ **b** $9 - (1 + 2)^2$ **c** $9 \div (1 + 2)^2$

Exam-style question

11 Luke is asked to find the value of $5^2 + 30 \div 5$.

Here is his working:

$$5^2 + 30 \div 5 = 25 + 30 \div 5$$
$$= 55 \div 5$$
$$= 11$$

Luke's answer is wrong. Explain what Luke has done wrong.   **(1 mark)**

> **Exam tip**
>
> Giving the correct answer to the calculation is not enough. You must write a sentence to explain what Luke has done wrong.

---

12 Work out the calculation on both sides, then write $=$ or $\neq$ between each pair of calculations to make the statements true.

a $8 - 4 \times 2 \;\square\; (8 - 4) \times 2$ **b** $(5 + 2) \times 4 \;\square\; 5 \times 4 + 2 \times 4$

c $(3 + 4)^2 \;\square\; 3^2 + 4^2$ **d** $(8 - 3) - 4 \;\square\; 8 - (3 - 4)$

e Rewrite any pair(s) of equations in **a–d** using $<$ or $>$ to make the statements true.

f **Reasoning.** Sarah says all the pairs of calculations could have the symbol $\leqslant$ between them. Is this correct? Explain.

> **Q12f hint**
>
> $\leqslant$ means 'less than or equal to'.

Example

Work out $(14 \times 12) \div (2 \times 3)$.

$$(14 \times 12) \div (2 \times 3) = \frac{14 \times 12}{2 \times 3}$$

         Write as a fraction.

$$= \frac{14}{2} \times \frac{12}{3}$$

         Split into fractions.

$$= 7 \times 4$$

         Simplify using division.

$$= 28$$

         Work out the calculations.

13 Work out

a $(6 \times 9) \div 2 = \dfrac{6 \times 9}{2} = \dfrac{6}{2} \times 9 = \square \times 9 = \square$

b $(11 \times 12) \div 3 = \dfrac{11 \times 12}{3} = 11 \times \dfrac{12}{3} = 11 \times \square = \square$

c $(7 \times 5) \div 5$ 　　　　　**d** $(16 \times 12) \div (8 \times 3)$

e $(24 \times 21) \div (6 \times 7)$ 　　**f** $(28 \times 18) \div (9 \times 4)$

Q13f hint

$\dfrac{28 \times 18}{9 \times 4}$ is the same as $\dfrac{18 \times 28}{9 \times 4}$

Key point

A function is a rule. The function $+2$ adds 2 to a number.

The **inverse** function is -2 because it *reverses* the effect of the function $+2$.

14 Write down the inverse for each function machine.

a

$3 \longrightarrow \boxed{+5} \longrightarrow 8$

$3 \longleftarrow \boxed{\square} \longleftarrow 8$

b

$5 \longrightarrow \boxed{\times 2} \longrightarrow 10$

$5 \longleftarrow \boxed{\square} \longleftarrow 10$

c

$12 \longrightarrow \boxed{\div 3} \longrightarrow 4$

$12 \longleftarrow \boxed{\square} \longleftarrow 4$

d

$9 \longrightarrow \boxed{-4} \longrightarrow 5$

$9 \longleftarrow \boxed{\square} \longleftarrow 5$

15 Write down a calculation to check each of these.

a $687 - 598 = 89$ 　　**b** $506 \div 46 = 11$ 　　**c** $264 \times 12 = 3168$

16 Use a function machine to decide which is the correct calculation to check $96 \div 6 + 4 = 20$.

A $20 - 4 \times 6$ 　　**B** $20 \times 6 - 4$ 　　**C** $(20 - 4) \times 6$ 　　**D** $20 \times (6 - 4)$

17 Complete these calculations. Check the answers using inverse operations.

a $15 \times 5 - 3 = \square$ 　　**b** $(18 + 7) \div 5 = \square$ 　　**c** $30 \times 6 \div 9 = \square$

Exam-style question

18 Caspar has three times as many chickens as Amy.
Amy has two more chickens than Theo.
Theo has nine chickens.
How many chickens does Caspar have? 　　**(2 marks)**

Exam tip

You can use a picture to help you to answer exam questions. You could start by drawing a function machine

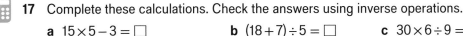

Key point

Finding the **square root** is the inverse of finding the square.
Finding the **cube root** is the inverse of finding the cube.

Use the $\sqrt{}$ and $\sqrt[3]{}$ keys on your calculator.

19 Complete these calculations. Check your answers using an inverse operation.

a $\sqrt{625} = \square$ 　　**b** $7^3 = \square$ 　　**c** $45^2 = \square$ 　　**d** $\sqrt[3]{1331} = \square$

1.2 Decimal numbers

- Round to a given number of decimal places.
- Multiply and divide decimal numbers.
- Use pictures to help you solve problems.

Active Learn
Homework

Warm up

1 **Fluency** Which of these numbers is rounded to

  a 1 decimal place   **b** 2 decimal places   **c** 3 decimal places?

  37.68   0.376   376.8

2 Copy and complete.

  a $1\,cm = \square\,mm$ **b** $1\,m = \square\,cm$ **c** $1\,km = \square\,m$ **d** $£1 = \square\,p$

3 Work out

  a $1.9 \div 10$ **b** 1.9×10 **c** 1.9×100 **d** 1.967×1000

Key point

To round a number to 1 decimal place (1 d.p.), look at the digit in the 2nd decimal place.
If it is 5 or more, round up. For example, 35.23 is 35.2 (1 d.p.) and 35.27 is 35.3 (1 d.p.).

4 Round these numbers to 1 decimal place.

  a 7.25   **b** 3.462   **c** 0.539   **d** 12.082   **e** 8.973

Key point

To round a number to 2 decimal places (2 d.p.), look at the digit in the 3rd decimal place.
To round a number to 3 decimal places (3 d.p.), look at the digit in the 4th decimal place.

5 Round these numbers to **a** 2 decimal places   **b** 3 decimal places

  i 4.0258   **ii** 16.1723   **iii** 0.1349   **iv** 11.862   **v** 703.4289

6 Round

  a 3.53 cm to the nearest mm

  b 6.846 m to the nearest cm

  c 25.3254 km to the nearest m

> **Q6a hint**
>
> 3.53 cm = 35.3 mm
> rounds to $\square$ mm.

7 **Problem-solving** Jamie cuts a 3.5 m length of wood into 6 equal pieces.
What is the length of each piece of wood correct to the nearest mm?

8 **a** **Problem-solving / Future skills** Four friends share a £35.90 taxi fare equally between them.
How much should they each pay?

  b **Reflect** How many decimal places should you round money answers to? Explain.

  c **Reflect** What calculation can you do to check your answer?

---

9 Work out

    a $0.4 \times 20 = 0.4 \times 2 \times 10$
        b $0.6 \times 30 = 0.6 \times 3 \times 10$
        c 0.3×500

               $= 0.8 \times \square$
                        $= 1.8 \times \square$

               $= \square$
                        $= \square$

10 Work out   **a** 3×2 **b** 3×0.2 **c** 0.3×2 **d** 0.3×0.2

    e Reflect Write a sentence explaining what you notice.

11 Work out

    a 0.8×9 **b** $0.8 \times 0.9 = 0.8 \times 9 \div 10$ **c** 0.8×0.09

                                  $= \square \div 10$

    d 0.8×-0.09 **e** -0.8×-0.09

12 **Problem-solving** Kelly multiplies two decimal numbers less than 1 and gets the answer 0.006. What two decimal numbers did she multiply? Is there more than one answer?

Example

Work out 3.4×5.6.

Estimate $3 \times 6 = 18$ ◂———— Estimate your answer by rounding.

$$\begin{array}{r} 3\ 4 \\ \times\quad 5\ 6 \\ \hline 2\ 0\ _2 4 \\ 1\ 7\ _2 0\ 0 \\ \hline 1\ 9\ 0\ 4 \end{array}$$

◂———— Use a standard method to work out 34×56.

$3.4 \times 5.6 = 19.04$ ◂———— Use your estimated answer to see where to put the decimal point.

13 Work out

    a 8.5×3 **b** 4.62×8 **c** 71×0.3 **d** 39.6×1.2 **e** 16.2×1.81

    f Reflect Count the number of digits after the decimal point in each calculation and the number of digits after the decimal point in each answer.
    Write a sentence explaining what you notice.
    Then check that your idea works for the worked example.

14 **Reasoning** Sarah multiplies 14.3 by 0.96 on her calculator. Her answer is 14.688.
    Without finding the exact value of 14.3×0.96, explain why her answer must be wrong.

Exam-style question

15 The table shows the cost of two different diaries.

Small	£2.40
Large	£3.50

Eleanor buys 10 small diaries and 5 large diaries.
She pays with a £50 note.
She thinks she will get more than £10 in change.
Is Eleanor correct? You must show how you get your answer. **(3 marks)**

Exam tip

Show your working by writing down every calculation you need to do. Make sure you clearly state your final answer.

16 Work out   **a** $\frac{8}{4}$ **b** $\frac{0.8}{4}$ **c** $\frac{8}{0.4}$ **d** $\frac{0.8}{0.4}$

      e Reflect Write a sentence explaining what you notice.

17 Work out   **a** $24.5 \div 5$ **b** $29.4 \div 3$ **c** $34.3 \div 7$

> **Q17a hint**
>
> Write as $5\overline{)24.5}$

      d Reflect How could you estimate the answer to use as a check?

Key point

To divide by a decimal, multiply both numbers by a power of 10 (10, 100, ...) until you have a whole number to divide by. Then work out the division.

Example

Work out $35.1 \div 1.5$.

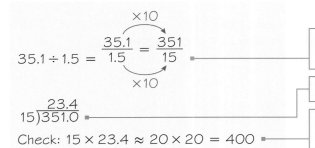

1.5 has 1 decimal place, so multiply both numbers by 10.

Divide.

Check using an inverse operation and estimation. $\approx$ means 'approximately equal to'.

$$35.1 \div 1.5 = \frac{35.1}{1.5} = \frac{351}{15}$$

$$15\overline{)351.0}^{\,23.4}$$

Check: $15 \times 23.4 \approx 20 \times 20 = 400$

---

18 Work out

    a $6.24 \div 0.4$ **b** $22.5 \div 0.15$

    c $280 \div 0.07$ **d** $234 \div 1.8$

> **Q18b hint**
>
> 0.15 has 2 decimal places, so multiply by 100.

19 Problem-solving / Future skills Lex pays £5.60 for some new pens. The pens cost 35p each. How many pens did Lex buy?

> **Q19 hint**
>
> Always work in one unit. 35p = £☐.☐☐

Example

It costs £5.95 to cut a front door key and £4.10 to cut a garage key.
Dan cuts some keys and pays £32. How many of each kind of key did Dan cut?

Draw a picture to represent different options.

○   ○ ○   ○ ○ ○   ○ ○ ○ ○

£5.95   £11.90   £17.85   £23.80

●   ● ●   ● ● ●   ● ● ● ●

£4.10   £8.20   £12.30   £16.40

○ = front door key
● = garage key

£23.80 + £8.20 = £32

Look for options that give the total you are looking for.

Dan cuts 4 front door keys and 2 garage keys.

Write a sentence answering the question.

Exam-style question

20 The weight of one brick is 3.4 kg. The weight of one bag of mortar is 20.5 kg.
Dave carries a weight of 58 kg in a wheelbarrow.
How many bricks and how many bags of mortar is he carrying?   **(3 marks)**

Exam tip

When using a picture to help you answer a problem, keep the picture very simple. For example, ☐ can represent a brick.

1.3 Place value

- Convert metric measures.
- Write decimal numbers of millions.
- Round to a given number of significant figures.
- Estimate answers to calculations.
- Use one calculation to find the answer to another.

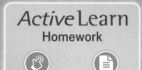

Active Learn
Homework

Warm up

1 **Fluency** Which of these calculations is the same as **a** $\div 5$ **b** $\div 50$?

$\times 10 \div 2 \quad \div 100 \times 2 \quad \div 100 \div 2 \quad \div 10 \times 2$

2 Work out
a 1.3×1000 **b** $2.6 \times 1\,000\,000$
c $34\,500\,000 \div 1\,000\,000$ **d** $1.5 \times 1000 \times 1000$

3 Work out
$$\frac{15 \times 40}{20} = 15 \times \frac{\square}{\square} = 15 \times \square = \square$$

4 Copy and complete $7.6 \times 2.1 = \square$, so $\square \div 7.6 = 2.1$ and $\square \div 2.1 = 7.6$

5 Write these numbers in figures.
a 7 million   **b** 4.6 million   **c** 10.1 million
d 2.45 million   **e** 3.125 million   **f** 5.5 million

> **Q5b hint**
> 4.6 million is $4.6 \times 1\,000\,000$.

6 Write these numbers in millions.
a 62 000 000   **b** 34 100 000   **c** 4 250 000
d 58 420 000   **e** 16 325 000   **f** 74 300 000

> **Q6b hint**
> 34 100 000 in millions is
> $34\,100\,000 \div 1\,000\,000$.

7 Work these out by calculating the number of millions. Then write your answers in figures.
a 0.8 million + 10.25 million   **b** 11.3 million − 0.75 million

Key point

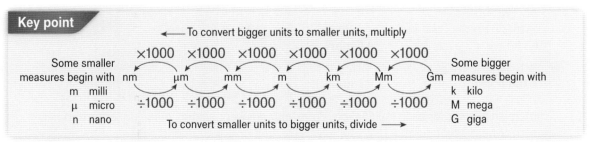

To convert bigger units to smaller units, multiply

Some smaller measures begin with
m milli
μ micro
n nano

$\times 1000 \quad \times 1000 \quad \times 1000 \quad \times 1000 \quad \times 1000 \quad \times 1000$

nm   μm   mm   m   km   Mm   Gm

$\div 1000 \quad \div 1000 \quad \div 1000 \quad \div 1000 \quad \div 1000 \quad \div 1000$

To convert smaller units to bigger units, divide →

Some bigger measures begin with
k kilo
M mega
G giga

8 Convert the units.
a 6437 m to km   **b** 431 000 mm to m   **c** 0.65 Gm to km
d 0.024 mm to μm   **e** 0.000 006 Mm to mm

You can round numbers to a number of significant figures (s.f.). The 1st significant figure is the one with the highest place value. It is the first non-zero digit in the number, counting from the left. Rounded numbers must have the same place value as the original number. For numbers greater than zero, this means you may need to put in zeros as 'place fillers'.

9 Round

   a 561.837 to 4 s.f.

> **Q9a hint**
>
> The 4th significant figure of 561.837 is 8. The next digit is 3, so leave the 8 as it is.

   b 0.003 468 to 3 s.f.

> **Q9b hint**
>
> The 3rd significant figure is 6.

   c 48 725 to 2 s.f.

> **Q9c hint**
>
> 48 725. Write in zeros to keep the place value.

10 Round these to the number of significant figures shown.

   a 67.14 (1 s.f.)   **b** 76 432 (3 s.f.)   **c** 342 510 (2 s.f.)   **d** 534 999 (2 s.f.)

11 Work out

   a $700 \times 400 = 7 \times 100 \times 4 \times 100$

                   $= \square \times 100 \times 100$

                   $= \square$

   b 8000×300

   c $2000 \div 500 =$ **d** $60 000 \div 30$

   e $51 000 \div 50 = 51\ 000 \div 100 \times 2$

                   $= \square \times 2$

                   $= \square$

> **Q11c hint**
>
>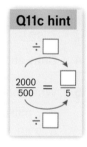
> $$\frac{2000}{500} = \frac{\square}{5}$$

To estimate the answer to a calculation, you can round every number to 1 s.f.

Exam-style question

12 Jamie's house has a value of £230 000 to 2 significant figures.

   a Write down the least possible value of the house. **(1 mark)**

   b Write down the greatest possible value of the house. **(1 mark)**

Exam tip

Sometimes it can be useful to draw a number line to help you to answer exam questions.

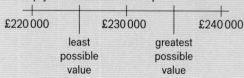

13 Estimate an answer for each calculation. Write your answer using ≈, meaning 'approximately equal to', where estimating, or = where exact. The first one is started for you.

   a $275 \times 421 \approx 300 \times \square =$

   b $\dfrac{876}{29}$

   c $\dfrac{41 \times 482}{1182}$

   d $\dfrac{284 \times 10.34}{4.52}$

   e $\dfrac{5.21 \times 3.84}{6.72}$

   f $\dfrac{9.83 \times 3.24}{7.65}$

> **Q13 hint**
>
> You may need to estimate the final division, for example $20 \div 7 \approx 3$.

Key point

For $\dfrac{\text{calculation 1}}{\text{calculation 2}}$ work out (calculation 1) ÷ (calculation 2) using priority of operations.

14 Estimate an answer for each calculation.

   a $\dfrac{19.37+6.1}{5.63-2.09}$ **b** $\dfrac{14.6-10.4}{2.7+0.9}$

Q15 hint

15 Check your answers to **Q13** and **Q14** using the fraction key on your calculator.

16 **Problem-solving / Future skills**

   a X is 8 million. Y is 2 million. How many times larger is X than Y?

   b At the end of 2013, the population of Beijing was 21.15 million. The population of London was 8.308 million. Estimate the number of times greater the population of Beijing was than the population of London.

Example

Use the information that $282 \times 56 = 15\,792$ to work out the value of

   a 28.2×5.6

   b $15.792 \div 5.6$ | The digits in the questions are the same.

a $282 \times 56 = 15\,792$ | Use the estimate to decide where to put the decimal point.

   Estimate: $28.2 \times 5.6 \approx 30 \times 6 = 180$

   $28.2 \times 5.6 = 157.92$ | Write the related division.

b $15\,792 \div 56 = 282$

   Estimate: $15.792 \div 5.6 \approx 15 \div 5 \approx 3$ | Estimate the answer.

   $15.792 \div 5.6 = 2.82$ | Use the estimate to decide where to put the decimal point.

17 **Reasoning** Use the information that $45 \times 127 = 5715$ to work out the value of

   a 4.5×1.27 **b** $57.15 \div 12.7$ **c** $5.715 \div 0.127$

18 **Reasoning** Use the information that $148 \times 39 = 5772$ to work out the value of

   a 148×40 **b** 147×39 **c** 148×38

Exam-style question

19 Calculate the value of $\dfrac{23.5-4.43}{18.40-3.22}$

   a Give your answer as a decimal.

   b Write down all the figures on your calculator display.

                                    (2 marks)

Exam tip

The calculation includes decimals up to and including 2 decimal places, so write your answer to part **a** to 2 decimal places.

1.4 Factors and multiples

- Recognise 2-digit prime numbers.
- Find factors and multiples of numbers.
- Find common factors and common multiples of two numbers.
- Find the HCF and LCM of two numbers by listing.

Active Learn
Homework

Warm up

1 Fluency What is the smallest prime number?

2 Find
   a the first ten multiples of   **i** 3   **ii** 4
   b the first even multiple of 9
   c a multiple of 8 between 50 and 60

3 Here is a list of numbers.
   4   5   6   18   24   36
   From the list write down
   a a factor of 12   **b** a prime number   **c** the product of 4 and 9

4 a Copy and calculate all the factor pairs of 36: $1 \times 36 = 36, 2 \times \square = 36, ...$
   b Write down all the factors of 36.

5 Write down all the factors of   **a** 28   **b** 32   **c** 40

6 Reasoning Adam is thinking of a number. He says his number is odd. It is a factor of 30 and a multiple of 5. There are two possible numbers Adam can be thinking of. Write down the two numbers.

Key point

A prime number has exactly two factors, itself and 1.

7 a List the prime numbers between 30 and 40.
   b Reasoning Alice says there are fewer prime numbers between 20 and 30 than between 30 and 40. Is Alice correct? You must show your working.

8 a Reasoning The product of two prime numbers is always even. Write down an example to show this statement is *not* correct.
   b Reflect How did you decide which example to use?

Exam-style question

9 Write down two 2-digit prime numbers with a difference of 10.   **(2 marks)**

Exam tip

When asked for a type of number (for example, a 2-digit prime number), it can help to write a systematic list. Therefore, begin by listing the first few 2-digit prime numbers in order.

The highest common factor (HCF) of two numbers is the largest number that is a factor of both numbers.

Find the HCF of 18 and 24.

$18: 1 \times 18 \quad 2 \times 9 \quad 3 \times 6$ ⟵ Work out the factors.

$24: 1 \times 24 \quad 2 \times 12 \quad 3 \times 8 \quad 4 \times 6$

18: 1, 2, 3, 6, 9, 18

24: 1, 2, 3, 4, 6, 8, 12, 24 ⟵ Ring the common factors.

The HCF is 6.

10 Find the highest common factor (HCF) of

    a 16 and 40   **b** 35 and 63   **c** 24 and 84   **d** 16, 24 and 40

The lowest common multiple (LCM) of two numbers is the smallest number that is a multiple of both numbers.

11 Find the lowest common multiple (LCM) of

    a 3 and 4   **b** 7 and 5   **c** 4 and 12

    d 8 and 6   **e** 12 and 30   **f** 5, 8 and 10

    g Reflect Write a sentence describing how you decided how many multiples to write for each pair of numbers.

> **Q11a hint**
>
> Look at the lists of multiples of 3 and 4 you wrote for **Q2a**. Ring the common multiples. Which is the lowest?

12 Find the HCF and LCM of

    a 12 and 18   **b** 40 and 60   **c** 30 and 75

13 Reasoning Write two numbers with an HCF of 16.

14 Reasoning Write a pair of numbers with an LCM of 40.

15 Problem-solving Matt prepares lucky dip bags for a pet shop. He wants each to be the same, with no items left over. Matt has 18 chews and 27 toys. What is the greatest number of lucky dips he can prepare?

> **Q15 hint**
>
> Find the HCF of 18 and 27.
> $18 = \square \times 2 \qquad 27 = \square \times 3$
> Use a diagram to check your answer.

16 Problem-solving One lighthouse flashes its lights every 20 seconds. A second lighthouse flashes its lights every 25 seconds. They both flash together at 6 pm. After how long will they next flash together?

> **Q16 hint**
>
> Find the LCM of 20 and 25.

17 Problem-solving Three sets of Christmas tree lights flash every 0.4, 0.6 and 0.8 seconds. They all flash together at midnight. How long does it take before they next flash together?

1.5 Squares, cubes and roots

- Find square roots and cube roots.
- Recognise powers of 2, 3, 4 and 5.
- Understand surd notation on a calculator.

Active Learn
Homework

Warm up

1 **Fluency** What are the first three

    a square numbers   **b** cube numbers?

2 Here is a list of numbers.

    18   32   56   81   104   125

From the numbers in the list, write down

    a a square number   **b** a cube number

3 Round these numbers to the number of decimal places or significant figures given.

    a 4.632 to 2 d.p.   **b** 0.8477 to 3 d.p.   **c** 52.36 to 3 s.f.   **d** 112.3 to 2 s.f.

 4 Use the power and root keys on your calculator to work out

    a 5.4^2 **b** $\sqrt{79}$ to 1 d.p.

    c 6.1^3 **d** $\sqrt[3]{112}$ to 3 s.f.

> **Q4 hint**
>
>

5 Work out

    a $\sqrt{0.25}$ **b** $\sqrt[3]{0.008}$

     c Use a calculator to check your answers.

> **Q5a hint**
>
> $\square \times \square = 25$
> $\square \times \square = 0.25$

6 Work out

    a 6^2 and $(-6)^2$ **b** $(-9)^2$ and 9^2 **c** $(-11)^2$ and 11^2

    d **Reflect** Write a sentence explaining what you notice about these answers.

7 The symbol $\pm$ shows that you are being asked for the positive and negative square root. Use your answers to **Q6** to work out

    a $\pm\sqrt{36}$ **b** $\pm\sqrt{81}$

8 Evaluate

    a 4^3 **b** $(-4)^3$ **c** 5^3 **d** $(-5)^3$

    e **Reflect** Write a sentence explaining what you notice about the cube of a negative number.

> **Q8 hint**
>
> Evaluate means 'work out the value'.

9 Use your answers to **Q8** to work out

    a $\sqrt[3]{-64}$ **b** $\sqrt[3]{-125}$

10 Work out
  a $(-4.2)^2$ **b** $(3.7)^3$ to 3 s.f.   **c** $\sqrt{481}$ to 1 d.p.
  d $\sqrt[3]{-564}$ to 3 s.f.   **e** $\sqrt[3]{122}$ to 3 d.p.

11 Write down
  a the positive square root of 2.56   **b** the cube root of 103.823

12 **Problem-solving** A square rug is 3.6 m by 3.6 m.
  What is the area of the rug?

> **Q12 hint**
> Write the correct measure unit in your answer.

13 **a** **Problem-solving** The area of another square rug is 49 m^2. What is the length of one of the sides?
  b **Reflect** Write a sentence explaining why the negative square root would not be correct.

14 Evaluate   **a** $2^2 \times 3$
  b $2^3 - 5^2$ **c** $3^3 - \sqrt{64}$ **d** $2^2 + 3^2 - 4^2$
  e $20 + 2^3 \div 4$ **f** $(2-8)^2 - 5^2$ **g** $(5 - \sqrt{36})^2 + 7$

> **Q14g hint**
> When using priority of operations, roots (for example $\sqrt{\ }$ and $\sqrt[3]{\ }$) are considered to be indices (powers).

15 **Reasoning** Adam is calculating $(1 + 7^2) \div \sqrt[3]{8} + 3$
  Here is his working.

> Brackets: $1 + 7^2 = 1 + 49$
> $\qquad\qquad\quad = 50$
> Indices: $\sqrt[3]{8} = 2$
> $(1 + 7^2) \div \sqrt[3]{8} + 3 = 50 \div 2 + 3$
> $\qquad\qquad\qquad\qquad = 10$

  a What mistake has Adam made?
  b What is the correct answer?

> **Q16c hint**
> Work out the answer to $8^2 - 3.4^2$ first. Then square root.

16 Work out
  a $2.7^2 + 1.5^3$ **b** $6.3^3 - 7.2^2$ **c** $\sqrt{8^2 - 3.4^2}$ to 3 s.f.

17 Use the fraction key on your calculator to work out $\sqrt[3]{\dfrac{9.4}{5.2}}$ to 3 s.f.

Example

Between which two numbers does $\sqrt{60}$ lie?

60 is between 49 and 64. ◂—————— Find the two square numbers 60 lies between.

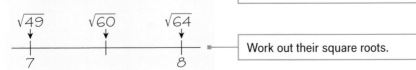

————— Work out their square roots.

So $\sqrt{60}$ lies between 7 and 8.

18 Reasoning Between which two numbers do these square roots lie?

a $\sqrt{55}$ b $\sqrt{41}$
c $\sqrt{90}$ d $\sqrt{14}$

Q18 hint

Go through the square numbers in order.

19 Reasoning Work out

a $\sqrt{4} \times \sqrt{16}$ b $\sqrt{4 \times 16}$

c **Reflect** Write a sentence explaining what you notice about your answers.

20 Reasoning $9 \times 25 = 225$. Use this fact to work out $\sqrt{225}$.

21 Work out

a $\sqrt{3} \times \sqrt{3}$ b $\sqrt{5} \times \sqrt{5}$

c **Reflect** Write a sentence explaining what you notice about your answers.

Q21 hint

$\sqrt{3} \times \sqrt{3} = \sqrt{3 \times 3}$

22 Use a calculator to work out these calculations. Give your answers to 2 decimal places where necessary.

a $\sqrt{25 + 9}$ b $\sqrt{25 \times 9}$

c **Reflect** Does your answer to part **a** give you the same answer as $\sqrt{25} + \sqrt{9}$?
Does your answer to part **b** give you the same answer as $\sqrt{25} \times \sqrt{9}$?

Exam-style question

23 Use a calculator to work out $\dfrac{\sqrt{214 + 316.1}}{(4.1 + 0.7)^2}$

Write down all the figures on your calculator display.

(2 marks)

Exam tip

Do not just write the answer. You should always show your working in an exam. For example, work out and write down the values of the numerator and denominator separately.

24 Give your answer to **Q23** correct to 2 significant figures.

Key point

Expressions with square roots like $3\sqrt{2}$ are in **surd** form. $3\sqrt{2}$ means $3 \times \sqrt{2}$. An answer in surd form is an exact value (it has not been rounded up or down).

25 Work out $\sqrt{12}$. Give your answer

a in surd form

b as a decimal

Q25 hint

Use the $S \Leftrightarrow D$ button on your calculator to switch your answer between decimal form and surd or fraction form.

26 Work these out. Give your answers in surd form.

a $\sqrt{6^2 - 5^2}$ b $\sqrt{\dfrac{5 \times 16}{25 - 15}}$ c $\sqrt{3^2 + 6^2}$

d $\sqrt{\dfrac{4 \times 12}{6}}$ e $\sqrt{10^2 - 5^2}$

1.6 Index notation

- Use index notation for powers of 10.
- Use index notation in calculations.
- Use the laws of indices.

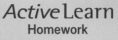

Active Learn
Homework

Warm up

1 Fluency What is

a 3^2 **b** 2^3 **c** $3^2 \times 2^3$

2 Work out

a $2 + -3$ **b** $2 - 5$ **c** $3 - -1$ **d** $\dfrac{2 \times 2 \times 2}{2}$ **e** 30×30

3 Convert   **a** 65 mm to cm   **b** 275 cm to m

Key point

In **index notation**, the number that is being multiplied by itself is called the **base**.
The number written above the base is called the **index** (plural: indices) or the **power**.
The index tells you the number of times that the base must be multiplied by itself.

Index or power

Base $\longrightarrow 10^{11} = 10 \times 10 \times 10 \times 10 \times 10 \times 10 \times 10 \times 10 \times 10 \times 10 \times 10$

4 Copy and complete the pattern of powers of 10.

$10^1 = 10$

$10^2 = 10 \times 10 = \square$

$10^3 = 10 \times 10 \times 10 = \square$

$10^4 = 10 \times 10 \times 10 \times 10 = 10\,000$

$10^5 = \ldots$

$10^6 = \ldots$

Exam-style question

5 List all the even numbers up to 40 that are powers of 2.

(1 mark)

Exam tip

You could list all the even numbers up to 40, and circle powers of 2. Or you could list powers of 2 in a pattern

$2^1 = \square$

$2^2 = \square \ldots$

and stop before going above 40.

6 Copy and complete.

a $3 \times 3 \times 3 \times 3 \times 3 \times 3 = 3^\square$

b $5 \times 5 \times 5 = 5^\square$ **c** $4 \times 4 \times 4 \times 4 \times 4 = 4^\square$

7 Write as a product.

a 2^6 **b** 5^4 **c** 7^5

Q7 hint

A product is a multiplication.

8 Write each product using index notation (powers).

    a $4 \times 4 \times 4 \times 4 \times 4$ **b** $2 \times 2 \times 2 \times 3 \times 3$ **c** $2 \times 2 \times 5 \times 5 \times 5$

9 Work out each pair of calculations. Then copy and complete using = or ≠.

    a $20^2 \square (2 \times 10)^2$ **b** $30^2 \square 3^2 \times 10^2$

    c $2^2 \times 5^3 \square 2^3 \times 5^2$ **d** $2^2 \times 2^3 \square 2 \times 2 \times 2 \times 2 \times 2 \times 2$

10 a Use your calculator to work out

      i 3^8 **ii** $3^2 \times 3^6$ **iii** $3^3 \times 3^5$ **iv** 3×3^7

> **Q10a iv hint**
>
> $3 = 3^1$

    b What do you notice about the index in part **a i** and the indices in part **a ii**?

    c Copy and complete this calculation using different indices to parts **a ii**, **iii** and **iv**.

      $3^8 = 3^\square \times 3^\square$

> ### Key point
>
> To multiply powers of the same number, add the **indices**.

11 Write these expressions as a single power.

    a $3^3 \times 3^4$ **b** $5^2 \times 5^3$ **c** 7×7^4 **d** $8^4 \times 8^5$

12 Reasoning Ross writes $2^4 \times 2^5$ as the single power 2^{20}. Write a sentence explaining what Ross has done wrong.

13 a Work out $\dfrac{3 \times 3 \times 3 \times 3 \times 3 \times 3 \times 3}{3 \times 3 \times 3 \times 3}$ by cancelling.

    b Write your answer to part **a** as a power of 3.

    c Copy and complete.

      i $\dfrac{3 \times 3 \times 3 \times 3 \times 3 \times 3}{3 \times 3 \times 3} = \dfrac{3^\square}{3^\square} = 3^\square$ **ii** $3^5 \div 3^3 = \dfrac{3^5}{3^3} = \dfrac{\square \times \square \times \square \times \square \times \square}{\square \times \square \times \square} = 3^\square$

    d Reflect Write a sentence explaining what you notice.

> ### Key point
>
> To divide powers of the same number, subtract the indices.

14 Write as a single power.

    a $4^6 \div 4^2$ **b** $5^4 \div 5$ **c** $7^5 \div 7^3$ **d** $2^8 \div 2^2$

15 Evaluate

    a $\dfrac{2^4 \times 2^2}{2^3}$ **b** $\dfrac{4^3 \times 4^5}{4^6}$ **c** $\dfrac{5^4 \times 5}{5^3}$ **d** $\dfrac{3^2 \times 3^3 \times 3^4}{3^6}$

16 Reflect Freddie is asked to write $(2^2)^3$ and $(6^3)^4$ as single powers. He writes

> $(2^2)^3 = 2^2 \times 2^2 \times 2^2$ $(6^3)^4 = 6^3 \times 6^3 \times 6^3 \times 6^3$
> $= 2^6$ $= 6^{12}$

Explain a quicker way to evaluate expressions like this.

17 Write as a single power.   **a** $(3^2)^5$ **b** $(5^4)^2$ **c** $(6^5)^3$ **d** $(7^3)^6$

18 Copy and complete.

a $100 = 10^{\square}$ **b** $100^4 = (10^{\square})^4 = 10^{\square}$ **c** $100^6 = (10^{\square})^{\square} = 10^{\square}$

d **Reflect** Write a sentence explaining what you notice about the powers of 10 in your answers and the powers of 100 in the question.

e **Problem-solving** How can you write 1000^2 as a single power of 10?

19 Copy and complete the pattern.

$10^3 = 1000$

$10^2 = 100$

$10^1 = 10$

$10^{\square} = 1$

$10^{-1} = \dfrac{1}{10} = 0.1$

$10^{\square} = \dfrac{1}{100} = \dfrac{1}{10^2} = 0.01$

$10^{\square} = \square = \square = \square$

20 Write as a single power.

a $10^3 \times 10^{-1}$ **b** $10^2 \times 10^{-3}$ **c** $\dfrac{10^2}{10^5}$ **d** $\dfrac{10^3}{10^{-1}}$

21 Copy and complete.

a 1 million $= \square = 10^{\square}$ **b** 1 billion = a thousand million $= \square = 10^{\square}$

c 1 trillion = a thousand billion $= \square = 10^{\square}$

Key point

Some powers of 10 have a name called a **prefix**. Each prefix is represented by a letter.

---

22 Copy and complete the table of prefixes.

Prefix	Letter	Power	Number
tera	T	10^{12}	1 000 000 000 000
giga	G	10^9	
mega	M		1 000 000
kilo	k		1000
deci	d	10^{-1}	
centi	c		0.01
milli	m	10^{-3}	
micro	μ		0.000 001
nano	n	10^{-9}	
pico	p		0.000 000 000 001

23 Copy and complete.

a 1 kilogram (kg) $= \square$g **b** 1 megabyte (MB) $= \square$B

c 1 microsecond (μs) $= \square$s **d** 1 picometre (pm) $= \square$m

Q23 hint

Refer to your completed table in **Q22**.

24 **Reasoning** What is 10^{12} bytes in gigabytes?

1.7 Prime factors

*Active*Learn
Homework

- Write a number as the product of its prime factors.
- Use prime factor decomposition and Venn diagrams to find the HCF and LCM.

Warm up

1   Fluency Which factors of 15 are prime factors?

> **Q1 hint**
>
> These are the factors of 15 that are prime numbers.

2 Write using index notation.

$$2 \times 2 \times 2 \times 3 \times 3 \times 3 \times 3 = 2^{\square} \times 3^{\square}$$

3 Evaluate $2^2 \times 3 \times 5$.

4   a Find the HCF of 16 and 36.　　　　**b** Find the LCM of 4 and 9.

Example

Write 180 as a product of its prime factors using index notation.

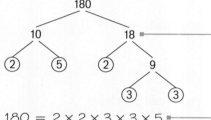

Make a factor tree using pairs of factors. Circle the prime factors.

$180 = 2 \times 2 \times 3 \times 3 \times 5$

Write the factors in order of size, smallest first.

$180 = 2^2 \times 3^2 \times 5$

Write their product using index notation.

5   a Complete these factor trees for 24.

b Write 24 as a product of its prime factors using index notation.

c   Reflect Does it matter which two factors you choose first?

Key point

All numbers can be written as a product of prime factors. This is called **prime factor decomposition**.

6 Write these numbers as products of their prime factors using index notation.

    a 18　　　　　　**b** 30　　　　　　**c** 56　　　　　　**d** 72

7 What is 300 as a product of its prime factors?

    A 3×100　　**B** $2^2 \times 3 \times 25$　　**C** $4 \times 3 \times 25$　　**D** $2^2 \times 3 \times 5^2$　　**E** $4 \times 3 \times 5^2$

8 $2^4 \times 3 \times 5$ is the prime factor decomposition of

   A 48   **B** 80   **C** 120   **D** 240   **E** 480

Example

a Find the highest common factor of 36 and 60.

b Find the lowest common multiple of 36 and 60.

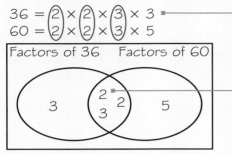

Write the products without powers.

Draw a Venn diagram. Put the common factors in the intersection.

HCF = product of numbers in the intersection.

a   HCF $= 2 \times 2 \times 3 = 12$

b   LCM $= 3 \times 2 \times 2 \times 3 \times 5 = 180$

LCM = product of all the numbers in the diagram.

9   Draw Venn diagrams to find the HCF and LCM of each of these pairs of numbers.

    a 60 and 96   **b** 24 and 108   **c** 120 and 150

Exam-style question

10 $A = 2^3 \times 5^2$ $B = 2^2 \times 5 \times 3$

    Write down

    a the highest common factor (HCF) of A and B **(1 mark)**

    b the lowest common multiple (LCM) of A and B.

                                     (1 mark)

Exam tip

Sometimes it can be useful to draw Venn diagrams to answer exam questions on the HCF and LCM.

11 **Reflect** Which is easier when working with large numbers: finding the LCM by prime factor decomposition or by listing multiples? Explain.

12 **Reasoning** $13 \times 17 = 221$ Use this information to find the LCM of 39 and 17.

Q12 hint

What type of numbers are 13 and 17? What is the LCM of 13 and 17?

$39 = 13 \times \square$

13 **Reasoning** 360 as a product of its prime factors is

    $360 = 2^3 \times 3^2 \times 5$

    Show that 90 is a factor of 360.

Q13 hint

Write 90 as a product of its prime factors. Does this divide into $2^3 \times 3^2 \times 5$?

14 **Reasoning / Problem-solving** Lea makes 225 milk chocolates, 165 dark chocolates and 180 white chocolates. She wants to put the chocolates into boxes.

    Each box will contain only one type of chocolate.

    She wants to use the largest possible box so that each box can be filled leaving no spaces.

    How many chocolates should each box hold?

1 Check up

Active Learn
Homework

Calculations

1 At a café, customers can order these different coffees:

Americano

latte

espresso

in one of the cups shown.

List all the possible combinations of ordering one cup of coffee.

2 Work out

a $8 \div 2 + 3 \times 9$ b $4 + (8 - 3)^2$ c $20 \div (2 - 6)$

d $(2 - 5)^2 - 4^2$ e $\dfrac{5 \times 9}{3}$ f $(15 \times 4) \div (16 \times 5)$

3 Write down a calculation you could do to check each of these.

a $840 \div 24 = 35$ b $13 \times 3 - 7 = 32$

4 Work out

a 4.2×3.8 b $3.5 \div 0.7$

5 Estimate an answer for each calculation.

a $\dfrac{491}{52.1}$

b $\dfrac{764 \times 96}{38}$

c $\dfrac{88.91 - 32.7}{8.7 + 2.62}$

6 Use the information that $436 \times 178 = 77\,608$ to work out the value of

a 43.6×1.78

b $776.08 \div 4.36$

7 Work out

a -0.7×0.05

b $8 \div 0.25$

 8 Use a calculator to work out $\sqrt{\dfrac{6700 - 2.38^2}{3.6^2}}$

a Write down all the figures on your calculator display.

b Give your answer to part **a** correct to 3 significant figures.

 9 Use a calculator to work out $2.6^3 - \sqrt[3]{5.4}$

Give your answer to 3 decimal places.

Powers and roots

10 **a** $5^{\square} = 125$

    b $6 \times 6 \times 6 \times 6 \times 6 = 6^{\square}$

    c $10^{\square} = 1000$

    d $\sqrt[3]{64} = \square$

11 Evaluate

    a $2^3 \times 3^2$

    b $\sqrt{81} - 2^2$

    c $\sqrt{4} \times \sqrt{4}$

12 Write as a single power of 7

    a $7^3 \times 7^4$

    b $7^5 \div 7^3$

    c $(7^4)^2$

    d $\dfrac{7^2 \times 7^3}{7}$

Factors, multiples and primes

13 Write two prime numbers that add together to give the answer 30.

14 **a** Find the HCF of 12 and 30.

    b Find the LCM of 8 and 10.

15 **a** Write 72 and 96 as products of their prime factors.

    b Find the highest common factor of 72 and 96.

    c Find the lowest common multiple of 72 and 96.

16 **Reflect** How sure are you of your answers? Were you mostly

Just guessing 😞   Feeling doubtful 😐   Confident 🙂

What next? Use your results to decide whether to strengthen or extend your learning.

Challenge

17 A birthday algorithm.

Step 1 Start with the number for the month of your birthday. For example, for December use 12.

Step 2 Multiply this by 5.

Step 3 Add 7.

Step 4 Multiply by 4.

Step 5 Add 13.

Step 6 Multiply by 5.

Step 7 Add the date of your birthday. For example, for 18 January add 18.

Step 8 Subtract 205.

If you have done this correctly you will be left with the month and day of your birthday!

1 Strengthen

Active Learn
Homework

Calculations

1 A game uses tokens that are four different shapes.

○ ▯ ▢ △

Each token can be one of these two colours: grey or white.
Draw all the different tokens for the game.

2 Round these numbers to 1 decimal place.

    a 6.32   **b** 15.07   **c** 0.438   **d** 11.972

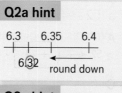

Q2a hint

6.3   6.35   6.4

6.32 ← round down

3 Round these numbers to 2 decimal places.

    a 11.257   **b** 9.072
    c 0.6352   **d** 28.983

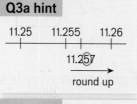

Q3a hint

11.25   11.255   11.26

11.257 → round up

4 Round these numbers to 3 decimal places.

    a 8.0462   **b** 14.1732
    c 0.0568   **d** 21.8139

Q4a hint

8.046   8.0465   8.047

8.0462 ← round down

5 Copy these numbers. Circle the first significant figure (the first non-zero digit starting from the left). Write its value.

    a 47.823   **b** 0.005 72
    c 432 650   **d** 0.6718

6 Copy these numbers. Circle the first significant figure then round the numbers to 1 significant figure (1 s.f.).

    a 51.3   **b** 487.2
    c 6234   **d** 8753

Q6a hint

The first significant figure is in the tens column, so you are rounding to the nearest 10.

7 Round these numbers to the number of significant figures shown.

    a 14.08 to 3 s.f.
    b 7.192 to 2 s.f.
    c 0.043 18 to 3 s.f.
    d 0.006 052 to 2 s.f.

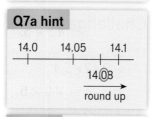

Q7a hint

14.0   14.05   14.1

14.08 → round up

8 Here is the priority of operations.

    ① brackets   ② indices (powers)   ③ ÷ and ×   ④ + and −

Use priority of operations to work out

    a $(4+6) \times 2$ **b** $(4+6)^2 \times 2$ **c** $(4+6) \times 2^2$

    d $4+6 \times 2^2$ **e** $(4+6) \times 2+3$ **f** $(4+6) \times (2+3)$

    g $6 \div 3+4 \times 2$ **h** $6 \div 3+\sqrt{4} \times 2$ **i** $6 \div 3-\sqrt{4} \times 2$ **j** $-6 \div 3-\sqrt{4} \times 2$

Q8 hint

i and **j** give negative answers.

9 Split into fractions. Then simplify.

a $\dfrac{7 \times 12}{4}$ b $\dfrac{15 \times 3}{5}$ c $\dfrac{25 \times 6}{5 \times 2}$

d $(16 \times 7) \div (14 \times 4) = \dfrac{16 \times 7}{14 \times 4}$

Q9a hint

$\dfrac{7 \times 12}{4} = \dfrac{7}{4} \times 12$ or $= 7 \times \dfrac{12}{4} = \square$

does not simplify   simplifies

10 Choose the inverse operation for

a subtract $-$ b divide $\div$

A add   **B** subtract $-$ **C** multiply $\times$ **D** divide $\div$

11 Choose the function machine that shows the correct inverse for each calculation.

a $52 \div 13 = 4$ b $52 \times 2 - 100 = 4$

Q11 hint

Check your answer by estimating

A
$52 \longleftarrow \boxed{+ 13} \longleftarrow 4$

B
$52 \longleftarrow \boxed{\times 13} \longleftarrow 4$

C
$52 \longleftarrow \boxed{\div 2} \longleftarrow \boxed{+ 100} \longleftarrow 4$

D
$52 \longleftarrow \boxed{+ 100} \longleftarrow \boxed{\div 2} \longleftarrow 4$

12 Work out

a $\times 10 \left(\begin{array}{ccc} 28 & \div & 1.4 \\ 280 & \div & 14 \end{array} \right) \times 10$

b $4.44 \div 2.4$ c $18.9 \div 1.2$

Q12a hint

It is easier to divide by a whole number.

d $7.2 \div 0.06$ e $8.1 \div 0.09$ f $71.5 \div 0.11$

13 Rewrite each calculation with the numbers rounded to 1 significant figure to work out an estimated answer.

a $546 \times 372 \approx 500 \times \square$

b $\dfrac{618}{34.6}$

c $\dfrac{291 \times 42}{59.3}$

d $\dfrac{21.8 - 5.3}{3.1 + 2.2}$

Q13c and d hint

When there is $\dfrac{\text{calculation 1}}{\text{calculation 2}}$ work out each calculation and then divide.

14 Reasoning $\boxed{4.2 \times 5.4 = 22.68}$

Copy and complete to use this fact to write down the value of

a 42×5.4

$\times 10 \left(\begin{array}{c} 4.2 \times 5.4 = 22.68 \\ 42 \times 5.4 = \square \end{array} \right) \times 10$

b 42×54

$\times 10 \left(\begin{array}{c} 4.2 \times 5.4 = 22.68 \\ \downarrow \times 10 \\ \square \times \square = \square \end{array} \right) \times 100$

c 0.42×5.4

$\div 10 \left(\begin{array}{c} 4.2 \times 5.4 = 22.68 \\ 0.42 \times 5.4 = \square \end{array} \right) \div 10$

d $226.8 \div 5.4$

$4.2 \times 5.4 = 22.68$

So $22.68 \div 5.4 = 4.2$

$\times 10 \left(\begin{array}{c} \\ 226.8 \div 5.4 = \square \end{array} \right) \times 10$

15 Use your calculator to work out the value of $\sqrt{(4.5^2 - 0.5^3)}$

   a Write down all the figures on your calculator display.

   b Write your answer correct to 2 decimal places (2 d.p.).

Powers and roots

1 Copy and complete.

   a $2^3 = 2 \times \square \times \square = \square$ **b** $\square^{\square} = 4 \times 4 \times 4 = \square$

   c $2 \times 2 \times 2 \times 2 = 2^{\square}$ **d** $2^2 \times 3^3 = 2 \times 2 \times 3 \times \dots$

2 **a** Copy and complete.

     i $\sqrt[3]{64} = 4$ because $4^3 = 4 \times 4 \times 4 = \square$

     ii $\sqrt[3]{1000} = \square$ because $10^3 = \square \times \square \times \square = 1000$

     iii $\sqrt[3]{\square} = 2$ because $2^3 = \square$

     iv The cube root of 27 is $\square$

   b Write the value of each root.

     i $\sqrt[3]{125}$ **ii** $\sqrt{81}$ **iii** $\sqrt[3]{1}$ **iv** $\sqrt{144}$

3 **a** Work out

     i 4^2 **ii** $(-4)^2$ **iii** $4 + (-4)^2$

> **Q3a ii hint**
>
> $(-4)^2 = -4 \times -4$

   b Copy and complete.

     The two square roots of 16 are $\square$ and $\square$.

   c Write the two square roots of

     i 9   **ii** 25   **iii** 144

4 Write each product as a single power.

   a $2^3 \times 2^4 = 2^{\square + \square} = 2^{\square}$

   b $5^4 \times 5 = 5^{4+1} = 5^{\square}$

   c $4^2 \times 4^3 = 4^{2\square 3} = \square$

   d $3^4 \times 3^5$

   e $6^2 \times 6^3 \times 6^4 = 6^{\square + \square + \square} = 6^{\square}$

   f $10^5 \times 10^4 \times 10$

> **Q4a hint**
>
> $2^3 \times 2^4 = \underbrace{2 \times 2 \times 2}_{3} \times \underbrace{2 \times 2 \times 2 \times 2}_{4}$
>
> How many 2s are multiplied together?

5 Write each division as a single power.

   a $7^6 \div 7^3 = 7^{\square - \square} = 7^{\square}$

   b $4^8 \div 4^2 = 4^{8 \square 2} = \square$

   c $3^5 \div 3^2$

   d $6^4 \div 6$

> **Q5a hint**
>
> $$7^6 \div 7^3 = \frac{7^6}{7^3} = \frac{\overbrace{7 \times 7 \times 7 \times 7 \times 7 \times 7}^{6}}{\underbrace{7 \times 7 \times 7}_{3}}$$
>
> How many 7s are left after cancelling 3 pairs of them?

6 Work out

   a $(6^2)^3 = 6^2 \times \square^{\square} \times \square^{\square} = 6^{\square}$

   b $(2^5)^2 = \square^{\square} \times \square^{\square} = \square$ **c** $(3^4)^3$

7 Work out

   a $\sqrt{2} \times \sqrt{2}$ **b** $\sqrt{11} \times \sqrt{11}$ **c** $\sqrt{7} \times \sqrt{7}$

8 Write the value of

   a $\sqrt{4} \times \sqrt{4}$ **b** $\sqrt{6} \times \sqrt{6}$ **c** $\sqrt{57} \times \sqrt{57}$

Factors, multiples and primes

1 **a** Write out the numbers 4 to 24.

   b Cross out the numbers that are divisible by 2 and the numbers that are divisible by 3.

   c What is the name for the numbers that are not crossed out?

2 **a** Write down two prime numbers that add together to give the answer 24.

   b Is there more than one answer to **2a**? Show working to support your answer.

3 Copy and complete.

   a Factors of 20: 1, 2, ...   **b** Factors of 30: 1, 2, ...

   c *Common* factors of 20 and 30: 1, ...

   d *Highest* common factor of 20 and 30: ...

   e Highest common factor of 28 and 42: ...

Q3c hint

Circle the numbers that appear in both lists. These are the common factors.

4 Copy and complete.

   a First 10 multiples of 6: 6, 12, ...

   b First 10 multiples of 7: 7, 14, ...

   c Lowest common multiple of 6 and 7: ...

   d Lowest common multiple of 5 and 8: ...

5 **a** Copy and complete the factor tree for 360.

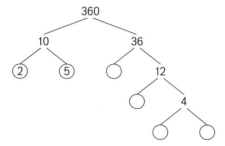

Q5a hint

The numbers in the circles are the prime factors.

   b Copy and complete 360 as the product of its prime factors.

   $$360 = 2 \times 5 \times \square \times \square \times \square \times \square$$
   $$= 2^{\square} \times 3^{\square} \times 5$$

   c Draw a factor tree then write each number as the product of its prime factors.

   i 144   **ii** 396   **iii** 450   **iv** 72   **v** 84

6 Write 72 and 84 as products of their prime factors.

   a Copy and complete the Venn diagram of their prime factors.

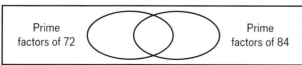

   b Multiply the common prime factors to find the HCF.

   c Multiply all the prime factors to find the LCM.

Q6b hint

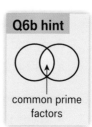

common prime factors

1 Extend

1 Copy and complete the pattern.

$2^3 = 2 \times 2 \times 2 = \square$

$2^2 = 2 \times \square = \square$

$2^1 = \square$

$2^0 = 1$

$2^{-1} = \dfrac{1}{2^1} = \square$

$2^{-2} = \dfrac{1}{2^2} = \square$

$2^{-3} = \dfrac{\square}{\square} = \square$

2 Work out

a $0.6 \times 0.4 \times 0.02$

b $0.6 \times 0.4 \div 0.02$

c $0.6 + 0.4 \div 0.02$

d $0.6 - 0.4 \div -0.02$

Exam-style question

3 Tim is selling tickets to climb a tower. In one day, he sells

    34 adult tickets at £2.50 each,

    11 senior tickets at £1.60 each,

    some children's tickets at 90p each.

Tim got a total of £153.

Work out how many children's tickets Tim sold.

(4 marks)

Exam tip

Think carefully about when you need to multiply, divide, add or subtract.

4 **Reasoning** Caroline writes down two numbers.
One number is a prime number less than 20.
The other is a square number.
She adds them together. Her answer is a cube number.

a What two numbers could Caroline have written down?

b What other two numbers could Caroline have written down?

5 $540 = 2^2 \times 3^3 \times 5$

Copy and complete.

a $540 = 2^2 \times \square$ b $540 = 3^3 \times \square$ c $540 = 3^{\square} \times 60$

Q5a hint

Work out $3^3 \times 5$.

6 **Reasoning**

a Given that $2304 = 36 \times 64$, work out $\sqrt{2304}$.

b Given that $216 = 8 \times 27$, work out $\sqrt[3]{216}$.

Q6a hint

$\sqrt{2304} = \sqrt{36 \times 64} = \sqrt{36} \times \sqrt{64}$

7 Sally used a spreadsheet to record the number of minutes she spent on social media for 10 days.
Her scores are in cells **A1** to **J1**.
Her mean score is in cell **K1**.

◢	A	B	C	D	E	F	G	H	I	J	K
1	72	41	18	19	19	21	49	53	82	11	28.5

a Use estimates to show that Sally's mean score is wrong.

b Work out Sally's correct mean. Give your answer in minutes and seconds.

8 Write brackets ( ) in this statement to make it true.
$$1.6 + 3.8 \times 2.45 = 13.23$$
Show your working.

9 Write down the whole number that is closest in value to $\sqrt{40}$.

Q9 hint

First find the two whole numbers $\sqrt{40}$ lies between.

Exam-style question

10 The cost of a return ticket for Alice to travel to work by train is £11.15.
Alice works out she will travel to work on 22 days this month.
She does the calculation $20 \times 11 = £220$ to estimate the cost.
Explain why Alice's calculation shows that the cost will be more than £220. **(1 mark)**

Exam tip

Giving the cost does not answer the question. You must write about Alice's calculation compared with the exact number of days and the exact cost of a ticket.

11 Reasoning Glen gets an answer on his calculator screen that begins 0.0674...
He rounds the answer to 3 significant figures, like this: 0.0675.
What digits could be after the 4 on Glen's calculator screen?

12 Problem-solving Trains from Manchester to Stockport leave every 15 minutes.
Trains from Manchester to London leave every 20 minutes.
A train to Stockport and a train to London leave together at 12 noon.
What is the next time a train will leave for Stockport and London at the same time?

13 Problem-solving Rhona wants to make up some party bags with identical contents.
She has 24 balloons and 30 party poppers.
What is the greatest number of party bags she can make up with no items left over?

14 Reasoning Rebecca divides 14.314 by 0.17 on her calculator.
Her answer is 8.42.
Without finding the exact value of $14.314 \div 0.17$, explain why her answer must be wrong.

15 Reasoning Given that $\dfrac{14.4}{(0.8)^2} = 22.5$, work out the value of $\dfrac{1.44}{(0.08)^2}$.

Q15 hint

$0.08 = 0.8 \div \square$
$(0.08)^2 = 0.08 \times \square$

16 The number 96 can be written in the form $2^n \times 3$. Find the value of n.

17 Problem-solving

a Alex says 1 trillion can be written as $(10^2)^2 \times 10^8$. Is Alex correct? Explain.

b Write at least two other ways of writing 1 trillion as a product of powers of 10.

1 Test ready

Summary of key points

To revise for the test:

Read each key point, find a question on it in the mastery lesson, and check you can work out the answer.

If you cannot, try some other questions from the mastery lesson or ask for help.

Key points

1 When there are m of one type of option and n of another type of option, the total number of options is $m \times n$. → **1.1**

2 The priority of operations is Brackets, Indices (powers and roots), Division and Multiplication, Addition and Subtraction. → **1.1**

3 $\neq$ means 'not equal to'. → **1.1**

4 $\approx$ means 'approximately equal to'. → **1.1**
A function is a rule that acts on a number (the input) to give an output number.
The inverse function reverses the effect of the original function.
The inverse of add $(+)$ is subtract $(-)$.
The inverse of multiply $(\times)$ is divide $(\div)$. → **1.1**

5 Finding the **square root** is the inverse of squaring.
Finding the **cube root** is the inverse of cubing. → **1.1**

6 To round a number to 1 decimal place (1 d.p.), look at the digit in the second decimal place. If it is 5 or more round up.
For example, to 1 d.p. 35.23 is 35.2 and 35.27 is 35.3. → **1.2**

7 To round a number to 2 decimal places, look at the digit in the third decimal place.
To round a number to 3 decimal places, look at the digit in the fourth decimal place. → **1.2**

8 To divide by a decimal, first multiply both numbers by a power of 10 (10, 100, ...) until you have a whole number to divide by. Then work out the division. → **1.2**

9 For any number, the first significant figure is the one with the highest place value. It is the first non-zero digit in the number, counting from the left. → **1.3**

10 Rounded numbers must have the same place value as the original number. For numbers greater than zero, this means you may need to put in zeros as 'place fillers'. → **1.3**

11 To estimate a calculation, round each number to 1 significant figure (1 s.f.) → **1.3**

12 A prime number has exactly two factors, itself and 1. → **1.4**

13 For $\dfrac{\text{calculation 1}}{\text{calculation 2}}$ work out (calculation 1) $\div$ (calculation 2) using priority of operations. → **1.3**

14 The **highest common factor (HCF)** is the highest factor that is common to two or more numbers. → **1.4**

15 The **lowest common multiple (LCM)** is the lowest multiple that is common to two or more numbers. → **1.4**

16 Expressions with square roots like $3\sqrt{2}$ are in **surd** form. $3\sqrt{2}$ means $3 \times \sqrt{2}$.
An answer in surd form is an exact value (it has not been rounded up or down). → **1.5**

17 In **index notation**, the number that is being multiplied by itself is called the **base**.
The number written above the base is called the **index** (plural: indices) or the **power**.
The index tells you the number of times that the base must be multiplied by itself. ⟶ **1.6**

18 To multiply powers of the same number, add the indices.
To divide powers of the same number, subtract the indices. ⟶ **1.6**

19 Some powers of 10 have a name called a **prefix**. Each prefix is represented by a letter. ⟶ **1.6**

20 All numbers can be written as a product of prime factors.
This is called **prime factor decomposition**. ⟶ **1.7**

21 When using a Venn diagram
  • to find the HCF, multiply the common prime factors
  • to find the LCM, multiply all the prime factors. ⟶ **1.7**

Sample student answers

Exam-style question

The diagram shows a rectangular noticeboard.

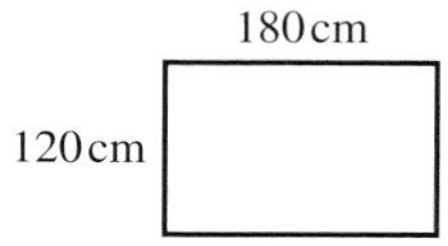

180cm

120cm

Chen wants to completely cover the noticeboard with postcards.
Each postcard is a rectangle

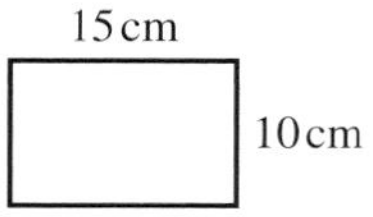

15cm

10cm

Each postcard costs 30p.
Chen has £50 to spend on the postcards.
Show that Chen has enough money to buy all the postcards he needs. **(4 marks)**

$180 \div 15 = 12$

$120 \div 10 = 12$

$12 \times 12 = 144$ postcards

$$
\begin{aligned}
\text{Cost} &= 144 \times 30 \\
&= 144 \times 3 \times 10 \\
&= £4320
\end{aligned}
$$

$$
\begin{array}{r}
144 \\
\times\ \ 3 \\
\hline
432
\end{array}
$$

Give two reasons why you think the student wouldn't gain full marks.

1 Unit test

1 Gavin buys 36 packets of raisins. Each packet of raisins costs £0.35.
Work out the total cost of all the packets that Gavin buys. **(2 marks)**

2 Three friends share the cost of a pizza equally between them.
The pizza costs £12.35. How much should they each pay? **(3 marks)**

3 Work out

  a $9 - (3 - 7)$ **(1 mark)**

  b $40 - 4^2 \div 8$ **(1 mark)**

  c $(8 - 2)^2 \div \sqrt{9}$ **(1 mark)**

4 Abdul has a '5p off per litre' voucher to use at his local petrol station.
The petrol normally costs 130.9p per litre.
Abdul fills his tank with 42 litres of petrol. How much does he pay? **(3 marks)**

5 Work out $\dfrac{20 \times 28}{7 \times 5}$ **(2 marks)**

6 Estimate the value of $\dfrac{70.1 \times 5.92}{0.19}$ **(3 marks)**

7 In the 2011 census the population of England was 53.0 million.
The population of York was 198 051.
Work out an estimate for the number of times greater the population of
England was than the population of York. **(3 marks)**

8 Work out

  a 25^2 **(1 mark)**

  b 0.5^3 **(1 mark)**

  c $\sqrt{0.16}$ **(1 mark)**

  d $\sqrt[3]{64}$ **(1 mark)**

9 Given that $32 \times 14 = 448$, write down the value of

  a 3.2×1.4 **(1 mark)**

  b $44.8 \div 1.4$ **(1 mark)**

  c $4.48 \div 320$ **(1 mark)**

10 Write each as a power of 9.

  a $9^3 \times 9^5$ **(1 mark)**

  b $9^6 \div 9^2$ **(1 mark)**

  c $\dfrac{9 \times 9^3}{9^2}$ **(1 mark)**

  d $(9^3)^5$ **(1 mark)**

11 In a shop postcards cost 65p each and greetings cards cost £2.10 each.
Sanjit buys some postcards and greetings cards in the shop.
He pays with a £10 note and receives £3.20 change.
Work out how many postcards and greetings cards Sanjit bought. **(3 marks)**

12 In a hospital, an alarm sounds every 4 hours. A second alarm sounds every 6 hours.
They sound together at 6 am.
How many times in the next 24 hours will they sound together? **(3 marks)**

13 a Express 75 and 90 as products of their prime factors. **(2 marks)**

  b For the numbers 75 and 90

   i find the highest common factor **(2 marks)**

   ii find the lowest common multiple **(2 marks)**

14 Work out the value of $\sqrt[3]{\dfrac{(3.3^2+4.2)}{5.1-2.02}}$

  a Write down all the figures on your calculator display. **(2 marks)**

  b Give your answer to 2 significant figures. **(1 mark)**

(TOTAL: 45 marks)

15 Challenge You can use Euclid's algorithm to find the highest common factor
(HCF) of two positive integers, M and N.

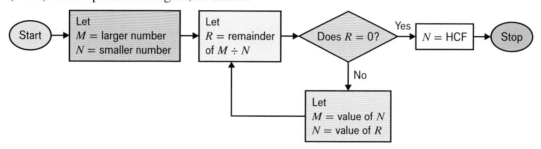

The table shows what happens if you input the numbers 36 and 8.

M	N	R	
36	8	4	$36 \div 8 = 4$ remainder 4 $(R \neq 0)$
8	4	0	$8 \div 4 = 2$ remainder 0
			HCF $= N = 4$

Use Euclid's algorithm to find the HCF of

a 45 and 18

b 1620 and 228

16 Reflect Look back at the work you did in this unit. Which topic did you find the most
difficult?
Choose 2 or 3 questions you found difficult to answer in that topic.
Work with a classmate (and/or your teacher) to better understand how to answer these
questions. (This is a good thing to do at the end of every unit.)

2 Algebra

Prior knowledge

2.1 Algebraic expressions

*Active*Learn
Homework

- Use correct algebraic notation.
- Write and simplify expressions.

Warm up

1 Fluency Simplify

  a $a+a$ **b** $h+h+h$ **c** $8x-7x$ **d** $4s-7s$

2 Simplify

  a $5a+3a-a$ **b** $9r-4r+6r$ **c** $8x-5x-7x$ **d** $10s-7s-4s$

3 What is the perimeter of the rectangle?

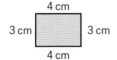

4 cm

3 cm   3 cm

4 cm

> **Q3 hint**
>
> Perimeter is the distance all around the outside.

Key point

A **term** is a number, a letter, or a number and a letter multiplied together.
Like terms contain the same letter to the same power (or do not contain a letter).
You can simplify an expression by collecting like terms.

$3x$ $7x$ These are 'like terms' as the **letters** are the same.

$3x$ $7y$ $2x^2$ These are not 'like terms' as the letters are different or the powers are different.

expression

$$\overbrace{3x + 1}$$

terms

Example

Simplify these expressions by collecting like terms.

  a $2a+3+a+4$ **b** $2x^2-2x+7x^2+4x$

> Add the letter terms: $2a+a$.
> Remember, $a=1a$. Add the numbers: $3+4$

  a $2a+3+a+4 = 3a+7$

  b $2x^2-2x+7x^2+4x = 9x^2+2x$

> x^2 and x are not like terms.

---

4 Simplify by collecting like terms.

  a $2y+4+7y+8$ **b** $5t+3-2t+4$ **c** $2p+3q+5p+q$

  d $7k+2t-3k+t$ **e** x^2+2+x^2+7 **f** $8m^2+5-7m^2-3$

  g $r^2+6r+8r^2-5r$ **h** $5x+3-2y-4x+y$ **i** $3c+2d-c+4d-10$

5 Collect like terms and simplify.

  a $3x+4x^2+4+6x^2+x+10$ **b** $5xy-6x+7xz-10xy-4xz$

6 **Reasoning** Sam and Ben simplify $10 + 4a + 7b - a$. Who is correct?

Sam's answer: $13a + 7b$ Ben's answer: $10 + 3a + 7b$

What mistake did the other student make?

> ### Key point
>
> When multiplying or dividing terms, you can simplify even if they are not like terms.
>
> $$a \times b = ab \qquad x \div y = \frac{x}{y}$$
>
> When multiplying:
> - write letters in alphabetical order
> - write numbers before letters.

> ### Example
>
> Simplify
>
> **a** $6 \times y$ **b** $5 \times 2p$ **c** $c \times b$ **d** $t \div 5$
>
> **a** $6y$ ————————————— 6 lots of y
>
> **b** $5 \times 2 \times p = 10 \times p = 10p$ ——— Multiply the numbers first.
>
> **c** bc ——————————
>
> **d** $\dfrac{t}{5}$ ———————— Write letters in alphabetical order.

7 Simplify

    a $3 \times m$ **b** $c \times 40$ **c** $4t \times 2$ **d** $d \times c$

    e $a \times b \times 5$ **f** $3s \times t$ **g** $h \div 3$ **h** $a \div b$

> ### Exam-style question
>
> **8**   **a**   Simplify $8t - 5t + t$.   **(1 mark)**
>
>   **b**   Simplify $x^2 + x^2$.   **(1 mark)**
>
>   **c**   Simplify $6 + 2a + 3b - 5a + b$.   **(2 marks)**

> ### Exam tip
>
> Write your answer using lowercase letters (e.g. t not T), as in the question.

> ### Key point
>
> You write an algebraic expression by using letters to stand for numbers. The letter is called a **variable** because its value can change or **vary**.

9 Write an expression for these.

    a   6 more than x

    b   7 less than x

    c   5 more than n **d**   3 less than y **e**   12 multiplied by y

    f   3 lots of m **g** k lots of 4   **h** y divided by 2

    i d halved   **j**   one quarter of t

10 **Reasoning** This stick is x cm long.

| x |

Match each description to an expression.

a A stick 3 cm longer than x

b A stick 3 times as long as x

c A stick 3 cm shorter than x

d A stick 3 times shorter than x

    A $x - 3$ **B** $x + 3$ **C** $3x$ **D** $\dfrac{x}{3}$

11 **Reasoning** Avinash is y years old.

a His brother is 2 years younger. Write an expression in y for Avinash's brother's age.

b His grandmother is 5 times as old as Avinash. Write an expression in y for Avinash's grandmother's age.

c His cousin is 3 years older. Write an expression in y for Avinash's cousin's age.

d Write and simplify an expression for the combined ages of all four.

Q11 hint

An 'expression in y' is an expression that contains the letter y.

12 Each bag holds n sweets.

Write an expression for the number of sweets in

a 2 bags $= 2 \times n =$

b 4 bags

c 10 bags

d x bags

Exam-style question

13 There are x cars in a queue.
There are 4 people in each car.
Write an expression, in terms of x, for the total number of people in the cars. **(1 mark)**

Exam tip

Use the letter given in the question in your expression.

---

14 **Problem-solving** Write an expression for the perimeter P of the rectangle.

b cm

h cm h cm

b cm

Exam-style question

15 Sweets are sold in bags and tubs.
There are 20 sweets in each bag.
There are 32 sweets in each tub.
Jay buys b bags of sweets and t tubs of sweets.
Write an expression, in terms of b and t, for the total number of sweets Jay buys. **(2 marks)**

2.2 Simplifying expressions

Active Learn
Homework

- Use the index laws.
- Multiply and divide expressions.

Warm up

1 **Fluency** Work out

a $-4 \times (-5)$ **b** $-2 \times 4 \times 3$ **c** $12 \div (-6)$ **d** $-8 \div 2$

2 Simplify

a $3 \times y$ **b** $l \times m$ **c** $h \div 4$ **d** $2n \times 4$

3 Write as a single power.

a $\dfrac{7^3}{7}$ **b** $\dfrac{4^5}{4^2}$ **c** $\left(5^2\right)^3$ **d** $\left(2^3\right)^2$

4 Copy and complete.

a $3 \times 3 = 3^{\square}$ **b** $x \times x = x^{\square}$ **c** $4 \times 4 \times 4 = 4^{\square}$ **d** $y \times y \times y = y^{\square}$

Key point

To multiply powers of the same letter, add the indices.

5 Copy and complete.

a $2^2 \times 2^4 = 2^{\square}$ **b** $x^2 \times x^4 = x^{\square}$ **c** $y^6 \times y^4 \times y = y^{\square}$ **d** $z^2 \times z^2 \times z^2 = z^{\square}$

6 **Reasoning** Write two terms that multiply together to give these answers.

a $\square \times \square = y^2$ **b** x^3 **c** y^3 **d** x^7

e **Reflect** Write another possible answer to part **d**.

Key point

To divide powers of the same letter, subtract the indices.

7 Copy and complete.

a $5^7 \div 5^4 = 5^{\square}$ **b** $y^7 \div y^4 = y^{\square}$ **c** $9^8 \div 9^3 = 9^{\square}$ **d** $x^8 \div x^3 = x^{\square}$

8 Simplify

a $\dfrac{a^6}{a^2} = a^6 \div a^2 = \square$ **b** $\dfrac{z^7}{z^3}$ **c** $\dfrac{x^4}{x}$ **d** $\dfrac{g^{10}}{g^5}$

9 Simplify

a $\dfrac{n^2}{n^5} = \dfrac{\not{n} \times \not{n}}{\not{n} \times \not{n} \times n \times n \times n} = \dfrac{1}{n^{\square}}$ **b** $\dfrac{x}{x^2}$ **c** $\dfrac{z}{z^3}$ **d** $\dfrac{t^2}{t^3}$

e $\dfrac{k^2}{k^5}$ **f** $\dfrac{m^5}{m^7}$ **g** $\dfrac{a^2}{a^5}$ **h** $\dfrac{s}{s^6}$

10 Simplify

a $\dfrac{e \times e \times f}{e}$ **b** $\dfrac{a \times a \times b}{a \times b \times b}$ **c** $\dfrac{c \times d \times d \times d}{c \times c \times d}$ **d** $\dfrac{g^2 h^4}{h}$

e $\dfrac{g^2 h^4}{h^3}$ **f** $\dfrac{g^2 h^4}{g h^3}$ **g** $\dfrac{x^2 y^3}{x^3}$ **h** $\dfrac{x^2 y^3}{x^3 y}$

Key point

To multiply algebraic terms, multiply the numbers first and then the letters.

Example

Simplify $2a \times 3b$

> Multiply the numbers first: 2×3.
> Then multiply the letters: $a \times b$

$2a \times 3b = 2 \times 3 \times a \times b$

$\qquad\qquad = 6ab$

> Put the number first, then the letters in alphabetical order.

---

11 Simplify

  a $4t \times w$ **b** $3k \times m$ **c** $2 \times s \times 3 \times t$ **d** $5 \times n \times q \times 2$

  e $3a \times 4b$ **f** $2r \times 7s$ **g** $6a \times 3b \times 2c$ **h** $9x \times 2y \times z$

  i $-6m \times 7n$ **j** $a \times 4a$ **k** $4t^2 \times t^3$ **l** $8s \times 3r \times 2s$

  m $4a \times -2a \times 2b$ **n** $4a \times -2a \times 3b$ **o** $4a \times -2a \times -2b$

12 Simplify

  a $n \times 3n^2$ **b** $2n \times 3n^2$ **c** $2n \times 3n^3$ **d** $2n^2 \times 3n^3$

  e $4a^2 \times 2a^3$ **f** $4b^3 \times 2b^6$ **g** $3cd \times 4c^2$ **h** $3cd \times 4c^2 d^3$

Key point

To divide algebraic terms, divide the numbers first and then the letters.

$\dfrac{10x}{2} = \dfrac{10}{2} \times x = 5x$

---

13 Simplify

  a $\dfrac{9b}{3} = \dfrac{\square}{\square} \times b = \square \times b =$ **b** $\dfrac{-40a}{10}$ **c** $\dfrac{26z}{13}$

  d $\dfrac{3m + 5m}{2}$ **e** $\dfrac{7n - 3n}{2}$ **f** $\dfrac{7p + 5p}{6}$

  g $\dfrac{2m}{4} = \dfrac{2}{4} \times m = \dfrac{\square}{\square} \times m = \dfrac{m}{\square}$ **h** $\dfrac{10p}{30}$ **i** $\dfrac{-8e}{4}$

  j $\dfrac{12t}{16} = \dfrac{12}{16} \times t = \dfrac{\square}{\square} \times t = \dfrac{\square t}{\square}$ **k** $\dfrac{6f}{9}$ **l** $\dfrac{-6d}{-12}$

Exam-style question

14 **a** Simplify $6x - x - 2x + 4x$.   **(1 mark)**

    b Simplify $5 \times r \times y \times 2$.   **(1 mark)**

15 **a** Simplify $4g \times 7h$. (1 mark)

    b Simplify $p \times p$. (1 mark)

    c Simplify $\dfrac{4x + 11x}{5}$. (1 mark)

Key point

$\dfrac{1}{2}x = \dfrac{x}{2}$ These fractions both mean 'half of x'.

16 **Reasoning** Lyn says $\dfrac{x}{x} = 0$. Jessie says $\dfrac{x}{x} = 1$. Who is correct?

Q16 hint

What is $\dfrac{4}{4}$? ... $\dfrac{6}{6}$?

17 Simplify

    a $\dfrac{3a^2}{a} = 3 \times \dfrac{a^2}{a} = 3 \times \square =$ **b** $\dfrac{8c^4}{2c^2}$ **c** $\dfrac{5x}{20x}$ **d** $\dfrac{-10z^2}{2z}$

    e $\dfrac{6p^2}{-36p}$ **f** $\dfrac{-15m^3}{-20m}$ **g** $\dfrac{8n^2}{2n^3}$ **h** $\dfrac{-8x^3}{2x^3}$

    i Explain why there is no 'x' in the answer to **Q17c** or **Q17h**.

18 **Reasoning** Insert the missing term in each question.

    Choose from $2x$ x^2 x $2xy$

    a $\dfrac{2x}{2} = \square$ **b** $\dfrac{\square}{x} = 2$ **c** $\dfrac{6x^2y}{30\square} = \dfrac{y}{5}$

    d $\dfrac{16x^2}{\square} = 8x$ **e** $\dfrac{3x^3y}{3xy} = \square$ **f** $\dfrac{6xy^2}{\square} = 3y$

19 Simplify

    a $(x^2)^3 = x^2 \times x^2 \times x^2 = x^{\square}$ **b** $(y^3)^3$ **c** $(z^5)^2$ **d** $(2x)^3 = 2x \times 2x \times 2x = \square x^{\square}$

    e $(3x)^2$ **f** $(5y)^2$ **g** $(5x^2)^2$ **h** $(5x^3)^2$

    i $(4a)^2$ **j** $(4a^2)^2$ **k** $(4a^2b)^2$ **l** $(4a^2b)^3$

20 Simplify

    a $\dfrac{8s^2t^3}{4st}$ **b** $\dfrac{9u^3v^3}{3u^2v}$ **c** $\dfrac{12a^5b}{3a^2b}$ **d** $\dfrac{12c^5d}{6c^4d^3}$

    e $\dfrac{12c^3d}{6c^4d^3}$ **f** $\dfrac{15e^3f^3}{3e^2f^2}$ **g** $\dfrac{15e^2f^2}{3e^3f^2}$ **h** $\dfrac{15e^2f^2}{30ef^3}$

21 **a** Simplify $n^2 \times n^4$. (1 mark)

    b Simplify $(2ab^2)^3$. (2 marks)

    c Simplify $\dfrac{18x^3y^5}{6x^3y^2}$. (2 marks)

2.3 Substitution

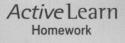

Active Learn
Homework

- Substitute numbers into expressions.
- Write more complex expressions.

Warm up

1 Fluency Work out

a $-9+11$ **b** $-6\times(-7)$ **c** $21\div(-3)$ **d** $4\div(-16)$

2 Match each statement to an expression.

$x-2$ $\dfrac{x}{2}$ $x+2$ $2x$

a x divided by 2   **b** double x **c** x subtract 2   **d** x add 2

3 Work out

a $3\times4+2$ **b** $2\times5+7\times3$ **c** $6\times2\div3$ **d** $2+3\times2^2$

Example

When $x=2$ and $y=5$ work out the value of

a $x+y$ **b** xy **c** $\dfrac{5x}{y}$ **d** $4x+3y$

a $2+5=7$ ———— | Replace x and y with the values given. |

b $2\times5=10$

c $5\times2\div5=10\div5=2$

d $4\times2+3\times5=8+15=23$ ———— | Use the priority of operations. |

4 Work out the value of these expressions when $a=5$ and $b=2$.

a $a+b$ **b** $a-2$ **c** ab **d** $\dfrac{8a}{b}$

e $4a-2b$ **f** $5ab$ **g** $-\dfrac{16}{b}$ **h** $\dfrac{3ab}{6}$

Exam-style question

5 $x=8$
$y=3$
Work out the value of $3x+4y$.   **(2 marks)**

Exam tip

Write down the calculation. Showing your working can gain marks.

---

6 Work out the value of each expression when $x=2$ and $y=3$.

a $x^2y=2^2\times3=$ **b** xy^2 **c** $(xy)^2$ **d** $4x^2y$

e $\dfrac{x^2y}{6}$ **f** $-5y^2$ **g** $\dfrac{-2xy^2}{12}$ **h** $\dfrac{2y^2}{3x}$

---

7 Find the value of each expression when $a = 3$, $b = -5$ and $c = 2$.

a $10a + 10b$ **b** $a + 2b$ **c** $\dfrac{10a}{b}$ **d** $\dfrac{ab}{b}$

e $3(a + c)$ **f** $2b^2$ **g** $b^2 - a$ **h** $b + a^2 c$

8 Write an expression for these statements.
Use n to represent the starting number.

a Tom thinks of a number and subtracts 20.

b Suzanne thinks of a number and multiplies it by 10.

c Ahmed thinks of a number and divides it by 5.

Q8a hint

Input → n → Subtract 20 → ☐ → Output

9 Write an expression for these statements.
Use n to represent the starting number.

a Christina thinks of a number, multiplies it by 4 then adds 5.

b Javed thinks of a number, doubles it then subtracts 6.

c Louisa thinks of a number, multiplies it by 3 then divides by 4.

d Matty thinks of a number, divides it by 2 then adds 4.

e Lisa thinks of a number, adds 4 then divides by 2.

f **Reasoning** Matty and Lisa both have starting number 6.
Show that their answers are not the same.

10 I think of a number, square it, and multiply by 6.
Write an expression for this statement.
Use n for the starting number.

11 A plate of biscuits has c chocolate biscuits and p plain biscuits.

a Write an expression in c and p for the total number of biscuits on the plate.

b Use your expression to work out the total number of biscuits on the plate when $c = 5$ and $p = 3$.

12 **Reasoning** Anna buys n cupcakes.

a She gives one to her mum. Write an expression in n for the number of cupcakes Anna has left.

b She gives half of the remaining cupcakes to her brother.
Write an expression in n for the number of cupcakes Anna gives to her brother.

c Use your answer to part **b** to work out how many cupcakes Anna gives to her brother when $n = 5$.

13 **a** **Reasoning** A pencil costs x pence. Write an expression for the cost of 3 pencils.

b A ruler costs y pence. Write an expression for the cost of 2 rulers.

c Josef buys 3 pencils and 2 rulers. Write an expression in x and y for the cost.

d Use your answer to part **c** to work out much Josef spends when $x = 35$ and $y = 90$.
Give your answer in pounds (£).

14 **Problem-solving** An airline charges passengers £55 for up to 20 kg of luggage, plus £4.50 for each kilogram over this limit.

a The expression used to work out how much each passenger will pay is £55 + £4.50k. Explain what k represents.

b Copy and complete the table.

kg over limit	$55 + 4.50k$
1	
3	
7	
10	

2.4 Formulae

- Recognise the difference between a formula and an expression.
- Write and use formulae.
- Use smaller numbers to help you see a pattern.

Active Learn
Homework

Warm up

1 **Fluency** Work out

   a 3^2 **b** 4^2 **c** $3^2 + 4^2$

2 Work out the value of these expressions when $a = 3$, $b = 6$, $c = 5$

   a $4b + 5$ **b** $\dfrac{b}{a}$ **c** $3c^2$ **d** $a^2 + c^2$

3 A red sweet costs r pence. A yellow sweet costs 4 pence more than a red sweet.
Write an expression for the cost of a yellow sweet.

Key point

A **formula** is a general rule that shows a relationship between variables.

For example, speed = distance ÷ time, which you can write as $s = \dfrac{d}{t}$

Speed, distance and time are variables. Although their values can vary, the rule stays true.
The plural of formula is formulae.

Exam-style question

4 Use this rule to work out the total hire charge, in pounds (£), for hiring a car for 2 weeks.

    Hire charge (£) = number of weeks × 120 + 50   **(2 marks)**

5 **a** $P = 3n + 5$
    Work out the value of P when $n = 4$.

   b $M = 17 - 5k$
    Work out the value of M when $k = 3$.

   c $A = 6x + y$
    Work out the value of A when $x = 2$ and $y = 5$.

   d $R = 5t + c$
    Work out the value of R when $t = 6$ and $c = 2$.

6 **Future skills** The formula to work out the force on an object is $F = ma$.
Work out the force on an object when

   a $m = 10$ and $a = 2$.

   b $m = 15$ and $a = 3$.

7 A formula to calculate a rough estimate for the area of a circle is $A = 3r^2$, where r is the radius.

Calculate an estimate for the area of a circle in cm^2 when

a $r = 3\,cm$ **b** $r = 5\,cm$ **c** $r = 10\,cm$

8 **Future skills** The formula for speed is $s = \dfrac{d}{t}$, where s = speed, d = distance and t = time

Work out the speed, s, in miles per hour when

a $d = 200$ miles and $t = 8$ hours

b $d = 70$ miles and $t = 2$ hours

c a car travels a distance of 240 miles in 4 hours

d a plane flies a distance of 1750 miles in 5 hours

9 **Future skills** The formula for the time taken for a journey is $t = \dfrac{d}{s}$, where t = time, d = distance and s = speed.

Work out the time, t, in hours a car takes to travel

a 180 km at 60 km/h

b 280 km at 70 km/h

c 60 km at 30 km/h

10 Choose words from the box to make each statement correct.

expression	factor	formula	multiple	term	letter

A _____ shows the relationship between variables, so it has an = sign.

$3x$ is a _____ in the _____ $3x + 5$ **(3 marks)**

11 Which of these are formulae and which are expressions?

a $2a + 3$ **b** $F = ma$ **c** $m^2 - 3pq$

d $\dfrac{t^3}{3}$ **e** $P = 2w + 2l$ **f** $A = bh$

Example

Sarah works h hours per week at an hourly rate of £m.

Write a formula to work out Sarah's total pay, P.

2 hours at £10 = 2 × 10 = £20 ⎤
3 hours at £10 = 3 × 10 = £30 ⎦   Use small numbers for h and m to see the pattern.
h hours at £10 = h × 10 = 10h
h hours at £m = h × m = hm
$P = hm$

12 **Reasoning** A red packet of hair bands costs n pence.

a Write an expression in n for a gold packet that costs 10 pence more.

b Write '$C=$' in front of your expression from part **a** to give a formula for the cost of a gold packet.

13 Reasoning There are 12 pencils in a box.

  a Write an expression for the number of pencils in x boxes.

  b Using your expression, write a formula for the number n of pencils in x boxes.

14 Reasoning A packet of sweets costs x pence.
Write formulae for the cost, C, of another packet which costs

  a 15 pence less

  b 3 times as much

  c half as much

15 Luke charges £x to hire a bicycle for 1 hour.
Write a formula to work out the charge for hiring a bicycle for h hours.

16 An electrician charges a call-out fee of £35 plus £20 per hour, h.

  a Write a formula for the cost, C, that a customer pays for an h-hour job.

  b Use your formula to work out the cost of a 3-hour job.

17 Write a formula for the number of sweets, N, each person receives when m people share y sweets.

18 Reasoning Darren is a tour guide. He works n hours per day and is paid a rate of £r per hour.

  a Write a formula for Darren's total pay per day, D, in terms of n and r.

  b Use your formula to work out D when $n = 8$ and $r = 7$.

  c At the end of each day Darren and a colleague share equally the tips, t, that they have earned.
  Write a formula for S, the amount Darren receives from the tips, in terms of t.

  d Use your formula from part **c** to work out Darren's share of the tips when $t = £30$.

  e Use your answers to parts **b** and **d** to work out Darren's total earnings for the day.

  f Reflect Would it matter if your formula used the letters x, y and z instead of D, n and r?

19 Reasoning A taxi firm charges £3 fixed charge plus £4 per kilometre.

  a Write a formula for the cost of a journey, C, of k kilometres.

  b Use your formula to work out the cost of travelling 5 km.

Exam-style question

20 Here are four pipes.

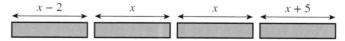

All measurements are in metres.
The total length of the four pipes is L m.
Find a formula for L in terms of x
Write your formula as simply as possible. **(3 marks)**

2.5 Expanding brackets

- Expand brackets.
- Simplify expressions with brackets.
- Write and use formulae with brackets.

Active Learn
Homework

Warm up

1 Fluency Work out

a -3×4 **b** -7×-6 **c** $3 \times b$ **d** $4 \times 3a$

e $x \times x$ **f** $a \times 2a$ **g** $-y \times 2y$ **h** $-3 \times 2y$

2 Work out the area of each rectangle.

a
```
    5
┌─────────┐
│         │ x
└─────────┘
```

b
```
 y
┌─┐
│ │ 4
└─┘
```

3 Simplify by collecting like terms.

a $3y - 6y$ **b** $2a + 4b - a - 7b$ **c** $12m^2 + 3m - 9m^2 - m$

4 Work out the area of each rectangle. Copy and complete the statements.

a $3 \times \square \quad + 3 \times \square \quad = 3(5 + \square)$

```
      5          4            5    4
  3 ┌────┐   + 3 ┌────┐  = 3 ┌────┬────┐
    └────┘       └────┘      └────┴────┘
```

b $2 \times \square + 2 \times \square \qquad = 2(\square + 7)$

```
    x           7              x    7
 2 ┌┐    + 2 ┌──────┐   = 2 ┌──┬──────┐
   └┘        └──────┘       └──┴──────┘
```

c $\square(y + \square) = 6\square + \square$

```
     y  3
 6 ┌──┬─┐
   └──┴─┘
```

d $3(\square + \square) = 3z + \square$

```
      z    4
 3 ┌────┬───┐
   └────┴───┘
```

Key point

Expanding single brackets means multiplying each term inside the brackets by the term outside.

$$5(x - 6) = 5x - 30$$

5 Expand

a $3(x + 2)$ **b** $5(a + 7)$ **c** $4(9 + t)$ **d** $6(y + 3)$

6 Expand

a $10(t - 6)$ **b** $4(x - 5)$ **c** $7(n - 2)$ **d** $5(3 - m)$

$\quad = 10 \times t - 6 \times 10 \qquad = 4 \times x - \square \times \square$

$\quad = 10t - \square \qquad\qquad\quad = \square - \square$

7 Expand

   a $-2(x+3)$ **b** $-3(a+5)$ **c** $-3(k-2)$ **d** $-2(y-1)$

Example

Expand $4(3a+2)$

$$4(3a + 2) = 12a + 8$$ ◄— Multiply each term inside the bracket by the term on the outside.

8 Expand

   a $3(2w+1)$ **b** $5(2x+3)$ **c** $4(3t+5)$ **d** $2(3m+1)$

   e $2(3m-1)$ **f** $5(2c-1)$ **g** $5(2d-3)$ **h** $-2(3p+1)$

   i $-3(4g+5)$ **j** $-3(4r-5)$ **k** $-2(1-3x)$ **l** $-2(1+3x)$

Exam-style question

9 Expand $6(2x-3)$ **(1 mark)**

10 **Reasoning** Hannah and Benton expand $3(2c-2)$.
Hannah's answer is $6c+6$. Benton's answer is $6c-6$.
Who is correct? What mistake has the other person made?

11 Expand

   a $x(x+1)$ **b** $r(r+4)$ **c** $g(g-2)$ **d** $h(3-h)$ **e** $m(2m-1)$

     $=x \times x + x \times 1$

     $=\square+\square$

   f $t(3t-2)$ **g** $-a(5a+4)$ **h** $-b(7-2b)$ **i** $-c(7+2c)$

Exam-style question

12 Expand
$5x(x+3)$ **(2 marks)**

13 Write an expression using brackets for each statement. Use n for the starting number.

   a I think of a number, add 3 and then multiply by 4.

     $(n+3) \times 4 = \square(n+3)$

   b I think of a number, add 5 and then multiply by 10.

   c I think of a number, subtract 5 and then multiply by 10.

   d I think of a number, add 1 and then double it.

   e I think of a number, subtract 1 and then double it.

14 **Problem-solving** Tallulah thinks of a number, n, adds 3 and then doubles the answer.

   a Write an expression in terms of n for Tallulah's result. Simplify your answer.

   b Work out another way for Tallulah to get the same result.

15 **Reasoning** A can of oil holds $3n+1$ litres.

   a Write and simplify an expression for the oil in 5 cans.

   b When $n=20$, how many litres are there in 5 cans?

16 Reasoning Brad is x years old and his sister is 2 years older. Their cousin is half the age of Brad's sister.

   a Write an expression for the age of Brad's sister.

   b Write an expression for the age of Brad's cousin.

   c Use your answer to part **b** to work out the age of Brad's cousin if Brad is 14.

17 Reasoning An entertainer charges £25 for a birthday party, plus £12 per child.

   a There are n children at a party. Write an expression in terms of n for the entertainer's charge.

   b Write a formula for the charge, C, in terms of n.

   c The entertainer does three parties. The same number of children, n, attend each one. Write a formula in n for her earnings, E.

   d When n is 30, how much does the entertainer earn for three parties?

18 Reasoning Water from a shower flows at $x + 2$ litres per minute.
In one shower, 70 litres of water are used, and then the shower is run for another 5 minutes.
The expression $70 + 5(x + 2)$ represents the total amount of water used.

   a Explain what the terms in the expression represent.

   b $x = 6$. How many litres in total were used during the shower?

19 Find the value of each expression when $a = 2$ and $b = 3$.

   a $3(b - 1)$ **b** $2a(b - 1)$ **c** $9(a + b)$

   d $(3b + a)^2$ **e** $a(b - 7)$ **f** $(5a)^2$

   g $(ab)^2$ **h** $-2(a - 5)$ **i** $4a(1 - b)$

20 Work out the value of these expressions.

   a $(3e)^2$ when $e = -2$ **b** $(2e - 2f)^2$ when $e = 9$ and $f = -1$

   c $\dfrac{e(f - 2)^2}{ef}$ when $e = -4$ and $f = 1$ **d** $(10e + 6f)^2$ when $e = 4$ and $f = -6$

21 Expand the bracket and collect like terms.

   a $3(t + 4) + 2$ **b** $5(m - 2) + 6$ **c** $a(a - 8) + 2a$

   d $5c - c(c + 4)$ **e** $-2(m + 3) + 7$ **f** $20 - 2(3 - 5x)$

   g $17e - (e + 2)$

> **Q21g hint**
>
> $-(e + 2) = -1(e + 2)$

22 Expand and simplify

   a $2(x + 1) + 3(x + 2)$ **b** $5(n + 7) - 2(n + 3)$

   c $4(k - 5) - 3(k - 1)$ **d** $4(2a + 3) - 3(a + 2)$

   e $2(2d - 3) + 3(d - 4)$ **f** $5(3r - 1) - 4(2r - 3)$

Exam-style question

23 Expand and simplify

   $6(x + 2) - 3(1 - 2x)$ **(2 marks)**

---

24 Reflect Brackets are used in text for information that isn't essential to the meaning of a sentence. For example: this is a GCSE maths book (for use in schools).
Write a short paragraph explaining how brackets are used in maths.

2.6 Factorising

- Factorise algebraic expressions.
- Use the identity symbol ≡ and the not equal to symbol ≠.

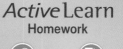

*Active*Learn
Homework

Warm up

1 **Fluency** Give the highest common factor (HCF) of each pair of numbers.

    a 2 and 8   **b** 9 and 21   **c** 20 and 30

2 Expand

    a $3(x+4)$ **b** $5(x-6)$ **c** $x(x+1)$ **d** $2x(x-1)$

3 **Problem-solving** Write the missing terms.

    a $2(a+\square) = 2a+6$ **b** $\square(a-4) = 4a-16$

    c $10a(\square+\square) = 10a^2+20a$ **d** $\square(7a-1) = 84a-12$

Key point

The factors of a term are all the numbers and letters that divide exactly into it.
A **common factor** is a factor of two or more terms.

---

4 Copy and complete to find the **highest common factor** (HCF) of $10t$ and 20.

    a The factor pairs for 20 are 1×20, $2 \times \square$, $\square \times 5$.

    b The factor pairs for $10t$ are $1 \times 10t$, $\square \times 5t$, $5 \times \square$, $10 \times \square$

    c The highest common factor is $\square$

5 Find the HCF of

    a $9x$ and 3   **b** $16y$ and 12   **c** $24r$ and 42

Key point

Factorise

$3m + 12 = 3(m + 4)$

Expand   HCF of
             $3m$ and 12

Example

Factorise $10y + 25$.

The HCF of $10y$ and 25 is 5.

$10y + 25 = 5(2y + 5)$ ←   Write the HCF of both terms outside the bracket. Work out the terms inside the bracket by dividing each term in the expression by the HCF.

$5(2y + 5) = 10y + 25$ ←   Check your answer by expanding.

---

6 Copy and complete. Check your answers by expanding the brackets.

a $8y + 16 = 8(y + \square)$

b $7m - 21 = \square(m - \square)$

c $6y + 24 = \square(y + 4)$

d $10t - 25 = 5(\square - 5)$

7 Phil and Emily factorise completely $8x + 12$.

Phil says the answer is $2(4x + 6)$. Emily says the answer is $4(2x + 3)$.

a Who is correct?

b What mistake has the other student made?

8 Factorise completely

a $9x + 18$ **b** $6w - 12$ **c** $15a + 10$ **d** $12 - 20t$

e $8m + 20$ **f** $10 - 4p$ **g** $6 - 9k$ **h** $6v + 24$

Exam-style question

9   **i** Expand $4(3x + 5)$.   **ii** Factorise $2x + 8$.   **(2 marks)**

10 Copy and complete.

a $8x + \square = 4(\square x + 3)$ **b** $\square y + 4 = 2(3y + \square)$ **c** $6z - \square = \square(3z - 5)$

11 Work out the HCF of

a cd and d **b** a^2 and a **c** bc and ab **d** x^2 and $2x$

e $3n^2$ and n **f** $3xy$ and y **g** $3xy$ and $3y$ **h** $6xy$ and $3y$

12 Work out the HCF of the numbers, then the letters, to find the HCF of each pair.

a $6xy$ and $36y$ **b** $15ab$ and $10a$ **c** $10a^2$ and $50a$

d $4x^2$ and $10x$ **e** $3cd$ and $6c^2$ **f** $6xy$ and $10x^2$

Example

Factorise

a $y^2 + y$ **b** $2ef + 4f^2$

          The HCF is y

a $y^2 + y = y(y + 1)$

b $2ef + 4f^2 = 2f(e + 2f)$ The HCF is $2f$

13 Factorise

a $n^2 + n$ **b** $k^2 - k$ **c** $r - r^2$ **d** $t + t^2$

e $4x^2 + x$ **f** $4x^2 + 3x$ **g** $4x^2 + 6x$ **h** $2st + 4t$

i $5ab - 20b$ **j** $5ab - 20b^2$ **k** $20ab - 5a^2$ **l** $15a^2 - 5ab$

14 Choose the correct factorisation for each expression.

a $a^2 + 7a$ **A** $a(a + 7)$ **B** $a(a + 7a)$

b $2ab - 3b$ **A** $b(2a - 3b)$ **B** $b(2a - 3)$

c $y^3 + y^2$ **A** $y(y^2 + y)$ **B** $y^2(y + 1)$

d $5d - d^2$ **A** $d(5 - d)$ **B** $5d(1 - d)$

15 Factorise these completely. Expand your answers to check them.

a $x^3 + x^2$ **b** $7x^3 - 21x$ **c** $9x^2 + 12x^3$ **d** $6y^3 - 2y$

e $4n - n^2$ **f** $4n + 2n^2$ **g** $3v^3 + v^2$ **h** $3v^3 + 2v^2$

i Reflect How do you know that your answer has been completely factorised?

Exam tip

Expand your answer to check it.

Key point

The $\equiv$ symbol shows an identity.

In an identity the two expressions are equal for all values of the variables.

$5(x+1) \equiv 5x + 5$ is an identity.

$5(x+1)$ has the same value as $5x + 5$ for all values of x.

17 Substitute different values for t in these expressions.

Which of these are true for *all* values of t rather than just *some* values of t?

Rewrite any identities that you find, replacing $=$ with $\equiv$.

a $t + 2 = 6$ **b** $2t + 4 = 2(t+2)$ **c** $t^2 = 6t$ **d** $5t + 7 = 7 + 5t$

18 Use $\equiv$ to write an identity for each of these expressions.

a $4a + a$ **b** $0.5a$ **c** $3(a+4)$ **d** $a + 2$

19 State whether each of these is an expression, formula or identity.

a $F = ma$ **b** $x^2 + 4x = x(x+4)$ **c** $9x - 3x^2 + 4$

d $4x^2 = (2x)^2$ **e** $y^2 + 2y$ **f** $E = mc^2$

Key point

The $\neq$ symbol is used to show that two expressions are not equal.

For example, $5x + 12 \neq 5(x+6)$

20 Put the sign $\neq$ or $\equiv$ in each box.

a $0.5x \ \square \ \dfrac{x}{2} - 1$ **b** $4(x+1) \ \square \ 4x$

c $16t + 4 \ \square \ 4(4t+1)$ **d** $p^2 - p \ \square \ -p(1-p)$

2.7 Using expressions and formulae

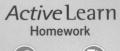
Active Learn
Homework

- Write expressions and simple formulae.
- Use maths and science formulae.

Warm up

1 Using x, write expressions for
  a 4 less than x **b** half of x **c** twice x **d** x divided by 3   **e** 10 lots of x

2 **Fluency** Work out the value of each expression when $a = 2$, $b = -3$, $c = 4$.
  a ac **b** $10a + 4b$ **c** $a - b$ **d** $b^2 - a$

3 A stapler costs r pence.
  a Write an expression for the cost of 3 staplers.
  b A hole-punch costs 20 pence more than a stapler.
     Write an expression for the cost of a hole-punch.
  c Beth buys 3 staplers and a hole-punch.
     Using your expressions, write a formula for the cost, C, simplifying your answer.

---

4 Write an expression using n as the unknown starting number.
  a I think of a number, multiply it by 5 then add 2.
  b I think of a number, subtract 1 and then multiply by 2.
  c I think of a number, multiply it by 3 and divide by 2.
  d I think of a number and multiply it by itself.
  e I think of a number, multiply it by itself and then multiply it by itself again.
  f I think of a number and square root it.

5 **Reasoning** Cakes are sold in packets or boxes. A packet contains x cakes and a box contains y cakes.
  a Write an expression for the number of cakes in 3 packets.
  b Write an expression for the number of cakes in 4 boxes.
  c Josh buys 3 packets and 4 boxes of cakes.
     Write an expression in terms of x and y for the number of cakes he buys.
  d Josh buys p packets and b boxes of cakes.
     Write a formula in terms of x, y, p and b for N, the number of cakes he buys.

6 **Reasoning** Alice scores m marks in her physics exam.
  a Write an expression in terms of m for
    i her biology marks, which are 20 marks less than in physics
    ii her chemistry marks, which are twice as many as her biology marks
  b Write and simplify an expression for Alice's total marks in physics, biology and chemistry.
  c Alice scored 60 marks in physics. Use your answer to part **b** to work out her total marks for all three science papers.

7 **Future skills** The formula to work out the mass, m, of an object is
$$m = dv$$
Work out m when $d = 2.7$ and $v = 4$.

Example

$v = u + at$
$u = 0$, $a = -4$ and $t = 5$
Work out the value of v.

$v = 0 + -4 \times 5$ ◀——— Substitute for u, a and t.
$v = -20$

---

8 **Future skills** Use the formula $v = u + at$ to work out v when
  a $u = 0$, $a = -2$, $t = 15$ **b** $u = 10$, $a = 5$, $t = 30$

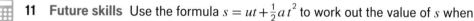

Exam-style question

9 $v = u + at$
  $u = 3$, $a = -2$ and $t = \frac{1}{2}$
  Work out the value of v.   **(2 marks)**

Exam tip

Write down the formula, replacing the letters with the numbers given.

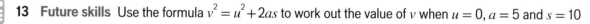

Exam-style question

10 $T = 3x + 5y$
  $x = 2$, $y = -1$
  Work out the value of T.   **(2 marks)**

---

11 **Future skills** Use the formula $s = ut + \frac{1}{2}at^2$ to work out the value of s when
  a $u = 20$, $t = 3$, $a = -10$ **b** $u = 30$, $t = 4$, $a = -5$

12 In a right-angled triangle $c^2 = a^2 + b^2$
  a Find c^2 when $a = 3$ and $b = 4$.
  b Square root your answer from part **a** to get c.

13 **Future skills** Use the formula $v^2 = u^2 + 2as$ to work out the value of v when $u = 0$, $a = 5$ and $s = 10$

14 A temperature in F (°Fahrenheit) can be converted to C (°Celsius) using the formula
$$F = \frac{9C}{5} + 32$$
  a Work out F when $C = 15°$.   **b** Work out F when $C = -5°$.
  c On Monday the temperature was 25 °Celsius and on Tuesday it was 75 °Fahrenheit.
    Which day was hotter?
    Give reasons for your answer.

15 **Problem-solving** To cook a chicken takes 40 minutes per kg plus an extra 20 minutes.
  a How long does it take to cook a 2.5 kg chicken?
  b Write a formula for the number of minutes, M, it takes to cook a chicken that weighs w kg.
  c Use your formula to find M when $w = 4$.
  d At what time should you put a 2 kg chicken in the oven to be ready for 7 pm?

2 Check up

Active Learn
Homework

Expressions and substitution

1   Simplify

   a $6a + 4b - a - 9b$

   b $10m^2 - 2m - 9m^2 - 6m$

   c $5m \times 4n$

   d $4a \times a \times b$

   e $2c \times 5c^2$

   f $-\dfrac{64\,x^2}{8x}$

   g $\dfrac{15ab}{b^2}$

   h $\left(3n^4\right)^2$

2   Work out the value of each expression when $x = 4$, $y = -2$, $z = 10$.

   a $2x + z$

   b $z - 2x$

   c y^2

   d $\dfrac{z}{y}$

   e $3x + 2z$

   f $5(x - y)$

   g $(xy)^2$

   h xy^2

3   Use x as the starting number to write expressions for these.

   a I think of a number and add 5.

   b I think of a number, multiply it by 4 and then divide by 5.

   c I think of a number, add 4 and then multiply by 5.

Expanding and factorising

4   Expand and simplify

   a $2(a + 1)$

   b $5(3f + 2)$

   c $y(6y - 2)$

   d $-2(3a + 5)$

5   Factorise completely

    a $6x+12$

    b $4x^2+16x$

    c $9y^3+21y$

    d $15xy-5y$

6   Choose the correct sign, $\neq$ or $\equiv$.

    a $10a+20ab\ \square\ 10a(a+2b)$

    b $4x(x+1)\ \square\ 4x^2+4x$

    c $36t+6\ \square\ 6(6t+1)$

    d $0.75x\ \square\ \dfrac{x}{4}$

Writing and using formulae

7   State whether each of these is an expression, a formula or an identity.

    a $y=mx+c$

    b $0.5x=\dfrac{x}{2}$

    c $2x+4$

8   **a** Use the formula $s=\dfrac{d}{t}$ to work out the value of s in km/h when $d=40$ km and $t=5$ hours.

    b $M=5n-2x$

       Work out the value of M when $n=3$ and $x=-4$.

9   There are b blue sweets and p pink sweets in a box.

    a Write a formula for the total number, T, of sweets in the box using b and p.

    b Use your formula to work out T when $b=20$ and $p=15$.

10 **a** Use the formula $v=u+at$ to work out the value of v when $u=0$, $a=-3$ and $t=10$.

    b Use the formula $v^2=u^2+2as$ to work out the value of v when $u=0$, $a=2$ and $s=16$.

11 **Reflect** How sure are you of your answers? Were you mostly

Just guessing 😦   Feeling doubtful 😐   Confident 🙂

What next? Use your results to decide whether to strengthen or extend your learning.

Challenge

12 **a** Multiply each of the four outer expressions by x and expand your answers.

    b Add together your answers and simplify the result.

    c Factorise your simplified expression.

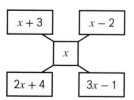

2 Strengthen

Active Learn
Homework

Expressions and substitution

1 Simplify by collecting like terms.

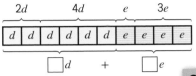

   a $2d + 4d + e + 3e$

   b $5x + 2y + 10x + 8y$

   c $7r + 6s - 5r + 2s$

   d $3p - 4q + p - 5q$

   e $6v - 10w - 9v + 4$

   f $8 + 3g + 4h - g + 2$

   g $2x^2 + 3x^2 + 5x^2$

   h $7a^2 + 3a + a^2 - a$

> **Q1c hint**
>
> The sign before a term belongs with the term: $7r - 5r = \square r$

> **Q1h hint**
>
> a^2 and a are not like terms.

2 Simplify

   a $a \times 6$

   b $4 \times n$

   c $-4 \times k$

   d $y \times (-2) = -\square y$

> **Q2a hint**
>
> 6 lots of a
>
> | a | a | a | a | a | a |

3 Simplify and write in alphabetical order.

   a $b \times a$ **b** $g \times f$ **c** $g \times f \times 2$ **d** $3 \times f \times h$

4 Simplify by multiplying the numbers, then the letters.

   a $5 \times 2a$

   b $6y \times 3$

   c $4s \times 10t = 4 \times 10 \times s \times t =$

   d $3s \times 2t$

   e $2p \times (-6q) = 2 \times (-6) \times p \times q =$

   f $3p \times (-4q)$

> **Q4a hint**
>
>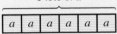
>
> | $2a$ | $2a$ | $2a$ | $2a$ | $2a$ |

5 Simplify by dividing the numbers first.

   a $\dfrac{20a}{10}$ **b** $14c \div 7$ **c** $18b \div -3$ **d** $\dfrac{5x}{10}$

6 Write an expression for these statements. Use n to represent the unknown starting number.

   a Lucy thinks of a number and adds 4.

   b Adam thinks of a number and subtracts 6.

   c Keisha thinks of a number and multiplies it by 2.

   d Bill thinks of a number and divides it by 3.

   e Jim thinks of a number, multiplies it by 2, then adds 4.

> **Q6a hint**
>
> ... a number and adds 4
>
>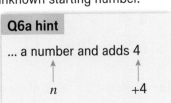
>
> n $+4$

7 Simplify

a $a \times a$

b $2a \times a$

c $a^3 \times a^2 = \overbrace{a \times a \times a}^{a^3} \times \overbrace{a \times a}^{a^2} = a^{\square}$

d $q^4 \times q$

e $2q^4 \times q$

f $2n^3 \times 3n$

g $2n^3 \times 3n^2$

h $5n^4 \times 3n^2$

8 Work out

a $\dfrac{4 \times 4}{4} = \dfrac{\cancel{4} \times 4}{\cancel{4}} = \square$

b $\dfrac{a \times a}{a} = \square$

c $\dfrac{s \times s \times t}{s} = \dfrac{\cancel{s} \times s \times t}{\cancel{s}} = \square$

d $\dfrac{3}{3 \times 3} = \dfrac{\cancel{3}}{\cancel{3} \times 3} = \dfrac{\square}{\square}$

e $\dfrac{x \times x \times y}{y \times y} = \dfrac{x \times x \times \cancel{y}}{y \times \cancel{y}} = \dfrac{\square}{\square}$

f $\dfrac{a \times a \times b}{a \times b \times b} = \dfrac{\square}{\square}$

9 Simplify

a $\dfrac{x^5}{x^2} = \dfrac{x \times x \times x \times x \times x}{x \times x} = \square$

b $\dfrac{a^7}{a^3}$

c $\dfrac{w^4}{w}$

d $\dfrac{t^4}{t^4}$

10 Simplify by dividing the numbers, then the letters.

a $\dfrac{6x^5}{2x} = \dfrac{6}{2} \times \dfrac{x^5}{x} = \square x^{\square}$

b $\dfrac{30f^6}{10f^2}$

c $\dfrac{15d^3}{5d}$

d $\dfrac{12s}{4s^2}$

11 Simplify

a $\left(2n^3\right)^2 = 2n^3 \times 2n^3 = \square n^{\square}$

b $\left(5t^2\right)^2$

c $\left(4x^3\right)^2$

d $\left(3p^5\right)^2$

12 Copy and complete to find the value of the expressions when $a = 2$ and $b = 5$.

a $2a = 2 \times \square = \square$

b $4b = 4 \times \square = \square$

c $2a + 4b = \square + \square = \square$

d $4b - 2a$

e $2b - a$

f $11a - 2b$

g $3b^2 = 3 \times \square \times \square = \square$

h $(5a)^2$

i $\dfrac{5a}{b} = \dfrac{5 \times \square}{\square} = \dfrac{\square}{\square} = \square$

13 Work out the value of these expressions when $a = 3$ and $b = 4$.

a $a + b$

b $3a - b$

c $7(a + b)$

d $\dfrac{8a^2}{b}$

14 Find the value of each expression when $f = -2$ and $g = 4$.

a $f + g$ **b** $g - f$ **c** $3g + 2f$ **d** f^2

e $4f$ **f** $\dfrac{2g}{f}$ **g** $2(f - g)$ **h** $\dfrac{-g^2}{f}$

Expanding and factorising

1 You can use a grid to expand brackets.

×	x	**+4**
5	$5x$	$+20$

$5(x + 4) = 5x + 20$

Expand, using a grid

a $3(x + 2)$ **b** $4(m + 1)$
c $5(n + 3)$ **d** $6(h + 4)$

2 Expand, using a grid

a $5(x - 3)$ **b** $2(x - 1)$
c $4(k - 5)$ **d** $3(n - 2)$

×	x	-3
2	$2x$	-6

$2(x - 3) = 2x - 6$

3 Expand, using a grid

a $3(2a + 1)$ **b** $4(3b - 1)$
c $5(2c + 3)$ **d** $5(3r - 2)$

×	$2x$	-3
4	$8x$	-12

$4(2x - 3) = 8x - 12$

4 Expand, using a grid

a $b(b + 4)$ **b** $t(t + 2)$
c $d(2d - 5)$ **d** $f(2f - 1)$

×	$3a$	$+2$
a	$3a^2$	$+2a$

$a(3a + 2) = 3a^2 + 2a$

5 Complete these factorisations.

a $2b + 6 = 2(\square + \square)$
b $3c - 9 = 3(\square - \square)$
c $8 - 2t = 2(\square - \square)$
d $5 + 15x = 5(\square + \square)$

6 **a** $bc = b \times c$
　　$b^2 = b \times b$
　　What is the HCF of bc and b^2?
b $5d^2 = 5 \times d \times d$
　　$10cd = 2 \times 5 \times c \times d$
　　What is the HCF of $5d^2$ and $10cd$?

7 Complete these factorisations.

a $3a^2 - 9a = \square(a - 3)$ **b** $16x^2 + 12x = \square(4x + 3)$
c $5a^2 + 15ab$ **d** $2q^3 + 8q$
e $84a - 12 = \square(\square - \square)$ **f** $5a^2 + ab = \square(\square + \square)$
g $y^3 + y^2 = \square(\square + \square)$

8 Choose the correct sign, $\neq$ (is not equal to) or $\equiv$ (is always equal to).

  a $4x \,\square\, 2x + 2x$

  b $6(x-3) \,\square\, 6x - 3$

  c $m^2 - m \,\square\, m(m-0)$

  d $y^2 + y \,\square\, y(y+1)$

Writing and using formulae

1 Use the formula $F = ma$ to work out F when

  a $m = 6$ and $a = 2$ $\qquad\qquad$ $F = m \times a$

  b $m = 20$ and $a = 3$ $\qquad\qquad$ $F = \square \times \square = \square$

  c $m = 100$ and $a = 5$

2 A film company pays extras an amount per day.

  a Work out the total amount paid to an extra who works for

   i 2 days at £50 per day

   iii 3 days at £a per day

   ii 10 days at £35 per day

   iv d days at £a per day

  b Write a formula for T, the total amount paid, for d days at £a per day.

  c Use your formula to work out the total amount paid to an extra who works for 3 days at £25 per day.

3 **a** Work out Amy's pay when she works for 4 hours at £5 per hour and earns £10 in tips.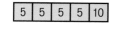

  b Write an expression for Amy's total earnings when she works h hours at £x per hour and gets £t in tips.

  c Complete this formula for Amy's total earnings E in terms of x, h and t.

   $E =$

  d Use your formula to work out how much Amy earns when $h = 10$ hours, $x = £5$ and $t = £20$.

4 **a** $A = bc$

   Work out the value of A when $b = 2$ and $c = 7$.

   $A = b \times c =$

  b $D = ef + 5$

   Work out the value of D when $e = 3$ and $f = 10$.

  c $G = \dfrac{h}{k}$

   Work out the value of G when $h = 12$ and $k = 3$.

  d $L = \dfrac{m}{n} + 8$

   Work out the value of L when $m = 6$ and $n = 2$.

  e $P = qr$

   Work out the value of P when $q = -2$ and $r = 3$.

2 Extend

1 **Problem-solving** Write four algebraic expressions that simplify to give $25x$.

2 The length of a line is x metres.
Write an expression, in terms of x, for the length of the line in
a centimetres   **b** millimetres

3 Work out the value of each expression when $x = -6$ and $y = 2$.
a $3(x - y)$ **b** $(x + y)^2$ **c** $2(x - y)^2$ **d** $5y + (y - x)^2$

4 Expand the brackets
a $x^2(xy + x)$ **b** $xy(x - y)$ **c** $an(n^2 + 1)$ **d** $2bc(b - c)$
e $3de(d^2 + e)$ **f** $3gh(2h + g)$ **g** $5pq(2p + 3q)$ **h** $4st(3t - 5s)$

5 Factorise fully
a $r^2s + rs$ **b** $r^2s + rs^2$ **c** $2r^2s - rs^2$ **d** $2r^2s + 4rs^2$
e $3jk - j^2k^2$ **f** $3jk + 6j^2k^2$ **g** $4g^2h - 10gh$ **h** $6g^2h + 10gh^2$

Exam-style question

6 **a** Factorise $7 - 21p$.   **(1 mark)**
  b Factorise fully $3xy^2 + 9x^2y$.   **(2 marks)**

---

7 Factorise
a $-7t - 21$ **b** $-a^2 - 5a$ **c** $-6f^2 - 12f$ **d** $-2r - 4r^2$

8 **Reasoning** In a café, coffee costs £x and tea costs £y.
a Write an expression for the total cost of 3 teas and 2 coffees.
b Write a formula for the total cost T of a teas and b coffees.

Exam-style question

9 Here are two straight lines, $ABCD$ and EF.

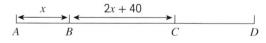

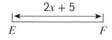

Diagrams not accurately drawn

In the diagrams all the lengths are in cm.
  $3EF = AD$

Find an expression, in terms of x, for the length of CD.

Give your answer in its simplest form.   **(4 marks)**

---

 10 Use the formula $k = \frac{1}{2}mv^2$ to work out the value of k when $m = 12$ and $v = 34$.

11 Reasoning An electrician uses this formula to calculate the charge in £ for n hours' work.

$$C = 30n + 40$$

a What is the electrician's hourly rate?

b Another electrician uses this formula.

$$C = 35n + 20$$

Which electrician is cheaper for a job that takes 3 hours?

12 Reasoning Simplify

a $a^6 \times a^3$ **b** $a^{10} \times a^4$ **c** $a^m \times a^n = a^{\square + \square}$

d $\dfrac{a^7}{a^4}$ **e** $\dfrac{a^8}{a^6}$ **f** $\dfrac{a^m}{a^n} = a^{\square - \square}$

Exam-style question

13 $x^2 \times x^y = x^7$

Work out the value of y.   **(1 mark)**

Exam-style question

14 $\dfrac{y^n}{y^3} = y^2$

Work out the value of n.   **(1 mark)**

---

15 Reasoning Simplify

a $(n^2)^3$ **b** $(n^3)^2$ **c** $(x^2)^4$ **d** $(2y)^3$

e $(2y^2)^3$ **f** $(a^m)^3$ **g** $(a^2)^m$ **h** $(a^m)^n$

Exam-style question

16 $(3^2)^t = 3^8$

Work out the value of t.   **(1 mark)**

---

 17 Future skills Use the formula $d = \dfrac{m}{v}$ to work out d when

a $m = 16$ and $v = 5.4$ **b** $m = 11.2$ and $v = 20$

Give your answers to 1 decimal place.

 18 Future skills Use the formula $s = ut + \frac{1}{2}at^2$ to work out s when

a $u = 0, t = 8.5, a = 3.7$ **b** $u = 5.2, t = 10, a = -2.5$

Give your answer to part **a** to 3 significant figures.

19 Work out

a $ab^2 - (ab)^2$ when $a = -5$ and $b = -2$ **b** $\dfrac{a^2 - b^2}{a + b}$ when $a = 7$ and $b = -2$

c $\sqrt{a^2 + b^2}$ when $a = -4$ and $b = -3$ **d** $\dfrac{a^2 b}{\sqrt[3]{c}}$ when $a = -3, b = 6$ and $c = 1000$

20 Use the values $q = 8, p = 3, r = -6, s = -2$ and no other numbers to write three expressions that will give each of the answers in the box. At least one of your expressions should involve a square or square root.

12	6	28	27

2 Test ready

Summary of key points

To revise for the test:

- Read each key point, find a question on it in the mastery lesson, and check you can work out the answer.

- If you cannot, try some other questions from the mastery lesson or ask for help.

Key points

1 A **term** is a number, a letter, or a number and a letter multiplied together. → **2.1**

2 **Like terms** contain the same letter to the same power (or do not contain a letter). You can simplify an expression by collecting like terms. → **2.1**

3 $3x$ and $7x$ are 'like terms' as the **letters** are the same; $3x$, $7xy$ and $2x^2$ are not 'like terms' as the letters are different or the powers are different. → **2.1**

4 When multiplying or dividing terms, you can simplify even if they are not like terms:
$$a \times b = ab \quad \text{and} \quad x \div y = \frac{x}{y}$$
→ **2.1**

5 When multiplying, write letters in alphabetical order and write numbers before letters. → **2.1**

6 You write an algebraic expression by using letters to stand for numbers. The letter is called a **variable** because its value can change or **vary**. → **2.1**

7 To multiply powers of the same letter, add the indices. → **2.2**

8 To divide powers of the same letter, subtract the indices. → **2.2**

9 To multiply algebraic terms, multiply the numbers first and then the letters. → **2.2**

10 To divide algebraic terms, divide the numbers first and then the letters. → **2.2**

11 $\dfrac{1}{2}x = \dfrac{x}{2}$ These fractions both mean 'half of x'. → **2.2**

12 A **formula** is a general rule that shows a relationship between variables.

 For example, speed = distance ÷ time, which you can write as $s = \dfrac{d}{t}$

 Speed, distance and time are variables. Although their values can vary, the rule stays true. The plural of formula is formulae. → **2.4**

13 **Expanding** single brackets means multiplying each term inside the brackets by the term outside. → **2.5**

14 The factors of a term are all the numbers and letters that divide exactly into it. A **common factor** is a factor of two or more terms. → **2.6**

15
Factorise
$$3m + 12 = 3(m + 4)$$
Expand   HCF of $3m$ and 12
→ **2.6**

Sample student answers

Exam-style question

Emma has x books.
Isabelle has 5 more books than Emma.
Laura has 3 fewer books than Emma.
Scarlett has twice as many books as Isabelle.

Write an expression in terms of x for the number of books the girls have altogether.
Simplify your answer.   **(3 marks)**

Emma	x
Isabelle	$x+5$
Laura	$x-3$
Scarlett	2 times $x+5$
	$= 2(x+5)$
	$= 2x+10$
Altogether	$x+x+5+x-3+2x+10$
	$= 5x+12$

a What is good about this layout?

b Why is it a good idea to use brackets for multiplying?

c Why is it a good idea to write out the whole calculation, not just the answer?

2 Unit test

Active Learn
Homework

1 Simplify

  a $5x - x + 2x - x$ **b** $4m \times 5m$ **c** $\dfrac{16b}{4}$ **d** $3 \times x \times y \times z$ **(4 marks)**

2 Expand and simplify

  a $4(p + 8)$ **b** $x(x - 3)$ **c** $5(4n - 2)$ **(3 marks)**

3 Factorise fully

  a $84a - 12$ **(1 mark)**
  b $3a^2 - 9a$ **(1 mark)**
  c $a^3 + a^2$ **(1 mark)**

4 **a** Expand
    $6(3x - 1)$ **(1 mark)**

  b Factorise
    $5m + 20$ **(1 mark)**

5 Work out the value of each expression when $a = 4$ and $b = -2$.

  a $a - b$ **(1 mark)**
  b $\dfrac{3a}{b^2}$ **(1 mark)**
  c $5a + (b + a)^2$ **(2 marks)**

6 Choose the correct sign, $\neq$ or $\equiv$.

  a $\dfrac{a}{a} \,\square\, 0$ **(1 mark)**
  b $2x - 4 \,\square\, 2(x - 2)$ **(1 mark)**

7 There are x sweets in a packet.
  Write an expression, in terms of x, for the number of sweets in

  a 5 packets   **(1 mark)**
  b 3 packets when 4 sweets are removed   **(1 mark)**

8 Write an expression using n as the unknown starting number.

  a I think of a number and add 9.   **(1 mark)**
  b I think of a number, add 5 and double the result.   **(1 mark)**

9 The mass of a box is t kg.
  Write an expression, in terms of t, for the mass of the box in grams.   **(1 mark)**

10 Tickets for a funfair are £a per adult and £c per child.
   Write an expression for the cost in terms of a and c for 2 adults and 3 children.   **(2 marks)**

11 State whether each of these is an expression, a formula or an identity.

  a $v = u + at$ **(1 mark)**
  b $u^2 - 2as$ **(1 mark)**
  c $4x(x - 2) = 4x^2 - 8x$ **(1 mark)**

12 Use the formula $E = kx^2$ to work out the value of E when $k = 100$ and $x = 4$. **(2 marks)**

13 The fuse size, I amps, of an electrical appliance can be found by dividing the power rating, P watts, by the voltage, V volts.

   a Write a formula for I in terms of P and V. **(2 marks)**

   b A hairdryer has a power rating of 1100 watts. Use your formula to work out the fuse size, I, when $P = 1100$ watts and $V = 220$ volts. **(2 marks)**

14 A paint-ball party company charges a fixed amount per guest plus an amount per hour for the venue hire.

   a Work out how much it would cost for a party of 10 guests at £25 each plus venue hire at £12 per hour for 2 hours. **(2 marks)**

   b Write an expression for the cost for a party of g guests at £p pounds per person plus venue costs for h hours at £v per hour. **(2 marks)**

   c Write a formula for the total cost, C, in terms of g, p, h and v. **(1 mark)**

   d Use your formula to work out C when $g = 15$ guests, $p = £20$ per person, $v = £8$ per hour, and $h = 4$ hours. **(2 marks)**

15 Expand and simplify
    $6(y+1) - 2(y-3)$ **(2 marks)**

16 **a** Simplify $t^2 \times t^4$ **(1 mark)**

   b Simplify $(3x^2)^3$ **(2 marks)**

   c Simplify $\dfrac{14a^5 b^2}{7a^3 b}$ **(2 marks)**

17 **a** $n^x \times n^2 = n^5$
    Find the value of x. **(1 mark)**

   b $(2^m)^2 = 2^6$
    Find the value of m. **(1 mark)**

(TOTAL: 50 marks)

18 Challenge When you add the expressions in each row, column or diagonal in this grid, the total is always the same.
Write each expression below into one of the cells
in a similar grid so that when each row, column or diagonal is added,
the simplified expression is the same.

$2a$	$7a$	$6a$
$9a$	$5a$	a
$4a$	$3a$	$8a$

| $8a + 3b$ | | $2a + b$ | | $7a + 6b$ | | $5a + 2b$ | | $a + 4b$ |

| $4a + 5b$ | | $6a + 9b$ | | $3a + 8b$ | | $7b$ |

19 Reflect Choose **A**, **B** or **C** to complete each statement about algebra.

In this unit, I did…   **A** well   **B** OK   **C** not very well

I think algebra is…   **A** easy   **B** OK   **C** hard

When I think about doing algebra I feel…   **A** confident **B** OK   **C** unsure

Did you answer mostly **A**s and **B**s? Are you surprised by how you feel about algebra? Why?

Did you answer mostly **C**s? Find the three questions in this unit that you found the hardest.

Ask someone to explain them to you. Then complete the statements above again.

3 Graphs, tables and charts

Prior knowledge

3.1 Frequency tables

ActiveLearn
Homework

- Designing tables and data collection sheets.
- Reading data from tables.

Warm up

1 Fluency How many tallies?
a ⵎ ||| b ⵎ ⵎ |

2 Eloise conducted a survey on what time Year 11 students get up on a Sunday morning.

a How many Year 11 students get up at 8 am or later?

b How many students did Eloise survey?

Time	Number of students
Before 7 am	6
7 to 7.59 am	12
8 to 8.59 am	35
9 to 9.59 am	25
10 am or later	62

3 A sports coach recorded the length of time (in seconds) John took to complete 15 races.

13.4, 13.0, 13.4, 13.1, 13.2, 13.2, 12.9, 13.3, 12.9, 13.1, 13.3, 13.4, 13.0, 12.9, 12.9

Draw a tally chart for this data.

Exam-style question

4 This tally chart shows information about the numbers of customers at a hairdresser.

	Tally	Frequency
Thursday	ⵎ |||	8
Friday	ⵎ ⵎ ⵎ ||	12
Saturday	ⵎ ⵎ ⵎ ||||	19

Write down **one** thing that is wrong with the tally chart.

(1 mark)

Exam tip

Write a sentence to explain what is wrong in the chart.

---

5 Chris works in a café. During lunch he records the number of customers sitting at each table.

a Work out the total number of tables in the café.

b Work out the total number of customers in the café.

Number of customers at table	Number of tables
0	4
1	5
2	10
3	7

Exam tip
Show all your working out clearly.

6   Jade buys some tubs of strawberries.
    There are 18, 19, 20 or 21 strawberries in each tub.
    The table gives some information about the number of
    strawberries in each tub.

Number of strawberries	Frequency
18	
19	5
20	8
21	2

The total number of strawberries is 351.
Complete the table.   **(3 marks)**

Key point

A **grouped frequency** table contains data sorted into groups called classes.

Key point

Discrete data can only have particular values. For example, shoe sizes are usually whole numbers.
For discrete data you can write groups like 1–5, 6–10.

7   A shop records the shoe sizes of 20 customers.
        3, 5, 11, 8, 9, 5, 6, 5, 2, 6, 6, 6, 8, 4, 4, 6, 7, 5, 9, 3
    Copy and complete the grouped frequency table.

Shoe size	Tally	Frequency
2–4		
5–7		

Key point

Continuous data is measured and can have any value, for example length or time.
Write inequalities for the groups, with no gaps between them.

8   For each inequality, list the numbers in the cloud that belong to it.
    a $25 \leqslant y < 26$
    b $25 < y \leqslant 26$
    c $26 \leqslant y \leqslant 27$

9   A college records the ages of 22 people taking a night class.
        22, 22, 28, 23, 26, 26, 18, 27, 19, 30, 26, 29, 28, 17, 25, 32, 34, 24, 17, 23, 20, 21

    a Copy and complete the grouped frequency table
           for this data.
    b What is the least common age group?
    c How many people are under 30?

Age (years)	Tally	Frequency
$15 \leqslant y < 20$		
$20 \leqslant y < 25$		
$25 \leqslant y < 30$		
$30 \leqslant y < 35$		

10 Ffion asked some students which country they would like to visit and recorded her results.
France, Spain, Spain, Greece, Peru, Spain, Ghana, Greece, France, France, Greece, Spain, Greece, India, Spain, France

a Which three countries are only mentioned once each?

b Design and complete a data collection sheet for Ffion's data.
Put the countries in part **a** together in a row marked 'other'.

Example

Edward recorded the time, in seconds, it took some Year 11 students to complete a task.

13, 14, 18, 21, 13, 18, 19, 13, 21, 20, 15, 15, 18, 13, 14

Design a suitable grouped frequency table for his data.

The smallest value is 13 seconds.

The largest value is 21 seconds. ————— Find the smallest and largest values.

There are 8 seconds between the smallest and largest values.

$8 \div 4 = 2$. A suitable number of classes is four. ————— Decide on a sensible number of classes. Four works here because the difference between the smallest and largest values is not very big.

Time (seconds)	Tally	Frequency
$13 \leqslant t < 15$		
$15 \leqslant t < 17$		
$17 \leqslant t < 19$		
$19 \leqslant t < 22$		

Use inequalities because the data is time, so it is continuous.

11 Aric measured the lengths, in cm, of 20 books.

13.2, 16, 18.5, 12, 16.2, 15, 19.4, 20, 15, 19.8, 14, 16.3, 14, 17, 12.5, 18, 16, 15.6, 19, 16.1

a Is this data discrete or continuous?

b Design and complete a suitable grouped frequency table for this data.

12 **Problem-solving** Ben records the number of leaves on some tomato seedlings.

2, 3, 3, 5, 3, 4, 3, 5, 3, 2, 4, 3, 4, 5, 5

Design and complete a data collection sheet for Ben's data.

13 **Problem-solving** A cyclist recorded the number of hours she trained each week for 15 weeks.

15, 12, 11.5, 26, 23, 21, 23, 27, 15.5, 14, 21, 24, 19, 22, 28

Design and complete a grouped data collection sheet for this data.

3.2 Two-way tables

- Use data from tables.
- Design and use two-way tables.

Active Learn
Homework

Warm up

1 Fluency a How many minutes in an hour? How many hours in a day?

b Change 4.25 pm to 24-hour time. Change 19:50 to 12-hour time.

2 Work out the time difference between

a 11.15 am and 1.25 pm   **b** 9.40 am and 11.45 am   **c** 09:15 and 20:30

3 a Alice works from 10.30 am to 2.45 pm. For how long is she at work?

b A plane takes $2\frac{1}{2}$ hours to get to Spain. It arrives at 15:45 UK time. At what time did it leave?

Exam-style question

4 Here is part of a train timetable.

Welwyn	09:53	09:59	10:22	10:29	10:52
Hatfield	09:57	10:04	10:26	10:34	10:56
Welham	—	10:07	—	10:37	—
King's Cross	10:20	10:41	10:49	11:11	11:19

a How many minutes should the 09:57 train from Hatfield take to reach King's Cross? **(1 mark)**

b Jo goes from Hatfield to King's Cross by train.
Jo takes 13 minutes to get from her house to Hatfield.
She takes 10 minutes to get from King's Cross to work.
She has to get to work at 11 am.
She leaves her house at 9.50 am.
Does Jo get to work by 11 am?
You must show all your working. **(3 marks)**

Exam tip

Circle the times you use for part **a**. Use a pencil so that you can rub out your circles. Then circle the times for part **b**.

Key point

A **distance chart** shows the distances between several places.

5 The distance chart shows distances in miles between four cities.

Leeds			
70	Lincoln		
42	87	Manchester	
35	45	42	Sheffield

Q5 hint

From Leeds to Sheffield is 35 miles.

How far is it from

a Leeds to Manchester   **b** Manchester to Sheffield   **c** Sheffield to Lincoln?

6 The chart shows the shortest distances, in kilometres, between cities.

London				
280	**Exeter**			
300	392	**Manchester**		
336	478	115	**York**	
648	725	351	336	**Edinburgh**

a Write down the distance between Exeter and York.

(1 mark)

b Luke drives from London to York by the shortest route. He drives 228 km and stops for a rest. Work out how many more kilometres he must drive.

(2 marks)

c Which two cities are the furthest apart? **(1 mark)**

A **two-way table** divides data into rows across the table and columns down the table. Calculate the totals across and down.

7 The table shows the numbers of medals won by a team in the Junior and Senior Games. How many more medals, in total, did the team win in the Senior Games than in the Junior Games?

	Gold	Silver	Bronze
Junior Games	29	17	19
Senior Games	34	43	43

8 Use the two-way table to work out how many females swim.

	Male	Female	Total
Swim	9		24
Run	24	12	36

9 **Reasoning** May wants to know if Rita, Sveta and Ali can teach squash, dance, football and tennis. Design a table to collect the information. Put sports down the left-hand side and names across the top.

10 **Reasoning** A teacher collects data on how late students are to school. She wants a table to record this information for each year group from Year 7 to Year 11. Design a two-way table to record the data.

11 **Reasoning** A factory makes three sizes of bookcase – small, medium and large. The bookcases are made from pine, oak or yew. The two-way table shows information about the bookcases the factory makes in a week.

	Small	Medium	Large	Total
Pine	7			23
Oak		6		24
Yew	3	8	2	13
Total	20		14	

a How many small oak bookcases does the factory make?

b How many large oak bookcases does the factory make?

c Copy and complete the two-way table.

Example

50 people chose one activity from swimming, squash or going to the gym.
   21 of the people were female.
   6 of the 8 people who played squash were male.
   18 of the people went to the gym.
   9 males went swimming.

a Put the data into a two-way table.

b Use the table to find the number of females who went to the gym.

> Put the activity down the left-hand side and male/female across the top. Include a 'total' column and row.

a

	Male	Female	Total
Swimming	9	24−9 = 15	24
Squash	6	8−6 = 2	8
Gym	29−9−6 = 14	21−15−2 = 4	18
Total	50−21 = 29	21	50

> Put the data you know into the table.

> Complete your table by working out the missing values.
> Start with a row or column that only has one missing value:
> Total male = 50−21
> = 29

b   4 females went to the gym.

> Read the value from your completed table.

12   The two-way table shows the numbers of students who play instruments.
Copy and complete the table.

	Piano	Guitar	Trumpet	Total
Year 7		11	6	37
Year 8				
Total	24	18		60

13   **Problem-solving** There are 40 people at a meeting.
Each person travelled to the meeting by car or by train.
   13 of the people are male.
   10 females travelled by train.
   8 males travelled by car.
Work out the total number of people who travelled by car.

14   **Problem-solving** Nadine asked 50 people which subject they like best from maths, English and science.
Here is some information about her results.
   19 out of the 25 males said they like science best.
   5 females said they like English best.
   Of the 7 people who said they like maths best, 4 were female.
Work out the number of people who like science best.

3.3 Representing data

- Draw and interpret comparative and composite bar charts.
- Interpret and compare data shown in bar charts, line graphs and histograms.

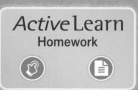

Active Learn
Homework

1 The bar charts show the temperature at midday in three different cities.

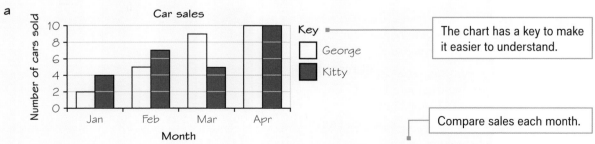

a Which bar chart shows midday temperature
   i decreasing each day   **ii** increasing each day?

b What was the highest midday temperature for city B?

Key point

A **dual bar chart** has bars side by side for each category. Dual bar charts are also called **comparative bar charts** because you can easily compare the data for each category.

Example

Kitty and George sell cars. The table shows the number of cars sold by each of them in the first four months of 2020.

	January	February	March	April
George	2	5	9	10
Kitty	4	7	5	10

a Draw a dual bar chart for this data.

b Compare Kitty and George's car sales over the 4-month period.

a

Car sales

Key
☐ George
■ Kitty

The chart has a key to make it easier to understand.

Compare sales each month.

b In January and February, Kitty sold more cars than George. George sold more than Kitty in March, and they both sold the same number in April.
George: 4 + 7 + 5 + 10 = 26   Kitty: 2 + 5 + 9 + 10 = 26
Over the 4 months, they sold the same total number of cars.

Work out and compare the total sales.

2 Jacob asks Year 7 and Year 8 students, "What is your favourite type of film?"
The dual bar chart shows the results.

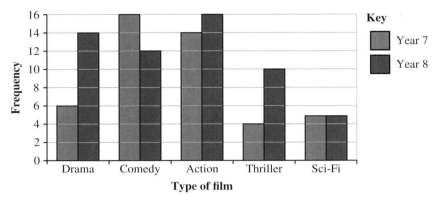

a What is the most popular type of film for Year 8 students? **(1 mark)**

b More students chose comedies than thrillers.
How many more? **(1 mark)**

c What is the total number of Year 7 students represented in the chart? **(1 mark)**

3 Kirsten records the number of mugs sold in a table. She divides the year into 'quarters' (periods of three months).

	Jan–Mar	Apr–Jun	Jul–Sep	Oct–Dec
2013	470	420	510	630
2014	490	540	770	820

a Draw a dual bar chart for the data.

b In which quarter of which year did she sell the most?

c Compare the sales of mugs in 2014 and 2013.

Key point

A multiple or **composite bar chart** compares features within a single bar.
A composite bar is also known as a compound bar chart.

4 Emma sells scarves through her website, in shops and at craft fairs. The composite bar chart shows information about sales over the last three years.

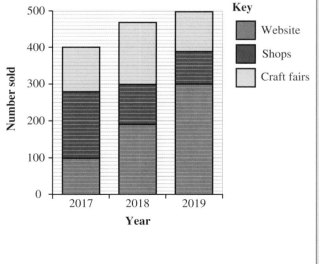

a Write down the number of scarves sold through her website in 2017. **(1 mark)**

b Work out the total number of scarves sold in shops in the 3 years. **(3 marks)**

c State whether her website, shop or craft fair sales had the biggest increase over the 3 years. Give a reason for your answer. **(2 marks)**

5 The table shows the number of gold, silver and bronze medals won by a team in 2012 and 2016.

	Gold	Silver	Bronze	Total
2012	15	17	15	47
2016	29	17	19	65

a Copy and complete the composite bar chart.

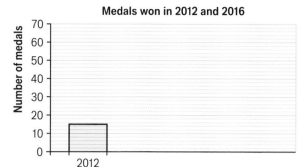

Medals won in 2012 and 2016

Key

Gold

Silver

Bronze

b Compare the performance of the team in 2012 and 2016.

6 **Reasoning** Here is a nursery's income for the first three months of 2014.

	January	February	March
Morning	£8500	£10 500	£9000
Afternoon	£12 000	£16 750	£14 000

a Draw a chart for this data.

b Draw a new set of axes. Draw two line graphs for this data on these axes.

---

7 **Future skills** An online book shop recorded the numbers of books and ebooks sold over 5 years.

Year	1	2	3	4	5
Books (1000s)	6	4.75	5.25	4.5	4.6
Ebooks (1000s)	3.5	3.75	3.5	4	4.8

a Draw a set of axes.
Put Year on the horizontal axis and Number sold on the vertical axis.

b Label your axes with the values in thousands.

c Draw a line graph for the books sold.
Use a different colour to draw a line graph for the ebooks sold.

d Describe the trends in the number of books sold and the number of ebooks sold.

---

8 The histogram shows the time taken by some students to complete a task.

a How many students took between 20 and 30 minutes?

b How many students took 30 minutes or less?

c How many students completed the task?

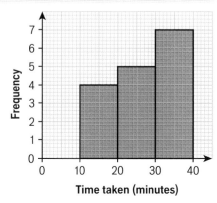

Time taken (minutes)

9 Ali and Jo asked a group of people how many minutes it took them to get to work.
Their results are shown in the table.
Draw a histogram to display this data.
Use axes like this:

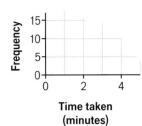

Time taken, t (minutes)	Frequency
$0 < t \leqslant 10$	4
$10 < t \leqslant 20$	7
$20 < t \leqslant 30$	9
$30 < t \leqslant 40$	5

Key point

To draw a **frequency polygon**, you can join the midpoints of the tops of the bars in a histogram with straight lines.

10 For the histogram you drew in **Q9**:
- mark the midpoint of the top of each bar
- join the points with straight lines to make a frequency polygon.

Exam-style question

11 The table shows information about the weights of 60 children.

Weight, w (kg)	Frequency
$20 < w \leqslant 30$	20
$30 < w \leqslant 40$	17
$40 < w \leqslant 50$	14
$50 < w \leqslant 60$	5
$60 < w \leqslant 70$	4

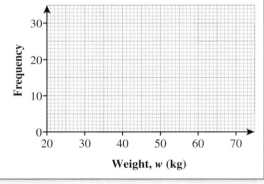

Draw a frequency polygon for the information in the table. **(2 marks)**

12 **Reflect** Look back at all of the graphs you have seen in this lesson.
Write down the types of graph you would you use to represent each of these sets of data.

a Percentage of males and females gaining GCSE geography grades in 2019.

Grade	No grade	1–3 only	4–6	7–9
Males (%)	6	23	25	46
Females (%)	4	15	25	56

b Students' heights.

Height, h (cm)	$120 \leqslant h < 130$	$130 \leqslant h < 140$	$140 \leqslant h < 150$	$150 \leqslant h < 160$
Frequency	18	24	30	27

3.4 Time series

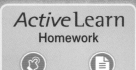

Active Learn
Homework

- Plot and interpret time series graphs.
- Use trends to predict what might happen in the future.

1 **Fluency** Write   **a** 1.6 million as a number   **b** 160 000 in millions.

2 The graph shows the height of a balloon at different times during a flight.

   a What is the height of the balloon at 35 seconds?

   b Estimate how long it took the balloon to reach a height of 40 m.

   c Did the balloon gain height steadily? Explain how you know.

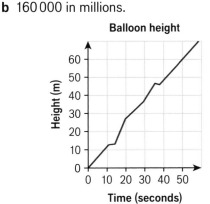

Balloon height

Key point

A **time series** graph is a line graph with time plotted on the horizontal axis.

Example

The table shows Mo's temperature over a 6-hour period.

Time	08:00	10:00	12:00	14:00
Temperature (°C)	36	37	36.5	36.8

a Draw a time series graph to represent this data.

b Estimate Mo's temperature at 11:00.

c Explain why you cannot predict Mo's temperature at 16:00.

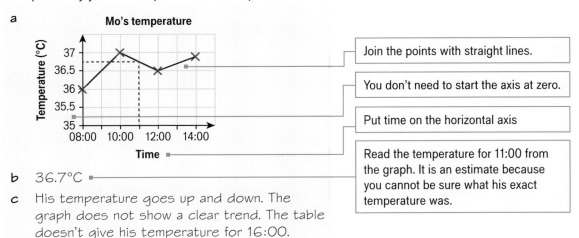

a Mo's temperature

Join the points with straight lines.

You don't need to start the axis at zero.

Put time on the horizontal axis

b 36.7°C

Read the temperature for 11:00 from the graph. It is an estimate because you cannot be sure what his exact temperature was.

c His temperature goes up and down. The graph does not show a clear trend. The table doesn't give his temperature for 16:00.

3 **Communication** Lucy's sales figures for the last three months are shown in the table.

Month	Oct	Nov	Dec
Sales figures (£)	12 340	14 500	20 090

a Draw a time series graph for this data.

b Lucy says, 'My sales figures are increasing at a steady rate.' Is Lucy correct? Explain how you know.

c Based on your graph, do you think her sales figures are likely to increase or decrease in January? Justify your answer.

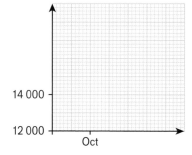

4 The table shows the temperature every two hours during one morning in January.

Time	4 am	6 am	8 am	10 am
Temperature (°C)	−2	2	6	9

a Draw a time series graph to represent the data.

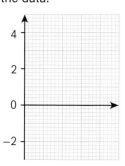

b Estimate the temperature at 9 am.

5 This time series graph shows the number of visitors, in thousands, to a museum.
Copy and complete this table:

Year	2016			
Number of visitors (thousands)		4.4		

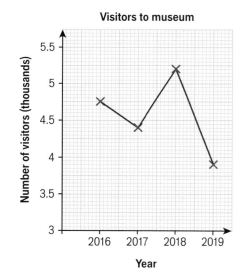

Visitors to museum

6 **Future skills** Jess is an electrician. She records the money she receives from her customers.

March: £280 from Mrs Jenkins, £1500 from Greens Garage, £4020 from Brants
April: £1340 from Fox Services, £1260 from Mr Cox
May: £4500 from ASA International

a Find the total amount of money Jess receives each month. Construct a time series table for Jess.

b Represent the data as a time series graph.

7 **Reasoning** The table shows the number of customers for an online business in 2018 and 2019. Q1 means the first quarter of the year (January to March).

	Q1	Q2	Q3	Q4
2018	860 000	1 300 000	1 500 000	1 100 000
2019	790 000	1 210 000	1 400 000	770 000

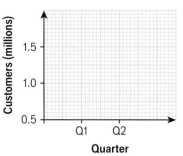

a Draw two time series graphs, one for each year's data, on the same axes.

b Describe how the number of customers changed during 2018.

8 **Reasoning** The table shows the numbers of students taking GCSE Physics and the numbers taking GCSE French in each of the years 2014 to 2019. The numbers are rounded to the nearest thousand.

Year	2014	2015	2016	2017	2018	2019
No. taking Physics	137 000	121 000	140 000	133 000	158 000	160 000
No. taking French	150 000	140 000	145 000	122 000	121 000	126 000

a Represent the data as two time series graphs on the same grid.

b Describe the overall trends in the number of students taking Physics and French between 2014 and 2019.

c Sally predicts that in 2020 and 2021 the number of students taking GCSE Physics will be greater than the number of students taking GCSE French. Is she correct? Explain your answer.

9 Lisa has drawn this time series graph to show the numbers, in thousands, of visitors to a seaside town.

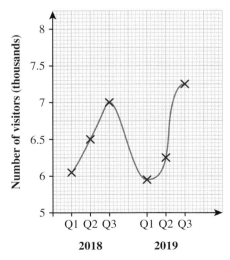

Write down two things that are wrong or could be misleading with this graph. **(2 marks)**

3.5 Stem and leaf diagrams

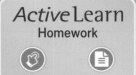

Active Learn
Homework

- Construct and interpret stem and leaf and back-to-back stem and leaf diagrams.

Warm up

1 Fluency Put these numbers in size order, starting with the smallest.

  a 132, 123, 125, 135, 143, 146, 125, 123   **b** 3.2, 6.7, 3.9, 4.1, 7.5, 1.4, 8.4, 3.9

Key point

A **stem and leaf diagram** shows numerical data split into a 'stem' and 'leaves'.
In a stem and leaf diagram the numbers are placed in order.

2 The stem and leaf diagram shows the length, in metres, of some bridges.

  a How many bridges were less than 20 m long?

  b What is the length of the longest bridge?

```
1 | 4 5 5 6 7 7
2 | 0 4 5 5 5 9
3 | 1 1 4
```
Key: 1 | 4 means 14 metres

3 Jason recorded the height (in mm) of some plants.
The information is shown in the stem and leaf diagram.

  a How many plants are there?

  b How many plants are more than 3 cm tall?

  c What is the difference between the tallest and the shortest height?

```
0 | 9
1 | 3 2 3 3
2 | 0 3 5 9 9
3 | 1 2 2 6 6 7
4 | 1 4 8
```
Key: 4 | 8 means 48 mm

Example

Here are the heights of some students (in cm).

  169, 163, 153, 173, 166, 178, 177

Construct a stem and leaf diagram for this data.

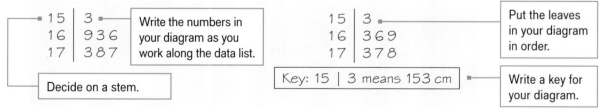

```
15 | 3
16 | 9 3 6
17 | 3 8 7
```
Write the numbers in your diagram as you work along the data list.

Decide on a stem.

```
15 | 3
16 | 3 6 9
17 | 3 7 8
```
Put the leaves in your diagram in order.

Key: 15 | 3 means 153 cm

Write a key for your diagram.

4 Here are the times some students took to complete a task (in seconds).

  220, 238, 220, 230, 235, 238, 205, 198, 238, 192

Draw a stem and leaf diagram to display the data.

5 Here are the times (in minutes) it took 21 teachers at a school to get to work.

  13, 18, 20, 35, 45, 34, 44, 23, 33, 12, 46, 21, 22, 17, 22, 31, 23, 8, 15, 22, 10

Construct a stem and leaf diagram to show this information. Use the tens digit as the stem.

6 Here are 15 students' marks in a test.

17, 16, 22, 20, 38, 14, 9, 26, 27, 30, 38, 33, 21, 16, 24

Show this information in a stem and leaf diagram.

(3 marks)

7 Daley measured the distance (in metres) that some Year 9 students jumped in a long jump competition.

4.6, 4.8, 5.1, 4.7, 3.9, 3.9, 3.4, 3.9, 4.7, 4.9

Draw a stem and leaf diagram to display this data. Use the whole number part as the stem.

A **back-to-back stem and leaf diagram** compares two sets of data.

8 **Reasoning** Some students watched a film. At the end of the film, the students' heart rates were recorded in beats per minute (bpm).

a What is the lowest female heart rate?

b What is the highest male heart rate?

c What is the difference between the lowest male heart rate and the highest female heart rate?

d Malek says, 'More males than females have a heart rate of more than 90 bpm'. Is Malek correct? Explain your answer.

e How does the shape of the back-to-back stem and leaf diagram show the answer to part **d**?

Females		Males
8 5	7	6 7 9
7 5 4 3 0	8	3 5 8
9 8 6 1	9	2 3 5 7 8 9
	10	1 3 7

Key: For females 5 | 7 means 75 bpm
For males 7 | 6 means 76 bpm

9 **Reasoning** A hotel chain records the age (in years) of the guests at two of its hotels.

Abbey Hotel: 2, 11, 15, 28, 32, 33, 19, 40, 45, 58, 39, 33, 35, 17, 21, 36, 23, 29, 36, 47, 47, 49, 39, 37, 39, 39, 48

Balmoral Hotel: 40, 45, 34, 37, 62, 64, 71, 63, 65, 50, 50, 50, 53, 56, 46, 26, 49, 40, 34, 51, 45, 63, 50, 75, 57, 67, 70, 56

a Draw a back-to-back stem and leaf diagram to display the data.

b Bronte says, 'The guests at the Abbey Hotel are younger than the guests at the Balmoral Hotel.' Is Bronte correct? Explain your answer.

10 Phoebe measures the heights of some Year 8 students to the nearest centimetre.

Boys	149	153	155	156	163	165	165	165	170	172
Girls	146	148	151	151	152	155	156	157	164	169

Draw a back-to-back stem and leaf diagram to display the data.
Use the 'hundreds' and 'tens' digits as the stem.

11 **Reasoning** Alex records the height (in cm) of tomato seedlings grown in the dark and in the light.

Grown in the dark (cm)	1.6	1.9	1.5	1.8	2.2	2.0	1.7	1.9
Grown in the light (cm)	2.8	2.7	3.5	4.4	3.8	4.1	4.5	4.3

a Draw a back-to-back stem and leaf diagram to display Alex's data.

b Did the tomato seedlings grow better in the dark or the light? Explain your answer.

3.6 Pie charts

- Draw and interpret pie charts.

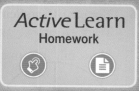

Warm up

1 **Fluency** What fraction or percentage of each of these circles is shaded?

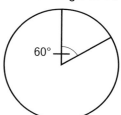

a

b

2 How many degrees are there at the centre of a circle?

3 a Draw a circle of radius 4 cm.

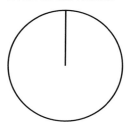

b Draw in the radius.

c Draw an angle of 60° on your radius.

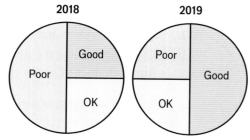

60°

Key point

A **pie chart** is a circle divided into sectors. Each sector represents a set of data.

4 **Reasoning** A council surveyed householders for their opinion on how well waste is collected. The pie charts show the results.

a What fraction of householders thought waste collection was good in 2018?

b Compare the percentage of householders that thought the waste collection was poor in 2018 and 2019.

c Do you think household waste collection is improving? Explain your answer.

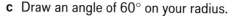

2018

2019

Good

Poor

Poor

Good

OK

OK

5 **Reasoning** A group of students were asked if they wanted more after school sports clubs.
The results are shown in the pie chart.

a What percentage of the students said 'Yes'?

b Measure the angle of the 'No' section with a protractor.

c Bethan says, 'More than twice the number of students said 'Don't know' than said 'Yes'.'
Explain why Bethan is correct.

Survey results

Yes

Don't Know

No

6 The pie charts show the match results for two netball teams. Team A played 20 games and Team B played 28 games.

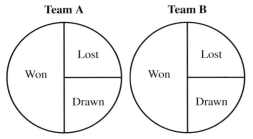

Team A   Team B

Lost   Lost

Won   Won

Drawn   Drawn

a Calculate the number of games that Team A lost.

(1 mark)

b The head coach says, 'Both teams won the same number of games.' Is he correct?
You must explain your answer. **(1 mark)**

Use calculations to explain your answer.

Example

The table shows the match results of a football team. Draw a pie chart to represent the data.

Result	Won	Drawn	Lost
Frequency	28	12	20

Total number of games = 28 + 12 + 20 = 60 — The total number of games is the total frequency.

$$\div 60 \left(\begin{array}{c} 60 \text{ games} : 360° \\ 1 \text{ game} : 6° \end{array} \right) \div 60$$

Work out the angle for one game.

Won: 28 × 6° = 168°

Drawn: 12 × 6° = 72° — Work out the angle for each result.

Lost: 20 × 6° = 120°

Check: 168 + 72 + 120 = 360 — Check that your angles total 360°.

Team results

Lost

Won

Drawn

Draw the pie chart. Give it a title and label each section, or make a key.

---

7 A café owner records the drinks sold in his café on one day. The information is shown in the table.

a Work out the total frequency.

b Work out the angle for one drink on a pie chart.

c Work out the angles for each type of drink on a pie chart.

d Draw a pie chart to show the information.

Drink	Frequency
hot chocolate	20
milkshake	15
coffee	25
tea	30

Exam tip

Label the pie chart sections with the categories, not the angles.

8 A group of people were asked where they went on holiday. The table gives information about their answers.

Country	Number of people
France	16
Spain	12
Germany	5
Italy	7

Draw an accurate pie chart for this information. **(3 marks)**

9 Problem-solving 30 people used a sports centre one evening. They each took part in one of four activities.

gym, swimming, squash, swimming, aerobics, swimming, aerobics, aerobics, aerobics, gym, aerobics, gym, gym, gym, squash, squash, gym, squash, gym, gym, gym, aerobics, aerobics, squash, gym, gym, aerobics, squash, gym, aerobics

Draw a pie chart to show this information.

10 Reasoning The pie chart shows the percentage of cake sales in a cake shop over one week.
The cake shop sold 30 chocolate cakes.

a How many cakes did it sell in total?

b How many

   i lemon   **ii** banana cakes did it sell?

Cake sales

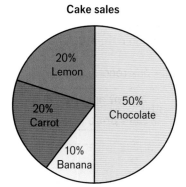

11 Margaret asked the students in a Year 8 English class their favourite type of book.
She drew a pie chart of her results.

a Measure the angle in the pie chart for 'Romance'.

b Margaret knows that eight students said 'Romance'. How many people are in the English class?

c What is the most popular type of book?

Favourite book type

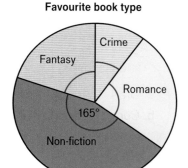

12 Reasoning Kato sells umbrellas.
Here are his sales figures for last week.

a Draw a pie chart for the data.

b Draw a bar chart for the data.

c Which chart best shows

   i the different numbers of each type sold

   ii the fraction of umbrellas sold that were spotty

   iii the most popular umbrella type?

Umbrella type	Number sold
spotty	24
plain	16
striped	32

3.7 Scatter graphs

- Plot and interpret scatter graphs.
- Determine whether or not there is a relationship between sets of data.

*Active*Learn
Homework

Warm up

1 Fluency The table shows the number of bedrooms and number of bathrooms in four houses.

Number of bedrooms	2	4	5	5
Number of bathrooms	1	2	3	5

  a Draw x- and y-axes on graph paper from 0 to 6.

  b Label the x-axis 'Number of bedrooms' and the y-axis 'Number of bathrooms'.

  c Plot the data, using crosses.

Key point

A **scatter graph** shows the relationship between two sets of data. Plot the points with crosses. Do not join them up.

2 Max recorded the sale price of 9 cars.

Age of car (years)	3	7	8	2	1	3	5	4	6
Sale price (£thousands)	7.5	1.5	1.2	8.0	9.2	7.4	3.6	6.3	2.1

  a Draw a scatter graph to display the data.
    Put age on the horizontal axis and sale price on the vertical axis.

  b Copy and complete As the age of the car increases, the value of the car _____

3 Beatrice recorded the ages of 9 people and the times they took to run 200 m.

Age (years)	35	17	28	23	19	37	51	43	60
Time (seconds)	32	26	29	27	25	33	36	34	38

> **Q3 hint**
>
> Scatter graphs are also called scatter diagrams.

  a Draw a scatter graph to display the data.

  b Copy and complete. As a person's age increases, their time to run 200 m _____

Key point

The relationship between the sets of data is called **correlation**.

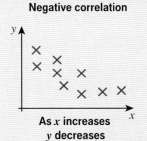

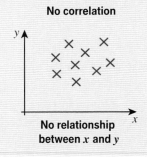

4 Look back at your graphs for **Q1–3** . Describe the type of correlation in each one.

Unit 3 Graphs, tables and charts   81

5 Match each real-life example to the correct type of correlation.

   a The relationship between height and arm span.   **A** No correlation

   b The relationship between age of car and price.   **B** Positive correlation

   c The relationship between eye colour and intelligence.   **C** Negative correlation

---

Exam-style question

6 The tread on a car tyre and the distance travelled by that tyre were recorded for a sample of cars. The data is shown in the scatter graph.

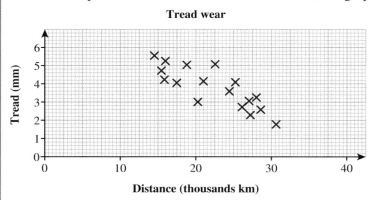

Tread wear

y-axis: Tread (mm); x-axis: Distance (thousands km)

Exam tip

'Describe the relationship' means write the type of correlation so, _____ correlation.

   a Describe the relationship between the tread on a car tyre and the distance travelled by that tyre. **(1 mark)**

   b A tyre with less than 1.6 mm of tread is illegal. If the law changed to less than 2.5 mm, how many tyres in this data set would be illegal? **(1 mark)**

---

7 The table shows the masses of 12 books and the number of pages in each book.

Number of pages	80	155	100	125	145	90	140	160	135	100	115	165
Mass (g)	160	330	200	260	320	180	290	330	260	180	230	350

   a Copy and complete the scatter graph to show the information in the table.

   b Copy and complete the sentences.
As the number of pages in a book increases, the mass _____
This is _____ correlation.

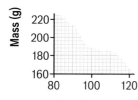

Key point

An **outlier** is a value that does not fit the pattern of the data.

You can ignore an outlier if it is due to a measuring or recording error.

---

8 The heights and lengths of seven sheep are given in the table.

Height (cm)	65	80	52	78	65	62	84
Length (cm)	100	110	86	106	80	95	115

   a Plot the information on a scatter graph, with height on the horizontal axis and length on the vertical axis.

   b One of the points is an outlier. Write down the coordinates of this point.

   c For all the other points, describe the correlation.

9 The scatter diagram shows information about 12 children.
   It shows the age of each child and their best times to tie their shoelaces.

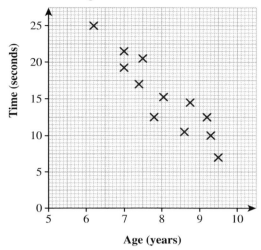

Exam tip

In part **b**, compare the position of the point to the other data points.

a Write down the type of correlation.   **(1 mark)**

b Ella is 6 years old. Her best time to tie her shoelaces is 10 seconds.
   The point representing this information would be an outlier on the scatter diagram.
   Explain why.   **(1 mark)**

10 The table shows the number of divorces and the amount of margarine eaten.

Number of divorces in a population of 1000	5.0	4.7	4.6	4.4	4.3	4.1	4.2	4.2	4.2	4.1
Amount of margarine eaten (lb)	8.2	7.0	6.5	5.3	5.2	4.0	4.6	4.5	4.2	3.7

a Plot the information on a scatter graph.

b Does 'Number of divorces' correlate with 'Amount of margarine eaten'?

c **Reasoning** Do you think eating more margarine causes the number of divorces to increase?

Key point

Correlation shows that there may be a link between two events. Correlation does not show causation, which means correlation does not show that one event caused the other.

11 Copy and complete the table. Tick the correct type of correlation for each set of data and whether you think one causes the other.

	Positive correlation	Negative correlation	No correlation	Possible causation
Arm length and leg length				
Exercise and weight				
Size of garden and running speed				
Hours of TV watched and shoe size				

3.8 Line of best fit

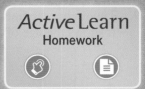
Active Learn
Homework

- Draw a line of best fit on a scatter graph.
- Use the line of best fit to predict values.

Warm up

1 Fluency Complete the statements.

a The more I work, the more I get paid.
This is _____ correlation.

b The more I watch TV, the less I revise for my exam.
This is _____ correlation.

2 Look at the graph.

a What is the value of x when y is 4?

b What is the value of y when x is 3?

Key point

A **line of best fit** is a straight line drawn through the middle of the points on a scatter graph. It should pass as near to as many points as possible and represent the trend of the points.

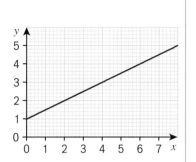

Example

Jayden records the time he spends watching TV and the time he spends revising one week.

Time spent watching TV (hours)	0.75	2.00	2.75	3.5	4.25	5.00	5.75
Time spent revising (hours)	5.25	4.75	3.75	3.25	3.00	2.00	1.5

a Show Jayden's results on a scatter graph. Draw a line of best fit.

b Jayden spends 4 hours watching TV. Estimate how much time he will spend revising.

a

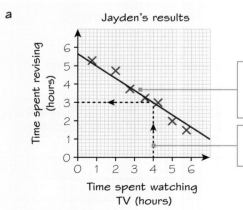

Position a transparent ruler over the scatter graph so it follows the overall trend. Move it until you have roughly the same number of points above and below the line.

Start at 4 on the 'Watching TV' axis, go up to the line of best fit and read off the answer on the 'Revising' axis.

b   3 hours

3 A scientist recorded the density of air and the speed of sound in each of six experiments.
The table shows her results.

Speed of sound (m/s)	355	340	337	320	317	340
Density of air (kg/m³)	1.1	1.2	1.3	1.2	1.4	1.3

  a Draw a scatter graph for the data.

  b Draw a line of best fit on your graph.
    There should be roughly the same number of points either side of the line.

4 **Reasoning** A school records the marks achieved by a group of students in a chemistry exam
and a music exam.

Chemistry mark	22	56	34	28	40	26	72	68
Music mark	46	62	82	30	70	62	38	88

  a Draw a scatter graph for the data.   **b** Describe the correlation.

  c Can you draw a line of best fit? Explain your answer.

5 The table gives the lengths and widths of
seven fossils.

Length (cm)	1.7	2.3	2.4	5.8	6.3	7.8	9.0
Width (cm)	0.9	2.3	3.3	6.0	6.2	7.0	8.5

  a Plot the points on a scatter graph.

  b Draw a line of best fit on the scatter graph.

  c Another fossil is 4.0 cm in length. Use your line of best fit to estimate its width.

6 **Reasoning** Some students took a French test and a Spanish test.
The table gives the marks for seven of the students.

French mark	20	25	27	50	64	80	90
Spanish mark	20	30	52	58	75	86	87

  a Plot the points on a scatter graph.

  b Are there any outliers? Explain your answer.

  c Describe the relationship between the French and Spanish marks.

  d Cherie's French mark is 60. Estimate her Spanish mark.

7 **Problem-solving** Jenny says, 'Richer
countries have a lower infant mortality
rate.' In the table, GDP per person
measures the wealth of a country.

  a Plot a graph using two appropriate
columns from the table to test Jenny's
statement.

  b Comment on Jenny's statement.

  c The table only has a sample of
countries of the world.
Does this affect your answer?
How?

Country	Infant mortality (deaths/1000 births)	GDP per person
Australia	4.43	$43 000
Brazil	19.21	$12 100
India	43.19	$4 000
Japan	2.13	$37 100
Morocco	24.52	$5 500
Norway	2.48	$55 400
South Africa	41.61	$11 500
Tonga	12.36	$8 200
Turkey	21.43	$15 300
United Kingdom	4.44	$37 300
United States	6.17	$52 800

8   **Reasoning** Toni sells used cars. The table gives the price and mileage of five cars sold.

Mileage	28 000	32 000	45 000	50 000	56 000
Price (£)	9200	8400	7500	6500	6200

a Plot the data on a scatter graph.

b Describe the correlation between mileage and price.

c Draw a line of best fit.

d Use your line of best fit to estimate the cost of a car with a mileage of

  i   40 000 miles

  ii 65 000 miles

e Which of your estimates in part **d** is more reliable? Explain.

Q8a hint

Plot mileage on the horizontal axis, from 20 000 to 70 000 miles. Plot price on the vertical axis, from £4000 to £10 000.

Exam-style question

9   The scatter graph shows the maths and science marks for 10 students.

Exam tip

Use a line of best fit to answer part **c**. Draw a line from the axis to the line of best fit to show your working.

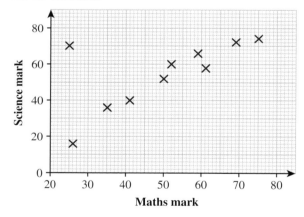

a One of the points is an outlier. Write down the coordinates of this point.   (1 mark)

b For all the other points, write down the type of correlation.   (1 mark)

c Another student has a science mark of 50.
   Estimate this student's maths mark.   (2 marks)

d Marek says 'Students with higher maths marks get higher science marks'.
   Does the scatter graph support what Marek says?
   Give a reason for your answer.   (1 mark)

3 Check up

*Active*Learn
Homework

Tables

1   Mandip measures the height of some students in metres.

   1.35, 1.28, 1.36, 1.42, 1.33, 1.29, 1.23, 1.41, 1.40, 1.34, 1.34, 1.46

   Use the information to copy and complete the table.

Height (m)	Tally	Frequency
$1.2 \leqslant h < 1.3$		
$1.3 \leqslant h < 1.4$		
$1.4 \leqslant h < 1.5$		

2   Here is part of a train timetable from Peterborough to London.

   a Which station should the train leave at 09:01?

   b The train arrives in Sandy at 09:12. How many minutes should it wait there?

   c The train should take 41 minutes to travel from Arlesey to London. At what time should the train arrive in London?

Station	Time of leaving
Peterborough	08:44
Huntingdon	09:01
St Neots	09:08
Sandy	09:15
Biggleswade	09:19
Arlesey	09:24

3   The two-way table shows numbers of tickets sold at a theatre.

   a Write the number of adults who chose luxury seats.

   b Copy and complete the two-way table.

	Budget seats	Standard seats	Luxury seats	Total
Adult		17	19	
Child	24		30	
Total	39			130

Graphs and charts

4   The table gives information about the numbers of fish in a lake. Draw an accurate pie chart to display this information.

Fish	Frequency
perch	10
bream	23
carp	39

5   The table shows the number of sunny days and windy days in a four-month period.

	May	Jun	Jul	Aug
Sunny days	14	22	18	23
Windy days	16	19	8	5

   a Draw a dual bar chart to display the data.

   b Write a statement comparing the number of sunny and windy days during the four-month period.

6 Caroline and Marc are in a darts team.
The pie charts show the number of games Caroline
and Marc won and lost last year.
Caroline played 52 games.
Marc played 160 games.
How many more games did Marc win than Caroline?

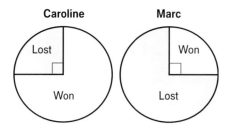

7 Amrita recorded the heart rate (in beats per minute) of
15 people. She asked them to walk up some stairs and
recorded their heart rates again. She showed her results
in a back-to-back stem and leaf diagram.

Before		After
9 8	5	
7 6 6 4 1 0	6	5 8 8 9
9 8 6 3 2	7	2 4 7 8
4 1	8	5 6 8
	9	1 3 7
	10	2

a What was the maximum heart rate after walking up
the stairs? Explain how you know.

b What was the difference in minimum heart rates
before and after walking upstairs?
Explain how you know.

Key: (Before) 8 | 5 means 58 bpm
(After) 6 | 5 means 65 bpm

Time series and scatter graphs

8 The graph shows the height of a candle as it burns.

a What is the height of the candle after 2 hours?

b Estimate the height of the candle at 13:30.

c How long does the candle take to burn down from
15 cm to 4 cm?

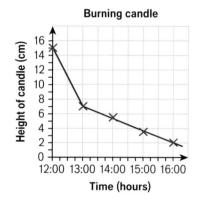

9 The scatter graph shows the number of ice creams sold and
the number of hours of sunshine on 9 days.

a One of the points is an outlier.
Write down the coordinates of this point.

b Describe the correlation shown.

c On another day there were 10 hours of sunshine.
Estimate the number of ice creams sold.

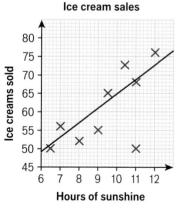

10 How sure are you of your answers? Were you mostly

Just guessing 😟   Feeling doubtful 😐   Confident 🙂

What next? Use your results to decide whether to strengthen or extend your learning.

Challenge

11 Design a data collection sheet for a traffic survey to be carried out near a primary school.
Decide what the purpose of the survey will be and the types of data you need to collect.

3 Strengthen

Tables

1 Erin measures the length of some runner beans from her garden.

14.7, 12.0, 13.8, 8.9, 14.9, 12.4, 9.0, 13.7, 7.0, 9.7, 11.5, 11.9, 14.2, 11.0, 12.5

a Which class does 9.0 belong in?

$7 \leqslant l < 9$ or $9 \leqslant l < 11$?

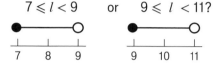

Length, l (cm)	Tally	Frequency
$7 \leqslant l < 9$		
$9 \leqslant l < 11$		
$11 \leqslant l < 13$		
$13 \leqslant l < 15$		

b Work through the lengths in order. Add a tally mark to your table for each one. Total your tallies to complete your frequency column.

2 **Future skills** Here is part of a train timetable from Dundee to London.
How long does it take the 06:30 train from Dundee to get to London?

Dundee	06:30
Peterborough	12:20
London	13:35

3 **Future skills** Here is part of a railway timetable.

a Work out how long the 10:13 train takes to travel from New Street to Coventry.

b Harry is at Birmingham International. He needs to be at Tile Hill by 10:50.

New Street	10:13	10:30	10:33
Marston Green	10:26	↓	10:41
Birmingham International	10:29	10:39	10:45
Hampton-in-Arden	10:32	↓	10:48
Tile Hill	10:40	↓	10:55
Coventry	10:47	10:49	11:00

i Which trains get to Tile Hill by 10:50?

ii What time is the latest train from Birmingham International he can catch?

4 The diagram shows some shapes. Copy and complete the two-way table to show the number of shapes in each category.

	White	Black
Circle		
Square		

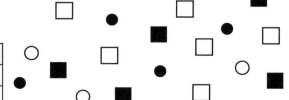

5 The two-way table shows how 100 students travelled to school on one day.

	Walk	Car	Bicycle	Total
Boys	15		14	54
Girls		8	16	
Total	37			100

  a Copy the table. Total the 'Girls' row and the 'Car' and 'Bicycle' columns.

  b Work out the number of girls who walked to school.

  c Work out the number of boys who travelled to school by car.

  d Complete the rest of the table.

6 80 students went on a school trip.
They went to either London or York.
23 boys and 19 girls went to London.
14 boys went to York.

	London	York	Total
Girls			
Boys			
Total			

  a Write this information into a copy of this two-way table.

  b Work out the missing values and complete the rest of the table.

Graphs and charts

1 The table shows the number of computer games and DVDs Ted bought in 2018 and 2019.

	Games	DVDs
2018	7	9
2019	10	6

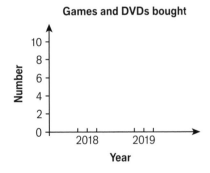

Games and DVDs bought

  a Draw a dual bar chart.

  b Copy and complete this statement for 2018 comparing the numbers of games and DVDs he bought.
In 2018 he bought more _____ than _____ .
Now write a statement for 2019.

2 The table shows the average daily hours of sunshine in Majorca and Crete over a five-month period.

	April	May	June	July	August
Majorca	9	9	11	11	10
Crete	6	8	11	13	12

  a Display the data in a dual bar chart.
Plot the month on the horizontal axis and the hours on the vertical axis.

  b Write a statement comparing the hours of sunshine in Majorca and Crete during the five-month period.

3 The table shows the number of sweets of each colour in a packet.

Colour	Number of sweets	Angle
red	10	
green	5	
yellow	6	
orange	9	
Total		360°

a Work out the total number of sweets.

b Work out the angle for one sweet in a pie chart.

$$\div \square \left(\underset{\text{1 sweet is } \square °}{\overset{\square \text{ sweets is } 360°}{}} \right) \div \square$$

c Work out the angle for the 10 red sweets.

d Work out the angles for the other sweets.

e Draw the pie chart. Give it a title. Label the sectors with the colours of the sweets.

4 In a chemistry experiment, Juan recorded the mass of chemical produced by a reaction (in grams). He repeated the experiment 15 times. Here are his results.

105, 112, 117, 127, 123, 103, 110, 125, 121, 108, 113, 125, 114, 119, 125

a Write the data in order, starting with the smallest.

b Copy and complete this stem and leaf diagram to show this information.

```
10 | 3
11 | 0
12 |
```

Time series and scatter graphs

1 The table shows the temperature in a greenhouse every 2 hours.

Time	11 am	1 pm	3 pm	5 pm
Temperature (°C)	12	20	29	25

a Draw a time series graph for the data. Join the points with straight lines.

b What is the time when the temperature is 25 °C?

c Read an estimate of the temperature at 4 pm from your graph.

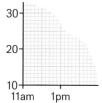

2 Draw two time series graphs on the same axes for the sunshine data for Majorca and Crete in **Graphs and Charts Q2** on p. 90.

3 The scatter graph shows the ages and weights of 8 lobsters.

a Find 'Age 3 years' on the correct axis.

b Draw a line for 'Age 3 years' up to the line of best fit and then across to the other axis.

c Read off the estimated weight of a lobster aged 3.

d Use the line of best fit to estimate the weight of

  i a lobster aged 5

  ii a lobster aged 2

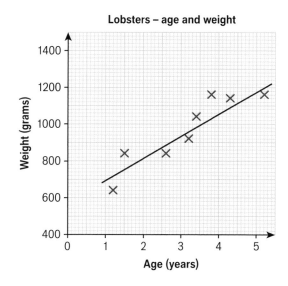

3 Extend

Exam-style question

1   The pictogram shows the numbers of text messages Katherine sent.

Monday	
Tuesday	
Wednesday	
Thursday	

Key ⬤ represents 8 messages

a   Write down the number of text messages Katherine sent on

   i   Monday   ii   Tuesday   **(2 marks)**

b   On Wednesday and Thursday, Katherine sent a total of 33 messages.
   The number she sent on Thursday was twice the number she sent on Wednesday.
   Draw the pictogram symbols for Wednesday.   **(2 marks)**

2   **Reasoning** The diagram gives information about the height of some buildings.

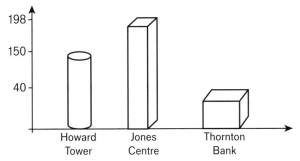

Q2 hint

What do you notice about the scale on the vertical axis? What about the shape of the bars?

Write two things that could be wrong or misleading in the diagram.

3   **Future skills** The graph shows the cost of the gas Charlie has used over the last 12 months.
   Quarterly means every quarter (3 months) of the year.
   Charlie wants to make monthly payments for his gas.
   He thinks his gas is going to cost the same for the next 12 months.
   Charlie works out he should pay exactly £40 per month to cover the cost over 12 months.
   Is he correct? Explain your answer.

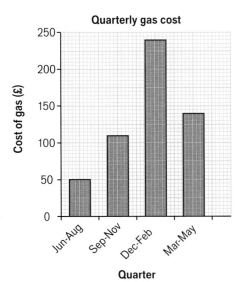

Exam-style question

4 Monica measures the body length and foot length of eight rodents, from one species.
The data is displayed in the table.

Body length (mm)	25	100	75	75	175	105	250	125
Foot length (mm)	14	25	18	24	35	19	43	28

  a Draw the scatter graph and a line of best fit. **(2 marks)**

  b Describe the type of correlation between body length and foot length. **(1 mark)**

  c An animal has a body length of 230 mm and a foot length of 20 mm.
    Is this animal likely to be one of this species of rodents?
    Explain your answer. **(2 marks)**

  d Another rodent has body length 300 mm. Monica says 'The scatter graph shows
    this rodent should have foot length 48 mm'. Comment on what Monica says. **(1 mark)**

Exam-style question

5 The bar chart shows the percentage of different types of mice found in samples from
homes near woodland and away from woodland.

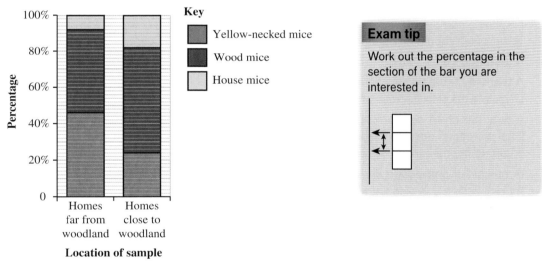

Exam tip

Work out the percentage in the section of the bar you are interested in.

  a What percentage of mice in homes close to woodland are wood mice? **(1 mark)**

  b What percentage of mice in homes far from woodland are not wood mice? **(2 marks)**

  c One of the pale bars is taller than the other. Does that mean there are more house mice
    in homes far from woodland than in homes close to woodland?
    Explain how you know. **(1 mark)**

6 **Future skills** This is a record of the money an electricity supplier receives in February.

Date	Amount (£)	Date	Amount (£)
5th	153 000	13th	468 000
7th	543 000	16th	38 000
9th	223 000	18th	540 000

Draw a graph or chart to represent this information. Choose a sensible scale.

3 Test ready

Summary of key points

To revise for the test:

- Read each key point, find a question on it in the mastery lesson, and check you can work out the answer.

- If you cannot, try some other questions from the mastery lesson or ask for help.

Key points

1 A **grouped frequency** table contains data sorted into groups called classes. → 3.1

2 **Discrete data** can only have particular values. For example, shoe sizes are usually whole numbers. For discrete data you can write groups like 1–5, 6–10. → 3.1

3 **Continuous data** is measured and can have any value, for example length or time. Write inequalities for the groups, with no gaps between them.

4 A **data collection sheet** is a table to record data as you collect it. → 3.1

5 A suitable number of classes for a grouped frequency table is four to six. The classes should be of equal width. → 3.1

6 A **distance chart** shows the distances between several places. → 3.2

7 A **two-way table** divides data into rows across the table and columns down the table. Calculate the totals across and down. → 3.2

8 A **dual bar chart** has bars side by side for each category. Dual bar charts are also called **comparative bar charts** because you can easily compare the data for each category. → 3.3

9 A multiple or **composite bar chart** compares features within a single bar. A composite bar is also known as a compound bar. → 3.3

10 A **line graph** is useful for identifying trends in data. The **trend** is the general direction of change. → 3.3

11 A **histogram** is a type of frequency diagram used for grouped continuous data. There are no gaps between the bars. → 3.3

12 To draw a **frequency polygon**, you can join the midpoints of the tops of the bars in a histogram with straight lines. → 3.3

13 A **time series** graph is a line graph with time plotted on the horizontal axis. → 3.4

14 A **stem and leaf diagram** shows numerical data split into a 'stem' and 'leaves'. In a stem and leaf diagram the numbers are placed in order. → 3.5

15 A **back-to-back stem and leaf diagram** compares two sets of data. → 3.5

16 A **pie chart** is a circle divided into sectors. Each sector represents a set of data. → 3.6

17 A **scatter graph** shows the relationship between two sets of data. Plot the points with crosses. Do not join them up. → 3.7

18 The relationship between the sets of data is called **correlation**. → 3.7

19 An **outlier** is a value that does not fit the pattern of the data. → 3.7

20 Correlation shows that there may be a link between two events. Correlation does not show that one event caused the other. → 3.7

21 A **line of best fit** is a straight line drawn through the middle of the points on a scatter graph.
    It should pass as near to as many points as possible and represent the trend of the points. → **3.8**

22 Using a line of best fit to predict data values within the range of the data given is called
    interpolation and is usually reasonably accurate.
    Using a line of best fit to predict data values outside the range of the data given is called
    extrapolation and may not be accurate.   → **3.8**

Sample student answers

Exam-style question

The temperature of a swimming pool is recorded over an 8-hour period.

Time	09:00	11:00	13:00	15:00	17:00
Temperature (°C)	26	27	28	27	28

Draw a suitable graph to show this information.   **(2 marks)**

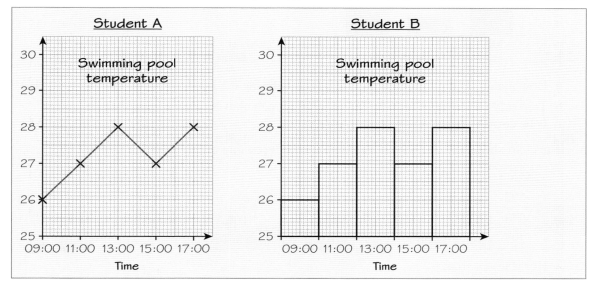

a Which student has drawn the more suitable graph?
   Explain why it is the more suitable type of graph for this data.

b Look at the graph that you chose in part **a**. Do you think this answer would gain full marks?
   Explain your answer.

3 Unit test

1 Here is part of a bus timetable.

Max gets to the bus stop in Banbury at 07:33.

| **Banbury** | 07:42 | 08:30 | 09:14 |
| **Oxford** | 09:00 | 09:45 | 10:06 |

   a Work out how many minutes he has to wait until 07:42. **(1 mark)**

   b Work out how long it will take the 07:42 bus to get to Oxford. **(2 marks)**

2 The bar chart shows how many students go to a French club.

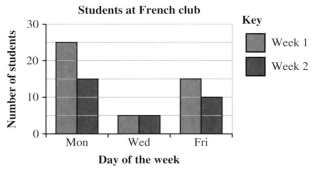

   a How many students attended the French club during the two weeks? **(2 marks)**

   b Tom said, 'The same students attended the French club on both Wednesdays.' Is he correct? You must justify your answer. **(1 mark)**

3 The chart shows the distances, in kilometres, between four cities.

Caen			
692	**Lyon**		
573	457	**Metz**	
295	662	705	**Nantes**

   a How far is it from

     i Caen to Metz   **ii** Lyon to Nantes? **(2 marks)**

   b Which two cities are the least distance apart? **(1 mark)**

4 This frequency table shows the lengths of grubs found in a garden.

Length, l (mm)	$0 < l \leqslant 5$	$5 < l \leqslant 10$	$10 < l \leqslant 15$	$15 < l \leqslant 20$	$20 < l \leqslant 25$
Frequency	3	7	14	4	1

   a How many grubs were found in total? **(1 mark)**

   b How many were greater than 15 mm in length? **(1 mark)**

   c Draw a frequency polygon for this data. **(2 marks)**

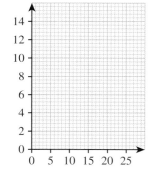

5 The incomplete table and composite bar chart show the percentage population in three age ranges, for two countries.

	Under 15	15 to 64	65 and over
UK	20%		
Mexico	36%	60%	4%

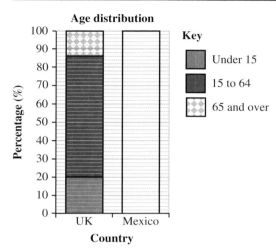

Age distribution

Key

Under 15

15 to 64

65 and over

a Use the information from the composite bar chart to complete the table. **(3 marks)**

b Use the information from the table to complete the Mexico bar on the chart. **(3 marks)**

6 A patient's temperature is recorded over an eight-hour period.

Time	6 am	8 am	10 am	12 noon
Temperature (°C)	38.4	38.3	38.9	37.3

a Draw a suitable graph to represent this data. **(2 marks)**

b Estimate the patient's temperature at 9 am. **(1 mark)**

c Describe the trend shown in the graph. **(1 mark)**

7 Amy asked each of 60 people to name their favourite pet.
Here are her results.

Pet	Frequency
cat	32
dog	18
rabbit	10

Draw an accurate pie chart for her results. **(3 marks)**

8 Joan records her dressage test scores for her two horses, Jigsaw and Percy.

Jigsaw	57	63	51	68	50	49	66	70
Percy	55	62	60	81	59	59	74	73

a Draw a back-to-back stem and leaf diagram for this data. **(3 marks)**

b Compare the dressage scores for the two horses. **(1 mark)**

9   120 students went on a school activities day.
They could go bowling, skating or to the cinema.
66 of the students were girls.
28 of the girls went bowling.
36 students went to the cinema.
20 of the students who went to the cinema were girls.
15 boys went skating.
Work out the number of students who went bowling.   **(3 marks)**

10   Lee has information about people's heights and the distances they can jump.
He is asked to draw a scatter graph and a line of best fit for this information.
Here is his answer.

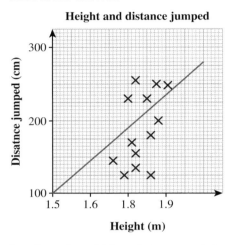

Height and distance jumped

Lee has plotted his points accurately.
Write down two things that are wrong with his answer.   **(2 marks)**

(TOTAL: 35 marks)

11 Challenge Look back at the questions where you drew a graph or chart.
Is there another type of graph you could draw for this data?
Draw it and explain which best represents the data.

12 Reflect The mathematics studied in this unit is called 'statistics'.
Choose **A**, **B** or **C** to complete each statement about statistics.

In this unit, I did…	**A** well	**B** OK	**C** not very well
I think statistics is…	**A** easy	**B** OK	**C** hard
When I think about doing statistics I feel…	**A** confident	**B** OK	**C** unsure

Did you answer mostly **A**s and **B**s? Are you surprised by how you feel about statistics? Why?
Did you answer mostly **C**s? Find the three questions in this unit that you found the hardest.
Ask someone to explain them to you. Then complete the statements above again.

4 Fractions and percentages

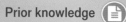

Prior knowledge

4.1 Working with fractions

- Compare fractions.
- Add and subtract fractions.
- Use fractions to solve problems.

Active Learn
Homework

Warm up

1 **Fluency** What is

    a the denominator of $\frac{2}{9}$ **b** the numerator of $\frac{5}{6}$?

2 Write $<$ or $>$ between each pair of fractions.

    a $\frac{3}{16}\square\frac{1}{16}$ **b** $\frac{4}{7}\square\frac{6}{7}$

3 Copy and complete these equivalent fractions.

    a $\frac{2}{3}=\frac{\square}{6}$ (×□ and ×2)   **b** $\frac{4}{9}=\frac{\square}{27}$ **c** $\frac{2}{5}=\frac{8}{\square}$ **d** $\frac{2}{3}=\frac{\square}{6}=\frac{8}{\square}$

4 Write each fraction in its simplest form.

    a $\frac{30}{40}=\frac{\square}{4}$ (÷10)   **b** $\frac{12}{18}$ **c** $\frac{9}{15}$ **d** $\frac{16}{24}$

5 Find the LCM (lowest common multiple) of these numbers.

    a 2 and 4   **b** 5 and 10   **c** 6 and 8   **d** 2, 3 and 4

6 Write each pair of fractions with a common **denominator**.

    a $\frac{1}{2}$ and $\frac{3}{4}$ **b** $\frac{7}{10}$ and $\frac{3}{5}$ **c** $\frac{1}{8}$ and $\frac{5}{6}$

> **Q6a hint**
>
> Use the LCM of the denominators as the common denominator.

Key point

To compare fractions, write them with a common denominator.

---

7 **Reasoning** Is $\frac{4}{9}>\frac{1}{3}$?

Show your working to explain your answer.

8 **Reasoning** Here are two fractions: $\frac{3}{5}$ and $\frac{2}{3}$. Explain which is the larger fraction.

9 **Problem-solving** Marcia has two spinners, A and B.
For spinner A, $P(\text{red}) = \frac{5}{8}$ For spinner B, $P(\text{red}) = \frac{7}{12}$
Which spinner is more likely to land on red?

Q9 hint

Write the fractions with a common denominator. Which is larger?

10 Here are some fractions: $\frac{12}{15}$ $\frac{16}{20}$ $\frac{4}{5}$ $\frac{20}{25}$ $\frac{8}{40}$

Which of these fractions is **not** equivalent to $\frac{24}{30}$? Show your working.

Example

Write $\frac{4}{5}, \frac{5}{8}$ and $\frac{3}{4}$ in order of size. Start with the smallest.

The LCM of 5, 8 and 4 is 40.

$$\frac{4}{5} = \frac{32}{40} \qquad \frac{5}{8} = \frac{25}{40} \qquad \frac{3}{4} = \frac{30}{40}$$

Write the fractions with a common denominator.

Answer: $\frac{5}{8}, \frac{3}{4}, \frac{4}{5}$

Write the original fractions in order.

11 Write these fractions in order of size. Start with the smallest fraction.
a $\frac{7}{12}, \frac{5}{6}, \frac{2}{3}$
b $\frac{2}{3}, \frac{3}{5}, \frac{3}{4}, \frac{1}{2}, \frac{1}{5}$

12 **Reasoning** Which of these fractions are greater than $\frac{1}{2}$?

$\frac{2}{5}$ $\frac{4}{7}$ $\frac{3}{8}$ $\frac{7}{10}$ $\frac{5}{11}$

Q12 hint

Is the numerator more than $\frac{1}{2}$ the denominator?

Exam-style question

13 Here are four fractions.

$\frac{7}{12}$ $\frac{1}{2}$ $\frac{3}{7}$ $\frac{5}{6}$

Write the fractions in order of size.
Start with the smallest fraction. **(2 marks)**

Exam tip

You can use a mix of strategies to answer questions. Here, you can consider whether the fractions are smaller or larger than $\frac{1}{2}$, then use equivalent fractions.

14 **Reasoning a** Write two fractions with the same numerator. Explain how you can tell which one is smaller.

b Explain which is larger, $\frac{7}{8}$ or $\frac{8}{9}$.

Key point

To add or subtract fractions, write them with a common denominator.

Example

Work out $\frac{2}{3} + \frac{1}{9}$.

$$\frac{2}{3} + \frac{1}{9} = \frac{6}{9} + \frac{1}{9} = \frac{7}{9}$$

The LCM of 3 and 9 is 9. Write the fractions with denominator 9 then add.

15 Work out these calculations. Simplify your answers if possible.

a $\frac{1}{2}+\frac{3}{8}$ **b** $\frac{3}{5}+\frac{2}{15}$ **c** $\frac{3}{4}-\frac{1}{2}$ **d** $\frac{2}{3}-\frac{1}{9}$

e $\frac{5}{8}-\frac{2}{16}$ **f** $\frac{1}{3}+\frac{3}{9}$ **g** $\frac{2}{5}-\frac{1}{15}$ **h** $\frac{17}{20}-\frac{1}{4}$

16 Work out

a $\frac{1}{4}+\frac{1}{3}$ **b** $\frac{1}{4}+\frac{2}{3}$ **c** $\frac{1}{2}-\frac{1}{5}$

d $\frac{1}{2}-\frac{2}{5}$ **e** $\frac{3}{4}-\frac{2}{3}$ **f** $\frac{3}{4}-\frac{2}{5}$

> **Q16a hint**
>
> Change both fractions so they have the same denominator. Use the LCM.

17 **Reflect** Look back at **Q15** and **Q16**. Write a sentence explaining when you have to change only one denominator, and when you have to change both denominators.

18 Work out these calculations. Simplify your answers where necessary.

a $\frac{1}{4}+\frac{1}{3}-\frac{1}{12}$ **b** $\frac{1}{4}+\frac{2}{3}-\frac{1}{6}$ **c** $\frac{3}{4}-\frac{2}{3}+\frac{1}{2}$ **d** $\frac{1}{2}-\frac{1}{5}+\frac{1}{10}$

19 Work these out. The first one is started for you.

a $1-\frac{1}{3}=\frac{3}{3}-\frac{1}{3}=$ **b** $1-\frac{3}{5}$ **c** $1-\frac{5}{7}$ **d** $1-\frac{7}{10}$

20 **Reasoning** A group of students went to a restaurant.
$\frac{1}{5}$ of them bought a chicken burger and $\frac{1}{2}$ of them bought a beef burger.
The remainder only bought a drink.
What fraction of the group bought a type of burger?

21 Work these out.

a $2-\frac{1}{4}$ **b** $2-\frac{3}{4}$ **c** $3-\frac{1}{4}$

> **Q21a hint**
>
> $2=\frac{\square}{4}$

Key point

A **unit fraction** has **numerator** 1. $\dfrac{1}{\square}$

---

22 **Problem-solving** The ancient Egyptians only used unit fractions.

For $\frac{3}{4}$ they wrote $\frac{1}{2}+\frac{1}{4}$.

For what fraction did they write $\frac{1}{3}+\frac{1}{5}$?

23 Work these out. Give each answer in its simplest form.

a $\frac{5}{12}+\frac{1}{8}$ **b** $\frac{3}{4}+\frac{1}{6}$ **c** $\frac{5}{6}-\frac{1}{4}$ **d** $\frac{3}{4}-\frac{1}{16}$

24 **Problem-solving** Work out the missing fractions.

a $\frac{3}{7}+\square=1$ **b** $\frac{3}{7}+\square=\frac{29}{35}$ **c** $\square-\frac{3}{7}=\frac{1}{63}$ **d** $\frac{3}{7}-\square=\frac{9}{28}$

4.2 Operations with fractions

- Find a fraction of a quantity or measurement.
- Use fractions to solve problems.
- Use bar models to help you solve problems.

Active Learn
Homework

Warm up

1 Fluency How many

  a cm in 1 m?   **b** g in 1 kg?   **c** minutes in 1 hour?

2 Work these out. Simplify your answers if possible.

  a $\frac{1}{8} + \frac{3}{8}$ **b** $\frac{1}{4} + \frac{3}{8}$ **c** $\frac{5}{12} - \frac{3}{8}$

3 Change these improper fractions to mixed numbers.

  a $\frac{7}{2}$ **b** $\frac{11}{9}$ **c** $\frac{50}{6}$ **d** $\frac{25}{12}$

4 Change these mixed numbers into improper fractions.

  a $2\frac{1}{2}$ **b** $4\frac{3}{4}$ **c** $7\frac{1}{6}$ **d** $10\frac{1}{2}$

Example

Work out $\frac{3}{5}$ of 40.   In mathematics, 'of' means multiply.

$$\times 3 \left(\begin{array}{l} \frac{1}{5} \text{ of } 40 = \frac{1}{5} \times 40 = \frac{40}{5} = 40 \div 5 = 8 \\ \frac{3}{5} \text{ of } 40 = 3 \times 8 = 24 \end{array} \right) \times 3$$

Multiply by 3 to find $\frac{3}{5}$.

5 Work out

  a $\frac{1}{9}$ of 18   **b** $\frac{2}{9}$ of 18   **c** $\frac{1}{8}$ of 32   **d** $\frac{7}{8}$ of 32   **e** $\frac{3}{7}$ of 35   **f** $\frac{4}{5}$ of 80

6 Work out

  a $\frac{3}{4}$ of 200 kg   **b** $\frac{4}{9}$ of 90 cm   **c** $\frac{5}{6}$ of 120 cm. Give your answer in metres.

  d $\frac{2}{3}$ of 1500 g. Give your answer in kg.   **e** $\frac{3}{5}$ of 1 hour. Give your answer in minutes.

7 Problem-solving A test has 60 marks. Monty gets $\frac{3}{4}$ of the marks.
How many marks does he get?

8 Problem-solving A parking meter takes £1, £2 and 50p coins.
At the end of the day, there are 96 coins in the meter.
$\frac{5}{12}$ of them are £2 coins. 26 of the coins are £1 coins.
The rest of the coins are 50p coins.
How much money did the parking meter collect that day?

> **Q8 hint**
>
> Read one sentence at a time.
> If you can, write a calculation.
> So you might begin by working
> out how many £2 coins there are.

$\frac{3}{10}$ of a number is 15. Find the number.

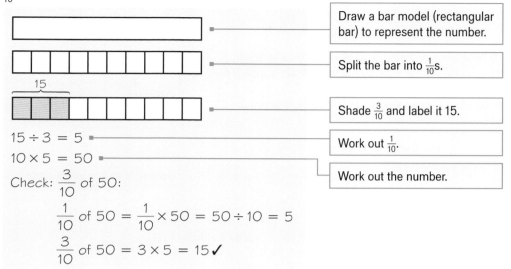

Draw a bar model (rectangular bar) to represent the number.

Split the bar into $\frac{1}{10}$s.

Shade $\frac{3}{10}$ and label it 15.

Work out $\frac{1}{10}$.

Work out the number.

$15 \div 3 = 5$

$10 \times 5 = 50$

Check: $\frac{3}{10}$ of 50:

$\frac{1}{10}$ of 50 $= \frac{1}{10} \times 50 = 50 \div 10 = 5$

$\frac{3}{10}$ of 50 $= 3 \times 5 = 15 ✓$

Exam-style question

9 $\frac{3}{5}$ of a number is 12. Find the number. **(2 marks)**

Exam tip

You can draw bar models to help you to answer exam questions.

10 **Future skills** Alicia and Gaby sell CDs on an internet auction site.
They split the money so that Alicia gets $\frac{2}{3}$ and Gaby gets the rest. They get £90 in total.

a Draw a bar model to represent the situation. **b** How much does Gaby get?

c Write working to check your answer.

Exam-style question

11 There are 700 students in a college.
All of the students are 16 years old, 17 years old, or 18 years old.
$\frac{1}{10}$ of the students are 16 years old. $\frac{1}{5}$ of the students are 18 years old.
Work out how many of the students are 17 years old.

(4 marks)

Exam tip

Check your answer gives a correct total number of students.

Animal	Frequency
hens	30
sheep	15
cows	5
goats	10

12 **Reasoning** The frequency table shows the numbers of animals on a farm. Leandra says, 'In a pie chart, the goats will be represented by $\frac{1}{4}$ of its area.' Explain why Leandra is wrong.

13 **Problem-solving** 600 children were asked to choose their favourite sport. This pie chart shows the results.

a How many children chose swimming?

b Estimate the number of children who chose football.

Favourite sports

Q13b hint

Estimate the fraction for football.

14 Work these out. Write your answers as mixed numbers.

a $\dfrac{2}{3}+\dfrac{3}{4}=\dfrac{\square}{12}+\dfrac{\square}{12}=\dfrac{\square}{12}=1\dfrac{\square}{12}$

b $\dfrac{4}{5}+\dfrac{3}{8}$ **c** $\dfrac{3}{5}+\dfrac{4}{7}$ **d** $\dfrac{5}{8}+\dfrac{5}{6}$

15 **Reflect** Check your answers to **Q14** using a calculator.
Write an example of how your calculator shows a mixed number.

16 **Problem-solving / Reasoning** Here are two fractions:

$\dfrac{7}{11}$ $\dfrac{11}{7}$

Work out which of the fractions is closer to 1.
You must show your working.

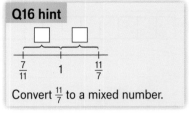

Q16 hint

Convert $\dfrac{11}{7}$ to a mixed number.

Key point

To add or subtract mixed numbers, convert them to improper fractions first.

Example

Work out $1\dfrac{1}{2}+2\dfrac{3}{4}$.

$1\dfrac{1}{2}+2\dfrac{3}{4}=\dfrac{3}{2}+\dfrac{11}{4}$ Convert to improper fractions.

$=\dfrac{6}{4}+\dfrac{11}{4}=\dfrac{17}{4}$ Add the fractions.

$=4\dfrac{1}{4}$ Convert back to a mixed number.

17 Work out

a $2\dfrac{1}{4}+3\dfrac{1}{2}$ **b** $7\dfrac{1}{2}+5\dfrac{3}{4}$ **c** $8\dfrac{1}{2}+3\dfrac{1}{4}$ **d** $2\dfrac{2}{3}+3\dfrac{1}{6}$

e $1\dfrac{1}{10}+2\dfrac{1}{4}$ **f** $3\dfrac{3}{4}+3\dfrac{1}{3}$ **g** $1\dfrac{1}{4}+2\dfrac{1}{6}$ **h** $3\dfrac{3}{10}+1\dfrac{2}{15}$

18 **Reasoning** Crista says, 'To answer **Q17**, I did not convert to improper fractions. I added the integers (whole numbers) together. Then I added the fractions together. Then I worked out the total.'
Use Crista's method to work out three calculations in **Q17**. Does her method work?

19 Work out

a $10\dfrac{1}{2}-5\dfrac{3}{4}$ **b** $6\dfrac{1}{10}-4\dfrac{3}{5}$ **c** $8\dfrac{2}{3}-6\dfrac{1}{6}$

d $3\dfrac{3}{4}-2\dfrac{1}{3}$ **e** $11\dfrac{1}{2}-8\dfrac{2}{5}$ **f** $3\dfrac{1}{10}-1\dfrac{3}{4}$

g $4-2\dfrac{4}{7}$ **h** $7-3\dfrac{3}{4}-2\dfrac{1}{2}$ **i** $9-1\dfrac{1}{2}-3\dfrac{2}{3}$

Q19g hint

$4=\dfrac{\square}{7}$

j **Reflect** Does Crista's method (in **Q18**) work for subtracting mixed numbers? Show working to explain.

20 **Problem-solving** In this diagram, the number in each box is the sum of the two numbers below it.
Find the missing numbers.

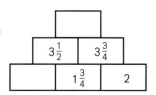

4.3 Multiplying fractions

Active Learn
Homework

- Multiply whole numbers, fractions and mixed numbers.
- Simplify calculations by cancelling.

Warm up

1 **Fluency** Which of these give the same answer?

a $\frac{1}{3}$ of 18 **b** $\frac{2}{3}$ of 18 **c** $\frac{1}{3}(6+12)$

d $\frac{4\times6}{2}$ **e** 3×18 **f** 18×3

2 Convert **a** $\frac{15}{2}$ to a mixed number **b** $3\frac{3}{5}$ to an improper fraction

3 Work these out. Write your answers as mixed numbers where necessary.
The first two are started for you.

a $7\times\frac{1}{2}=\frac{7\times1}{2}=\frac{\square}{2}=\square\frac{1}{2}$ **b** $18\times\frac{2}{3}=\frac{\square}{3}=\square$ **c** $2\times\frac{5}{2}$ **d** $\frac{5}{3}\times6$

4 **Future skills** Scott has to complete 5 tasks. Each task takes $\frac{2}{3}$ of an hour to complete.
Scott starts the tasks at 9 am.
Will he finish by 12 midday? You must show working.

5 **Reasoning** Ali is selling his bike for £460. He reduces the price by $\frac{1}{4}$.

a What is the new price?

b Ali works out the answer using this calculation: $\frac{3}{4}\times460$. Does he get the correct answer?

6 Use the priority of operations to work out the value of these expressions when $x=3$ and $y=6$.
a $\frac{1}{2}(x+5)$ **b** $\frac{1}{4}(y+10)$ **c** $\frac{1}{3}(x+y)$ **d** $\frac{1}{2}(y-x)$

Key point

To multiply fractions, multiply the numerators and multiply the denominators.

Example

Work out $\frac{2}{3}\times\frac{1}{5}$

$\frac{2}{3}\times\frac{1}{5}=\frac{2\times1}{3\times5}=\frac{2}{15}$ ◄—— Multiply the numerators and the denominators.

7 Copy and complete.

a $\frac{1}{2}\times\frac{1}{4}=\frac{1\times1}{2\times4}=\frac{\square}{\square}$ **b** $\frac{2}{5}\times\frac{3}{7}=\frac{2\times3}{5\times7}=\frac{\square}{\square}$

8 Work these out. Write answers in their simplest form where necessary.

a $\frac{1}{7}\times\frac{6}{7}$ **b** $\frac{2}{7}\times\frac{4}{9}$ **c** $\frac{3}{10}\times\frac{2}{3}$ **d** $\frac{5}{6}\times\frac{3}{5}$

9 **Reflect** Does the rule 'multiply the numerators and multiply the denominators' work for $7 \times \frac{3}{2}$ and $\frac{2}{5} \times 6$?
Write a sentence to explain.

Key point

Numerators can be cancelled with denominators if they are divisible by the same number.

Example

Work out $\frac{5}{6} \times \frac{2}{7}$

$$\frac{5}{6} \times \frac{2}{7} = \frac{5 \times 2}{6 \times 7}$$

Look for numbers in the numerator and the denominator with a common factor. 2 is a factor of 2 and 6.

$$= \frac{\cancel{2} \times 5}{\cancel{6}^3 \times 7} = \frac{1 \times 5}{3 \times 7}$$

Change the order to give $\frac{2}{6}$, which simplifies to $\frac{1}{3}$.

$$= \frac{5}{21}$$

10 Work out by cancelling numerators and denominators.

  a $\frac{4}{5} \times \frac{1}{2}$ **b** $\frac{7}{9} \times \frac{3}{10}$ **c** $\frac{5}{12} \times \frac{6}{7}$

  d $\frac{8}{9} \times \frac{1}{4}$ **e** $\frac{10}{21} \times \frac{7}{15}$ **f** $\frac{13}{16} \times \frac{4}{7}$

  g **Reflect** Which is easier – rearrange and simplify or simplify at the end? Explain why.

11 In a cycling club, $\frac{1}{4}$ of the members are female.
Of these, $\frac{2}{3}$ are under 16.
What fraction of the cycling club are female and under 16?

12 **Problem-solving** Garcia buys $\frac{3}{4}$ kg of cheese. He gives his friend $\frac{1}{3}$ of it.
Garcia needs 600 g of cheese for a recipe. Does he have enough?

13 Work out

  a $3 \times 1\frac{3}{4}$ **b** $6 \times 2\frac{1}{7}$

  c $2\frac{1}{2} \times 7$ **d** $3\frac{1}{5} \times 11$

14 **a** Check your answers to **Q13** using a calculator.

  b **Reflect** Describe how you enter mixed numbers on your calculator.

15 Uzma's paper round takes her $1\frac{1}{2}$ hours each day.
How long does she spend on her paper round each week (Monday to Friday)?

Exam-style question

16 Sam's weekday rate of pay is £15 per hour.
When Sam works on weekends she is paid $1\frac{1}{5}$ times her weekday rate of pay.
Last weekend Sam worked for 10 hours.
Work out the total amount of money Sam earned last weekend. **(4 marks)**

4.4 Dividing fractions

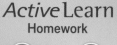
*Active*Learn
Homework

- Divide a whole number by a fraction.
- Divide a fraction by a whole number or a fraction.

Warm up

1 Fluency How many

   a twos in 8   **b** halves in 1 whole   **c** thirds in 1 whole?

2 Write

   a $3\frac{3}{4}$ as an improper fraction   **b** $\frac{19}{3}$ as a mixed number

3 Work out

   a $\frac{1}{5}$ of 60   **b** $\frac{2}{5}$ of 60   **c** $\frac{2}{3}\times\frac{9}{11}$ **d** $5\times\frac{2}{3}$

Key point

The **reciprocal** of a fraction is the 'upside down' fraction. The reciprocal of 2 (or $\frac{2}{1}$) is $\frac{1}{2}$.

4 Write down the reciprocal of

   a $\frac{3}{4}$ **b** $\frac{7}{2}$ **c** $\frac{1}{5}$ **d** 4

   e Reflect Multiply $\frac{3}{4}, \frac{7}{2}, \frac{1}{5}$ and 4 by their reciprocals.
     What do you notice when you multiply a number by its reciprocal?

Key point

Dividing by a number is the same as multiplying by its reciprocal. $12\div 2$ means $\frac{1}{2}$ of 12 or $12\times\frac{1}{2}$

Example

Work out $6\div\frac{2}{3}$

$$6\div\frac{2}{3} = 6\times\frac{3}{2}$$ Change to multiplication by the reciprocal.

$$= \frac{6\times 3}{1\times 2} = \frac{\overset{3}{\cancel{6}}\times 3}{\underset{1}{\cancel{2}}\times 1} = 9$$ Give the answer in its simplest form.

5 Work out

   a $4\div\frac{1}{3} = 4\times\square = \square$ **b** $5\div\frac{1}{5}$ **c** $12\div\frac{1}{8}$ **d** $4\div\frac{3}{4}$

6 Problem-solving How many $\frac{3}{4}$-litre cups can you fill from a 2-litre jug?

7 Work out these divisions. Write your answer as a mixed number if necessary.

   a $5\div\frac{3}{2} = 5\times\frac{2}{\square} =$ **b** $12\div\frac{4}{3}$ **c** $14\div\frac{11}{5}$ **d** $20\div\frac{9}{4}$

8   For each calculation write the mixed number as an improper fraction then work out the division. Write your answer as a mixed number if necessary.

   a $7 \div 2\frac{1}{2} = 7 \div \frac{\square}{2} =$ **b** $10 \div 2\frac{2}{5}$ **c** $8 \div 5\frac{1}{3}$ **d** $12 \div 2\frac{3}{4}$

9   **Reasoning** Work out the missing integers or mixed numbers.

   a $\square \times \frac{2}{3} = 4$

   b $\square \times \frac{7}{3} = 5$

   c $\square \times 1\frac{1}{2} = 8$

   d $\square \div \frac{3}{10} = 6$

   e $\square \div \frac{4}{5} = 8$

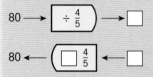

Q9a hint

$\square \longrightarrow \boxed{\times \frac{2}{3}} \longrightarrow 4$

$\square \longleftarrow \boxed{\div \frac{2}{3}} \longleftarrow 4$

Q9d hint

$\square \longrightarrow \boxed{\div \frac{3}{10}} \longrightarrow 6$

$\square \longleftarrow \boxed{\square \frac{3}{10}} \longleftarrow 6$

Exam-style question

10   Gunnar buys 80 m of fabric.
   He needs $\frac{4}{5}$ of a metre of fabric to make a cushion.
   Gunnar wants to make 30 cushions per week.
   Has Gunnar bought enough fabric to last 4 weeks?
   You must show how you get your answer.   **(3 marks)**

Exam tip

Check your answer using the inverse operation.

$80 \longrightarrow \boxed{\div \frac{4}{5}} \longrightarrow \square$

$80 \longleftarrow \boxed{\square \frac{4}{5}} \longleftarrow \square$

Exam-style question

11   Sally has to work out the exact value of $30 \div \frac{1}{3}$.
   She writes $30 \div \frac{1}{3} = 10$.
   Sally's reason is, 'There are 3 thirds in 1, so there are 10 thirds in 30.'
   Explain what is wrong with Sally's reason.   **(1 mark)**

Exam tip

Show the correct answer to the calculation $30 \div \frac{1}{3}$. Then write a sentence explaining what is wrong with Sally's reason.

---

12   Work these out. The first one is started for you.

   a $\frac{1}{3} \div 2 = \frac{1}{3} \times \frac{1}{2} =$ **b** $\frac{2}{3} \div 6$ **c** $\frac{6}{5} \div 6$

13   For each calculation, write the mixed number as an improper fraction.
   Then work out the division.

   a $1\frac{1}{4} \div 2$ **b** $2\frac{1}{3} \div 5$ **c** $3\frac{3}{4} \div 8$

   14   Check your answers to **Q13** using a calculator.

15   **Problem-solving** Kirsten has to walk $2\frac{3}{4}$ km to school.

   How far has she walked when she is halfway?
   Give your answer in metres. Check your answer using an inverse operation.

16   **Future skills** Sharnia takes $2\frac{4}{5}$ of an hour to do 7 tasks. Each task takes the same time.

   a What fraction of an hour does 1 task take?

   b How long is this in minutes?

17   Work out these divisions. Write your answer as a mixed number if necessary.

   a $\frac{1}{2} \div \frac{1}{3} = \frac{1}{2} \times 3 = \frac{\square}{\square} = 1\frac{\square}{\square}$ **b** $\frac{3}{4} \div \frac{1}{5}$ **c** $\frac{7}{10} \div \frac{1}{2}$

4.5 Fractions and decimals

- Convert fractions to decimals and vice versa.
- Use decimals to find quantities.
- Work out divisions with decimal answers.
- Write one number as a fraction of another.

Active Learn
Homework

Warm up

1 **Fluency** Which of these numbers have the same value?

7.07   7.00   7   7.77   7.000 000

2 Write

a $\frac{3}{10}$ as a decimal   **b** 0.4 as a fraction   **c** $\frac{48}{100}$ in its simplest form

3 Here are four decimals.   0.39   0.4   0.349   −0.43

a Write these decimals in order. Start with the smallest.

b Which decimal is closest to 0.5?

4 Work out

a $4\overline{)5.00}$ **b** $6\overline{)25.00}$ as a decimal   **c** 3×0.2 **d** 6×0.25

5 Write

a 5 as a fraction of 12   **b** 6 as a fraction of 9 in its simplest form

Key point

You can use short or long division to convert fractions to decimals. For example, $\frac{3}{5}$ means $3 \div 5$.

Example

Write $\frac{7}{8}$ as a decimal.

$$\frac{7}{8} = 8\overline{)7} = 8\overline{)7.^{7}0^{6}0^{4}0}^{\;0.\;8\;7\;5}$$

Rewrite as written division.

Keep putting zeros at the end of the decimal until there is no remainder.

$$\frac{7}{8} = 0.875$$

6 Use division to write $\frac{1}{8}$ as a decimal.

Exam tip

Think carefully about which number is the divisor and which is the dividend.

divisor   dividend

$$\Box\,\overline{)\Box}.0\,0$$

Exam-style question

7 Write $\frac{7}{25}$ as a decimal.   **(1 mark)**

8 Copy and complete this table.

Fraction	$\frac{1}{1000}$	$\frac{1}{100}$	$\frac{1}{10}$	$\frac{1}{8}$	$\frac{1}{5}$	$\frac{1}{4}$
Decimal			0.1			0.25

Unit 4 Fractions and percentages   109

9   Write these fractions as decimals. Use your table from **Q8** to help you.

a $\frac{2}{5} = 2 \times \square$ b $\frac{3}{8}$ c $\frac{5}{4}$

10 Work these out. The first one is started for you.

a $\frac{1}{4}$ of 10 $= 0.25 \times 10 =$

b $\frac{3}{8}$ of 100   c $\frac{1}{100}$ of 50

Q10 hint

Converting a fraction to a decimal can make some calculations easier.

11 Convert these fractions to decimals to work out which is closest to $\frac{1}{2}$.

$\frac{5}{8}$ $\frac{7}{10}$ $\frac{49}{100}$

Example

Write 0.723 as a fraction.

$$0.723 = \frac{723}{1000}$$

The smallest place value is 3 thousandths, so use denominator 1000.

12 Write these decimals as fractions in their simplest form.

a 0.61   b 0.78   c 0.229   d 0.450   e 0.096

13 **Problem-solving** Write these in order of size.
Start with the smallest and write the original values in order.

a $\frac{2}{3}$, 0.6, $\frac{5}{8}$, 0.628   b $\frac{5}{2}$, $-\frac{17}{4}$, -4.5, 2.8

Q13 hint

Convert the fractions into decimals.

14 **Future skills** Ed and Sam share a £225 car repair bill. Ed pays $\frac{3}{8}$ and Sam pays $\frac{5}{8}$.

a How much does each pay?

b Will the amounts in your answers to part **a** pay the whole bill?

c **Reflect** Should you round up or down when sharing a bill?

15 Here are four fractions.

$\frac{2}{3}$ $\frac{7}{12}$ $\frac{8}{15}$ $\frac{17}{25}$

Write the fractions in order of size.
Start with the smallest fraction.

Q15 hint

Use division on your calculator to convert the fractions to decimals.

Key point

You can write one quantity as a fraction of another.

16 **Reasoning** Zule draws a pie chart to display the meals bought in the school canteen.
Show that the fraction of the meals that are lasagne is $\frac{5}{12}$.

School meals bought

Q16 hint

$\frac{\square}{360}$

17 **Reasoning** Kim says, '18 minutes as a fraction of an hour is 0.18 hours.'
Write 18 minutes as a fraction of an hour in its simplest form to show that Kim is wrong.

4.6 Fractions and percentages

- Convert percentages to fractions and vice versa.
- Write one number as a percentage of another.

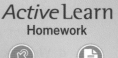

Active Learn
Homework

Warm up

1 **Fluency** How many **a** pence in £1 **b** ml in 1 litre?

2 Use a calculator to convert $\frac{7}{15}$ to a decimal. Round your answer to 1 decimal place.

3 Write these decimals and fractions as percentages.

a 0.35 **b** 0.2 **c** $\frac{3}{10}$ **d** $\frac{1}{4}$

4 Work out

a $\frac{8}{20} \times 100$ **b** $\frac{28}{40} \times 100$

> **Q4 hint**
>
> Cancel and then work out the answer.

5 Write as a fraction in its simplest form. The first one has been started for you.

a $8\% = \frac{8}{100} = \frac{\square}{\square}$ **b** 24% **c** 65% **d** 96%

6 35% of students are driven to school. Write this percentage as a fraction in its simplest form.

7 **Reasoning** Vakita says that 30% is $\frac{1}{3}$. Show that Vakita is wrong.

8 Write these fractions as percentages.
The first one has been started for you.

a $\frac{7}{50} = \frac{\square}{100} = \square\%$ **b** $\frac{3}{25}$ **c** $\frac{4}{5}$ **d** $\frac{18}{200}$

> **Q8d hint**
>
>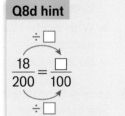
>
> $$\frac{18}{200} = \frac{\square}{100}$$

9 By comparing percentages, show that 16% is larger than $\frac{3}{20}$

Key point

To convert a fraction to a percentage, you can convert the fraction to one with the denominator 100, or you can multiply by 100%.

Example

There are 20 students in a class. 6 are male. What percentage of the class is male?

Method A: $\frac{6}{20} = \frac{30}{100} = 30\%$ ×5

> Write as a fraction then convert to a fraction with denominator 100.

Method B: $\frac{6}{20} \times 100\% = \frac{6 \times {}^5\cancel{100}}{{}^1\cancel{20}}\% = 30\%$

> Or multiply by 100%.

10 A class of 25 primary school children choose a musical instrument to learn.
10 choose the violin. The rest choose the recorder. What percentage choose

    a the violin   **b** the recorder?

    c **Reflect** Add together your answers to **a** and **b**. Write a sentence explaining what you notice.

    d Write a calculation to work out the percentage that choose the recorder, using the percentage that choose the violin.

11 Write

    a 15 as a percentage of 50

    b 24 as a percentage of 60

    c 50 as a percentage of 750

    d 70p as a percentage of £2

    e 500 ml as a percentage of 10 litres

    f **Reflect** Which of **a–e** did you work out by converting the fraction to denominator 100? Which did you work out by multiplying the fraction by 100? Write a sentence explaining how you chose which method to use.

> **Q11b hint**
>
> $\frac{24}{60} \times 100$

> **Q11d, e hint**
>
> Write both amounts in the same units.

12 **Reasoning** Elliot took a maths test and a spelling test.
He scored 35 out of 50 for maths and 48 out of 60 for spelling.
In which test did Elliot score the lower percentage?

13 Harry got 64 out of 80 in a recent test. Jill scored 84% in the same test.
Who achieved the higher score? Show your working.

14 There are 240 passengers on a train. 80 are getting off at the next station.

    a What fraction of the passengers are getting off at the next station?
Give your answer in its simplest form.

    b What percentage of passengers are getting off at the next station?
Round your answer to the nearest whole number.

> **Q14b hint**
>
> Use your calculator to multiply the fraction in part **a** by 100.

15 Lucy is on work experience at a garage.
She draws a bar chart to show the number of cars serviced.

    a What percentage of the cars serviced were Vauxhalls?

    b Lucy says, 'There were half as many Kias and Audis serviced as Vauxhalls.'
Show that Lucy is incorrect.

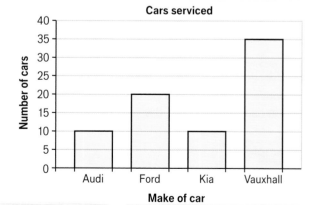

Cars serviced

Exam-style question

16 A factory line produces 260 chocolate bunnies in an hour.
17 are rejected.
What percentage of the chocolate bunnies are not rejected?
Give your answer to 1 decimal place.   **(3 marks)**

> **Exam tip**
>
> Always read the question carefully. Here, you are not working out $\frac{17}{260}$ as a percentage.

4.7 Calculating percentages 1

- Convert percentages to decimals and vice versa.
- Find a percentage of a quantity.
- Use percentages to solve problems.
- Calculate simple interest.

Active Learn
Homework

Warm up

1 **Fluency** What is

a $\frac{1}{4}$ of £160   **b** $\frac{3}{4}$ of £160   **c** 50% of £160   **d** 25% of £160?

e Which two parts **a–d** have the same answer? Explain why.

2 Work out

a $38 \div 100$ **b** 0.71×100 **c** 0.2×100 **d** 0.2×300

e 0.4×300 **f** $6.1 \div 100$ **g** $8.4 \div 100$ **h** $0.5 \div 100$

3 Work out

a $\frac{1}{5} \times 200$ **b** $\frac{3}{10} \times 150$ **c** $1\frac{1}{2} \times 60$ **d** $2\frac{4}{5} \times 80$

4 Write these lengths of time as a decimal number of years.

a 18 months   **b** 39 months

5 Write these percentages as decimals. The first one has been started for you.

a $65\% = \frac{65}{100} = 0.\square\square$ **b** 42%   **c** 7%

6 The diagram shows a fraction–decimal–percentage triangle.

Draw fraction–decimal–percentage triangles for

a $\frac{1}{4}$ **b** 80%   **c** 0.6

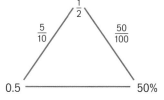

7 Greg draws this function machine for writing percentages as decimals.

percentage $\longrightarrow$ ⟨ $\div 100$ ⟩ $\longrightarrow$ decimal

a Draw a function machine for writing decimals as percentages.

b Use your function machine to write these decimals as percentages.

i 0.99   **ii** 0.09   **iii** 0.9   **iv** 0.009

8 Convert these to decimals to write them in order of size. Start with the smallest.

a 59%, 0.6, 65%, $\frac{1}{2}$, $\frac{11}{20}$ **b** 0.5%, $\frac{1}{100}$, 0.5, $\frac{26}{100}$

9 A student loan has an interest rate of 3.3%.
Write this percentage as a decimal.

10 a Write 20% as a fraction. **b** Work out 20% of 1500.

11 Work out
   a 90% of 400   **b** 70% of 300
   c 30% of 900   **d** 5% of 200

12 Work out 15% of 220 g.

Q12 hint

10% + 5% = 15%

Exam tip

After writing your answer, read the question again to check you have fully answered it. Have you only worked out the percentage or have you worked out the difference too?

Exam-style question

13 It rained on 20% of the days in September.
It rained on 9 days in October.
Work out the difference between the number of rainy days in September and October. **(3 marks)**

14 a Write 20% as a decimal.
   b Work out 20% of 300.

15 Work out these percentages by converting to a decimal first.
   a 60% of 375   **b** 35% of 600   **c** 5% of 280

16 Jo buys a £25 coat reduced by 35% in a sale. How much money does Jo save?

17 On Friday a shop sells 140 loaves of bread.
25% of the loaves are wholemeal bread.
Here are two ways of starting to work out how many of the loaves sold are **not** wholemeal.

    1)   25% of 140
    2)   100% − 25% = 75%

   a Copy and complete each method.
   b **Reflect** Write a sentence explaining which method you prefer.

18 **Problem-solving** Ava's grandmother receives a 25% discount on her council tax as she lives alone.
Her annual council tax bill, before the discount is applied, is £1050.60.
The council says Ava's grandmother must pay £87.50 each month for a year.
Check that the council are charging correctly. Show your working.

Key point

A **deposit** is a first payment towards the total cost of something.
The remainder that is owing is called the **balance**.

Exam-style question

19 Chris is buying a greenhouse.
The greenhouse costs £1200.
Chris will pay a deposit of 5%.
She will pay the balance in 4 equal monthly payments.
How much is each monthly payment? **(2 marks)**

Exam tip

Write each calculation you do on a separate line, so your working is clear.

20 Write as a decimal   **a** 130%   **b** 120%   **c** 350%

21 A loaf of bread costs $1\frac{1}{2}$ times as much today as it did
10 years ago.
What percentage has it increased by?

> **Q21 hint**
>
> $1\frac{1}{2} = 1.5 = \square\%$

22 Work out   **a** 150% of £80   **b** 125% of 200 ml   **c** 250% of £700

23 A used car dealer buys a car for £8000. She sells it at 115% of this cost.
How much did the car dealer sell the car for?

Example

Find the simple interest when £5000 is invested at 2.75% per annum over 2 years.

$2.75\% = 0.0275$ ◂——— | Convert the percentage to a decimal.

$5000 \times 0.0275 = £137.50$ ◂——— | principal investment × rate of interest = simple interest in 1 year

$£137.50 \times 2 = £275$ ◂——— | Multiply your answer by 2.

24 Find the simple interest when
   a £3000 is invested at 2.75% per annum (p.a.) over 1 year
   b £250 is invested at 3.25% p.a. over 3 years
   c £4000 is invested at 2.2% p.a. over 18 months.

> **Q24 hint**
>
> Per annum or p.a. means
> 'each year'.

25 John's grandmother gives him £2000.
He saves the money in a bank account with an annual rate of simple interest rate of 2.8%.
How much money will John have in the bank account after 30 months?

26 **Problem-solving** Alice invests £2000 for 4 years in a
savings account.
By the end of the 4 years she has received £80 simple interest.
   a How much does Alice receive each year?
   b Work out the annual rate of simple interest.

> **Q26 hint**
>
> principal investment × rate of
> interest (as a decimal) = simple
> interest in 1 year

27 Use the formula for simple interest to check your answers to **Q24**.

4.8 Calculating percentages 2

- Calculate percentage increases and decreases.
- Use percentages in real-life situations.
- Calculate VAT (value added tax).

Active Learn
Homework

Warm up

1 **Fluency** What is
   a 100% + 10%   **b** 100% − 40%   **c** 100% − 70%

2 Convert these percentages to decimals.
   a 20%   **b** 320%   **c** 3%   **d** 23.5%

 3 Work out
   a 35% of 60   **b** 125% of 120   **c** 250% of 70

Key point

To increase a number by a percentage, work out the increase and add this to the original number.

Example

Increase 30 by 10%.

10% of 30 = 3 ◄———— Work out 10%.

30 + 3 = 33 ◄———— Add your answer to the original number.

4 Increase
   a 33 by 20%   **b** 400 by 33%   **c** 500 by 1%   **d** 2000 by 0.1%

 5 **Reasoning** Daisy works part time for £40 a week. Her employer raises her weekly wage by 3%. What is Daisy's new weekly wage?

6 **Problem-solving** The cost of living increased by 30% from 2004 to 2014.
   In 2004, Sharnia's wage was £240 a week.
   In 2014, her wage was £300 a week.
   Explain if Sharnia's wage has increased more or less than the cost of living.

Key point

To decrease a number by a percentage, work out the decrease and subtract this from the original number.

---

 7 Decrease
   a 48 by 25%   **b** 600 by 45%
   c 450 by 6%   **d** 5000 by 0.6%

Q7a hint

Work out 25% of 48. Subtract the answer from 48.

---

8 **Future skills** A carpet company reduces its prices by 40%.
A carpet was £300. What is the new price of the carpet?

9 The Smith family spent £120 per week on food last year. This has gone up by 2% this year.
The Singh family spent £130 per week on food last year.
They have changed supermarket and this has reduced their weekly spend by 3.5% this year.
Which family's food bill is higher this year? You must show your working.

10 Train fares are increasing by 25%. A train fare was originally £18.
  a Work out 25% of £18.   **b** Work out the new price of the train fare.
  c Work out 125% of £18.
  d What do you notice about your answers to parts **b** and **c**? Explain.

11 Write the decimal multiplier you can use to work out the new
amount after an increase of

> **Q11a hint**
> $100\% + 37\% = \square\% = \square.\square$

  a 37%   **b** 20%   **c** 75%   **d** 12%
  e 50%   **f** 5%   **g** $87\frac{1}{2}\%$ **h** 7.5%

12 **a** Which of these multipliers is correct to work out an increase of 112%?
      1.12   11.2   112   212   2.12
  b **Reflect** Write a sentence explaining how you decided on the correct multiplier.

13 Increase
  a 120 by 20%   **b** 500 by 5%   **c** 5000 by 87.5% **d** 1.5 by 140%

14 Write the decimal multiplier you can use to work out the new
amount after a decrease of

> **Q14a hint**
> $100\% - 40\% = 60\% = \square.\square$

  a 40%   **b** 35%   **c** 6%   **d** 1.2%

15 Decrease
  a 200 by 40%   **b** 500 by 35%   **c** 130 by 6%   **d** 4000 by 1.2%
  e **Reflect** Write a sentence explaining why you cannot decrease a number by more than 100%.

16 **Problem-solving** A 5% service charge is added to the cost of a meal.
A meal costs £52. What is the total charge?

17 **Future skills** Harry buys a car for £9800.
The value of the car depreciates by 15% each year.
Work out the value of the car at the end of the year.

> **Q17 hint**
> **Depreciates** means that the value of the car decreases.

18 **Reasoning** Ruth wants to increase 170 by 4%. She writes down

$$170 \times 1.4 = 238$$

Ruth's method is wrong. Explain why.

19 **Future skills** A market stall has this offer:
    Cloths £1.10 each
    Buy 12 or more cloths, get 15% off the total cost
Work out the cost of buying 20 cloths using this offer.

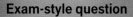

Exam-style question

20 Ashley wants to buy some tins of paint.
He finds out the costs of paint at two shops.

Paint R Us
Normal price £2.19 a tin
Special Offer
Buy 2 tins at the normal price and get the 3rd tin free

Deco Mart
Normal price £1.80 a tin
Special Offer
10% off the normal price

> **Exam tip**
>
> Make sure you make a statement at the end that includes the cost of paint for each shop (with units), from which shop Ashley should buy the paint, and why.

Ashley needs 9 tins of paint. He wants to get all the tins of paint from the same shop.
Which of the two shops is the cheapest for Ashley to buy his paint from?
You must show how you get your answer. **(3 marks)**

21 **Problem-solving** Mr Elliott and his five children are going to London by train.
An adult ticket usually costs £24. A child ticket usually costs £12.
Mr Elliott has a family rail card which gives $\frac{1}{3}$ off adult tickets and 60% off child tickets.
Work out the cost of the rail tickets.

22 **Future skills** Orlav is self-employed. Last year, he earned £18 940.
He does not pay income tax on the first £12 500 he earns.
He pays tax of 20% for each pound he earns above £12 500.
How much tax must he pay?

Key point

Value added tax (VAT) is charged at 20% on most goods and services.

23 **Future skills** Debbie gets two plumbers' quotes.
Quote 1 £660 including VAT
Quote 2 £500 excluding VAT
Which is the more expensive quote? Show your working.

> **Q23 hint**
>
> Add VAT at 20% to Quote 2.

24 **a Problem-solving** Ben buys three concert tickets. The cost of one ticket is £99 plus VAT.
Work out the cost of three tickets.

b Reasoning Sarah works out the cost of one ticket plus VAT and then multiplies by 3.
Peter works out the total cost of three tickets and then adds VAT.
Do Sarah and Peter get the same answer? Show working to explain.

25 **Problem-solving** A company plans a competition for its employees.
Each winner gets a prize of two tickets to a sporting event.
The cost of two tickets is £120 plus VAT. The company wants to spend up to £1000.
How many winners can there be?

26 **Future skills** Travis buys a campervan.
The cost of the campervan is £18 600 plus VAT at 20%.
Travis pays a deposit of £3000.
He pays the rest of the cost in 12 equal payments.
Work out the amount of each of the 12 payments.

4 Check up

Active Learn
Homework

Operations with fractions

1 Work these out. Simplify your answers if necessary.

a $\frac{1}{5}+\frac{1}{3}$

b $\frac{3}{8}+\frac{1}{4}$

c $\frac{3}{4}-\frac{1}{2}$

d $\frac{11}{12}-\frac{1}{8}$

2 Work these out. Simplify your answers if necessary.

a $\frac{2}{5}\times\frac{3}{5}$

b $\frac{2}{7}\times\frac{14}{5}$

c $4\times1\frac{3}{5}$

d $1\frac{2}{9}\times2$

3 Work out

a $4\div\frac{3}{8}$

b $\frac{3}{4}\div8$

c $\frac{3}{4}\div\frac{1}{2}$

d $1\frac{1}{3}\div4$

4 Work these out. Write your answers as mixed numbers.

a $\frac{7}{8}+\frac{1}{6}$

b $3\frac{3}{4}+2\frac{1}{8}$

c $5\frac{3}{4}-3\frac{1}{2}$

d $4\frac{1}{6}-2\frac{3}{4}$

Percentages, decimals and fractions

5 Write these decimals as fractions in their simplest form.

a 0.007   **b** 0.325

6 Write 36 minutes as a fraction of an hour.

7 Write $\frac{11}{25}$ as a decimal.

8 Write as a fraction in its simplest form.

a 5%   **b** 28%   **c** 150%

9 Write $\frac{3}{25}$ as a percentage.

10 Write these in order of size. Start with the smallest.

$\frac{3}{5}$, 0.62, 58%, $\frac{2}{3}$

11 Write 25 as a fraction of 60 in its simplest form.

 12 **a** A 450 g pot of yoghurt contains 60 g of fruit.
What percentage of the total is fruit? Give your answer to the nearest whole number.

b A 300 g pot of yogurt contains 45 g of fruit.
Does this have a greater or smaller percentage of fruit?

Calculating percentages

13 Work out 5% of £180.

 14 Work out
a 12% of 150 cm **b** 8% of 225 km **c** 130% of £82

 15 Lindy buys a bag priced at £22. She gets 15% off. How much does she save?

 16 A price of £65 is increased by 20%. Work out the new price.

 17 £500 is reduced by 35%. Work out the new price.

 18 Building work costs £8500 + VAT. VAT is 20%. Work out the total cost.

 19 Find the simple interest when £2500 is invested for 1 year at 1.75%.

20 **Reflect** How sure are you of your answers? Were you mostly

Just guessing 😞   Feeling doubtful 😐   Confident 🙂

What next? Use your results to decide whether to strengthen or extend your learning.

Challenge

 21 Sam earns £32 000 per annum.
He does not pay income tax on the first £12 500 he earns.
He pays 20% for each pound he earns above £12 500.

a How much does Sam receive each month, after paying tax?

b $\frac{7}{10}$ of Sam's pay each month goes on bills. 10% of the rest he saves.
How much does Sam have after paying bills and saving?

> **Q21b hint**
> Draw a bar model.

22 Shapes A, B, C, D and E fit together to make a large rectangle. A is a square.
B, C, D and E are rectangles.

a What percentage of the area of the large rectangle is

i A **ii** B

iii C **iv** A and D?

b Which three shapes together are 66% of the area of the large rectangle?

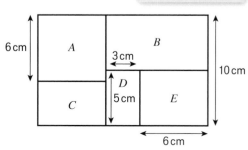

4 Strengthen

Operations with fractions

1 Copy and complete to work these out.

a i $\dfrac{1}{2} = \dfrac{\square}{4}$ (×□, ×2) **ii** $\dfrac{1}{2} + \dfrac{1}{4} = \dfrac{\square}{4} + \dfrac{1}{4} = \dfrac{\square}{4}$ **b i** $\dfrac{1}{4} = \dfrac{\square}{8}$ (×□, ×□) **ii** $\dfrac{1}{4} + \dfrac{3}{8} = \dfrac{\square}{8} + \dfrac{3}{8} = \dfrac{\square}{8}$

c i $\dfrac{1}{3} = \dfrac{\square}{9}$ **ii** $\dfrac{2}{9} + \dfrac{1}{3} = \dfrac{2}{9} + \dfrac{\square}{9} = \dfrac{\square}{\square}$ **d i** $\dfrac{1}{3} = \dfrac{\square}{6}$ **ii** $\dfrac{1}{3} - \dfrac{1}{6} = \dfrac{\square}{6} - \dfrac{1}{6} = \dfrac{\square}{\square}$

2 Work out

a $\dfrac{1}{8} + \dfrac{1}{6}$

b $\dfrac{1}{5} + \dfrac{1}{8}$

c $\dfrac{1}{2} - \dfrac{1}{3}$

d $\dfrac{1}{3} - \dfrac{1}{10}$

Q2a hint
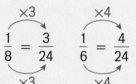
The lowest number that both 6 and 8 'go into' is 24.
$\dfrac{1}{8} + \dfrac{1}{6} = \dfrac{\square}{24} + \dfrac{\square}{24}$
$\dfrac{1}{8} = \dfrac{3}{24}$ (×3) $\dfrac{1}{6} = \dfrac{4}{24}$ (×4)

3 Work these out. Simplify your answer if necessary.

a $\dfrac{5}{6} - \dfrac{1}{4}$

b $\dfrac{3}{4} + \dfrac{1}{5}$

c $\dfrac{7}{12} + \dfrac{1}{8}$

d $\dfrac{2}{5} - \dfrac{1}{4}$

Q3a hint
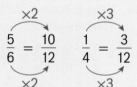
The lowest number that both 6 and 4 'go into' is 12.
$\dfrac{5}{6} - \dfrac{1}{4} = \dfrac{\square}{12} - \dfrac{\square}{12}$
$\dfrac{5}{6} = \dfrac{10}{12}$ (×2) $\dfrac{1}{4} = \dfrac{3}{12}$ (×3)

4 Work these out. Give each answer as a mixed number. The first one is started for you.

a $\dfrac{3}{4} + \dfrac{1}{2} = \dfrac{3}{4} + \dfrac{2}{4} = \dfrac{\square}{4} = 1\dfrac{\square}{4}$ **b** $\dfrac{2}{3} + \dfrac{5}{9}$ **c** $\dfrac{4}{5} + \dfrac{1}{4}$ **d** $\dfrac{6}{7} + \dfrac{1}{2}$

5 Work out

a $1\dfrac{5}{6}$ as an improper fraction: $\dfrac{\square}{6}$

b $2\dfrac{1}{2}$ as an improper fraction: $\dfrac{\square}{2}$ **c** $2\dfrac{1}{2} - 1\dfrac{5}{6} = \dfrac{\square}{6} - \dfrac{\square}{6} = \dfrac{\square}{6} = \dfrac{\square}{3}$

6 Follow the method in Q5 to work out

a i $5\dfrac{1}{8} + 2\dfrac{3}{4}$ **ii** $5\dfrac{1}{8} - 2\dfrac{3}{4}$ **b i** $4\dfrac{3}{4} + 3\dfrac{1}{2}$ **ii** $4\dfrac{3}{4} - 3\dfrac{1}{2}$

c i $4\dfrac{7}{8} - 3\dfrac{1}{2}$ **ii** $4\dfrac{7}{8} + 3\dfrac{1}{2}$ **d i** $3\dfrac{3}{4} - 2\dfrac{5}{8}$ **ii** $3\dfrac{3}{4} + 2\dfrac{5}{8}$

7 Work these out. The first one is started for you.

a $2 \times \dfrac{2}{3} = \dfrac{2 \times 2}{3} = \dfrac{\square}{3} = \square\dfrac{\square}{3}$ **b** $5 \times \dfrac{4}{7}$ **c** $15 \times \dfrac{3}{5}$

8 Work these out. The first one is started for you.

a $\dfrac{1}{2} \times \dfrac{1}{3} = \dfrac{1 \times 1}{2 \times 3} = \dfrac{1}{\square}$

b $\dfrac{4}{5} \times \dfrac{1}{2}$

c $\dfrac{7}{11} \times \dfrac{4}{5}$

d $\dfrac{4}{5} \times \dfrac{4}{5}$

9 Work these out. The first one is started for you.

a $\dfrac{2}{5} \times \dfrac{15}{4} = \dfrac{15 \times 2}{5 \times 4} = \dfrac{15}{5} \times \dfrac{2}{4} = 3 \times \dfrac{1}{2} = \square$

b $\dfrac{3}{4} \times \dfrac{8}{10}$

c $\dfrac{4}{5} \times \dfrac{5}{7}$

d $\dfrac{3}{4} \times \dfrac{12}{15}$

Q9 hint

Change the order so you can cancel the common factors.

10 Copy and complete.

a $1\dfrac{1}{3}$ as an improper fraction is $\dfrac{\square}{3}$

b $4 \times 1\dfrac{1}{3} = 4 \times \dfrac{\square}{3} = \dfrac{4 \times \square}{3} = \dfrac{\square}{3} = 5\dfrac{\square}{3}$

11 Follow the method in **Q10** to work out **a** $3 \times 2\dfrac{1}{4}$ **b** $5 \times 1\dfrac{1}{7}$ **c** $1\dfrac{2}{5} \times 7$

12 Copy and complete.

a $7 \div \dfrac{1}{5} = 7 \times \square = \square$
   ↑
   reciprocal of $\dfrac{1}{5}$

b $3 \div \dfrac{2}{3} = 3 \times \dfrac{\square}{\square} = \dfrac{3 \times \square}{\square} = \dfrac{\square}{\square} = 4\dfrac{\square}{\square}$
   ↑
   reciprocal of $\dfrac{2}{3}$

13 Work these out. The first one is started for you.

a $\dfrac{1}{2} \div 5 = \dfrac{1}{2} \times \dfrac{1}{5} = \dfrac{\square}{\square}$

b $\dfrac{1}{3} \div 4$

c $\dfrac{2}{5} \div 6$

d $\dfrac{3}{7} \div 6$

14 Work these out. Write your answer as a mixed number if required. They are started for you.

a $\dfrac{1}{4} \div \dfrac{2}{15} = \dfrac{1}{4} \times \dfrac{\square}{\square} =$
   ↑
   reciprocal of $\dfrac{2}{15}$

b $1\dfrac{1}{2} \div 3 = \dfrac{\square}{2} \times \dfrac{\square}{\square} =$
   ↑   ↑
   improper   reciprocal
   fraction   of 3

Percentages, decimals and fractions

1 Copy and complete to write these fractions as decimals.

a $\dfrac{1}{8}$ **b** $\dfrac{5}{8}$ **c** $\dfrac{8}{25}$ **d** $\dfrac{1}{25}$

$\dfrac{1}{8} = 8\overline{)1.^{1}0^{2}0^{\square}0}^{\,0.1\,\square\,\square}$

$\dfrac{5}{8} = 8\overline{)5.0\ \ 0\ \ 0}^{\,0.\square\square\square}$

$\dfrac{8}{25} = 25\overline{)8.0\ 0}^{\,\square.\square\square}$

$\dfrac{1}{25} = \square\overline{)\square.\square\square}^{\,\square.\square\square}$

2 Write these decimals as fractions. Simplify if necessary.

a $0.6 = \dfrac{6}{10} = \dfrac{\square}{5}$

b $0.03 = \dfrac{\square}{100}$

c 0.25

d $0.005 = \dfrac{\square}{1000} = \dfrac{\square}{\square}$

e 0.033

f 0.125

3 Write as a fraction in its simplest form.

$\div \square$

a $2\% = \dfrac{2}{100} = \dfrac{1}{\square}$ **b** 5% **c** 48% **d** 78% **e** $160\% = \dfrac{160}{100} =$ **f** 180%

$\div \square$

4 Write as a percentage.

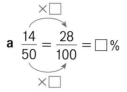

a $\dfrac{14}{50} = \dfrac{28}{100} = \square\%$

b $\dfrac{41}{50}$

c $\dfrac{3}{25}$

5 Write as a decimal, then a percentage.

Q5 hint
$$5\overline{)4.0}^{0.\square}$$

a $\dfrac{4}{5} = 0.\square = \square\%$

b $\dfrac{2}{5}$

c $\dfrac{3}{8}$

6 Write as a fraction in its simplest form. The first and third ones are started for you.

a 15 as a fraction of 40 $\dfrac{15}{40} = \dfrac{\square}{\square}$ ÷5

b 34 as a fraction of 50

c 6 minutes as a fraction of 1 hour $= \dfrac{\square}{60} = \dfrac{\square}{\square}$

d 9 minutes as a fraction of 1 hour

e **Reflect** When writing minutes as a fraction of 1 hour, why is the denominator 60? Write a sentence to explain.

7 Copy and complete these two methods to work out 24 as a percentage of 50.

$\dfrac{24}{50} = \dfrac{\square}{100} = \square\%$ $\dfrac{24}{50} \times 100 = \dfrac{24 \times 100^{\square}}{{}_1 50} = \square\%$

8 Use one of the methods in **Q7** to work out
a 8 as a percentage 50
b 12 as a percentage of 25
c 10 cm as a percentage of 40 cm
d 30 ml as a percentage of 150 ml

9 Use a calculator to work out 40 g as a percentage of 280 g.

10 Write these percentages as decimals.

a $40\% = \dfrac{40}{100} = \dfrac{4}{10} = 0.\square$ ÷10

b 70%

c $12\% = \dfrac{12}{100} = 0.\square\square$

d 47%

e $2\% = \dfrac{2}{100} = 0.\square\square$

f 2.5%

Calculating percentages

1 Work out
   a 10% of £80   **b** 10% of 70   **c** 10% of 220

Q1 hint

$10\% = \frac{10}{100} = \frac{1}{10}$

2 Work out
   a 30% of 80   **b** 20% of 70   **c** 60% of 220

Q2a hint

$30\% = 3 \times 10\%$

3 Work out
   a 5% of 60   **b** 5% of 200   **c** 15% of 1250

Q3a hint

$5\% = \frac{1}{2}$ of 10%

4 Use your answers to **Q10** in Percentage, decimals and fractions to help you to work out
   a 12% of 46
   b 47% of 700
   c 2% of 90
   d 2.5% of 150

5 **Future skills**
   a Find the simple interest when £1500 is invested for 1 year at 2.5%.
   b Find the simple interest when £10 000 is invested for 2 years at 1.75%.

Q5b hint

Simple interest for 2 years = $2 \times$ simple interest for 1 year.

6 **a** Work out 10% of 40.
   b Add 40 to your answer to part **a**. This is 40 increased by 10%.

7 Follow the steps in **Q6** to
   a increase 30 by 10%
   b increase 20 by 50%
   c increase 180 by 80%
   d increase £3000 by 5%

8 **Future skills** VAT is 20%. Add VAT to £1200.

9 **a** Work out 10% of 70
   b Subtract your answer to part **a** from 70. This is 70 decreased by 10%.

10 Follow the steps in **Q9** to
   a decrease 60 by 10%
   b decrease 90 by 20%
   c decrease 40 by 60%
   d decrease £1000 by 5%

Q10a hint

$100\% - 10\% = 90\% = 0.9$
$60 \times 0.9 = \square$

4 Extend

1   Work out

   a $-\frac{1}{4} \times \frac{2}{7}$ **b** $1\frac{1}{4} \times -\frac{2}{7}$ **c** $\frac{1}{4} \div -\frac{2}{7}$ **d** $\frac{1}{4} - \frac{2}{7}$

2   **Problem-solving** A class records the weather each day for a project.
   At the end of the project they report:

   $\frac{2}{5}$ of the days were rainy.

   $\frac{1}{3}$ of the remaining days were sunny.

   The rest of the days were dry and cloudy.
   There were 12 dry and cloudy days.

   For how many days did the project last?

3   Find the value of the reciprocal of 1.25. Give your answer as a decimal.

4   **Problem-solving** Scotland has an area of 30 000 square miles.
   $\frac{1}{6}$ of the area is woodland.
   15% of the area of woodland is covered with pine trees.
   Work out the area of woodland covered with pine trees.

5   **Reasoning** There are 21 questions in a test on biology, chemistry and physics.
   There are 7 biology questions and 8 chemistry questions.
   What percentage of the questions are on physics?

   Q5 hint
   Work out the fraction of questions that are on physics.

6   **Reasoning** The value of a car depreciates by 20% each year.
   The price when new was £7500.
   Work out the value of a two-year-old car.

   Q6 hint
   Work out the value after 1 year, then decrease that by another 20%.

7   **a** Work out $\frac{1}{4}$ of 20% of 300.

   b Which would you rather have: $\frac{1}{5}$ of 30% of £500 or 5% of $\frac{1}{2}$ of £800?
      You must show your working.

8   Millie is collecting small coins in a jar.
   She empties the jar and counts 120 coins.
   $\frac{2}{5}$ are 1p coins.
   25% are 2p coins.
   The rest are 5p coins.
   What percentage of the coins are 5p coins?

   Q8 hint
   Work out the percentage of 1p coins first.

9   **Problem-solving** Anwar and Bethany each earn the same weekly wage.
   Each week, Anwar saves 12% of his wage and spends the rest.
   Each week, Bethany spends $\frac{7}{8}$ of her wage and saves the rest.
   Who saves more money each week?
   You must show your working.

Unit 4 Fractions and percentages   125

10 **Reasoning** Gunnar is doing a project for his health and social care course.
He finds the following information.

Year	Percentage of patients unable to secure a GP or nurse appointment
2013	8.9
2014	10.3

Gunnar writes, 'The percentage of patients unable to secure an appointment has increased by more than 10%.'
Is Gunnar correct? Show your working.

11 **Problem-solving / Future skills** Riccardo's parents need a mortgage to buy a house.
They are offered two options.
Option 1 Three and a half times their joint salary.
Option 2 Six times the larger salary plus one and a half times the smaller salary.
Riccardo's dad's salary is £22 000. His mum's salary is £37 000.
Riccardo's parents need a mortgage greater than $\frac{1}{4}$ million pounds.
Show that only one of the options is suitable.

12 **Reasoning** Randell uses this formula to work out the time to cycle to his aunt's house.

$$\text{Time (hours)} = \frac{\text{distance (km)}}{18}$$

The distance is 42 km.
Work out the time. Give your answer in hours and minutes.

13 **Problem-solving** A supermarket buys chicken for £1.35 per kg and sells it at 250% of the cost price.
How much does the supermarket sell 750 g of chicken for?
Write your answer to the nearest penny.

14 A company shares its profits from £72 000.
The managing director (MD) receives 10%.
30% is shared among 5 directors.
The rest is shared among the remaining 54 staff.
One of the staff says, 'If the profits are shared equally between all 60 people at the company, I will get 50% more money.'
Is the member of staff correct? You must show how you get your answer. **(5 marks)**

> **Exam tip**
>
> Sometimes you have to show a lot of working. It can help to label your calculations so you know what your working is. For example, 'MD receives:'.

15 Jennie's council has a target for households to recycle $\frac{1}{5}$ of their waste each month.
In January, Jennie recycled $\frac{1}{10}$ of her household waste.
In February, she recycled 15 kg of her 120 kg of household waste.
Her result for March was 13% recycled out of 112 kg of household waste.
Has Jennie met the council's target?
In which month did she recycle the largest percentage of her household waste? **(4 marks)**

> **Exam tip**
>
> For each month, write the amount recycled in the same form, for example all as fractions or all as percentages.

4 Test ready

To revise for the test:

- Read each key point, find a question on it in the mastery lesson, and check you can work out the answer.

- If you cannot, try some other questions from the mastery lesson or ask for help.

Summary of key points

Key points

1 To compare, add or subtract fractions, write them with a common **denominator**. → **4.1**

2 To add or subtract **mixed numbers**, it often helps to convert to improper fractions first. For example, write $2\frac{2}{3}$ as $\frac{8}{3}$. → **4.2**

3 To multiply fractions, multiply the **numerators** and multiply the **denominators**. → **4.3**

4 Numerators can be cancelled with denominators if they are divisible by the same number. → **4.3**

5 The **reciprocal** of a fraction is the original fraction turned upside down. The reciprocal of 2 (or $\frac{2}{1}$) is $\frac{1}{2}$. → **4.4**

6 Dividing by a number is the same as multiplying by its reciprocal. For example, $12 \div 2$ means $\frac{1}{2}$ of 12 or $12 \times \frac{1}{2}$. → **4.4**

7 You can use short or long division to convert fractions to decimals. For example, $\frac{3}{5}$ means $3 \div 5$. → **4.5**

8 To convert a fraction to a decimal, divide the numerator by the denominator. → **4.5**

9 You can write one quantity as a fraction of another. → **4.5**

10 **Percentage** means 'out of 100'. For example $24\% = \frac{24}{100}$. → **4.6**

11 To convert a fraction to a percentage, you can convert the fraction to one with the denominator 100, or you can multiply by 100%. → **4.6**

12 A **deposit** is a first payment towards the total cost of something. The remainder that is owing is called the **balance**. → **4.7**

13 Percentages can be bigger than 100%. The cost of a loaf of bread has increased by more than 100% since 1975. → **4.7**

14 **Simple interest** is interest paid out each year by banks and building societies. → **4.7**

15 The formula for simple interest is
simple interest $=$ principal investment $\times$ rate $\times$ time
$$I = \frac{PRT}{100}$$
→ **4.7**

16 Value added tax (VAT) is charged at 20% on most goods and services. → **4.8**

17 To increase a number by a percentage, work out the increase and add this to the original number.
To decrease a number by a percentage, work out the decrease and subtract this from the original number. → **4.8**

Sample student answers

Exam-style question

1 Samantha wants to buy a new pair of trainers.
There are 3 shops that sell the trainers she wants.

Sports '4' All Trainers	Edexcel Sports Trainers	Keef's Sports Trainers
£5 plus 12 payments of £4.50	$\frac{1}{5}$ off usual price of £70	£50 plus VAT at 20%

From which shop should Samantha buy her trainers to get the best deal?
You must show all of your working. **(5 marks)**

$12 \times 4.50 = 54 + 5 = £59$
$\frac{1}{5}$ of $70 = 70 \div 5 = 14$
$70 - 14 = 56$
20% of $50 = \frac{1}{5} \times 50 = £10 = 50 + 10 = £60$
Edexcel Sports is the best deal.
Samantha should buy her trainers from Edexcel Sports.

The student has given the correct answer. However, the working is not accurately set out.
Explain what is wrong with the student's working.

Exam-style question

2 Sanjit paid £150 per year for car tax on his old car.
For his new car he pays £210 per year for car tax.
Write this increase as a fraction of the car tax on his old car.
Give your answer in its simplest form. **(2 marks)**

Increase = £210 – £150 = £60
Fraction: $\frac{60}{150} = \frac{6}{15}$

How could the student improve their answer?

4 Unit test

1   Samuel gets 24 out of 50 in a science test.
    Write 24 out of 50 as a percentage.   **(2 marks)**

2   Wayne took two tests.
    He got $\frac{12}{25}$ in the first test and $\frac{56}{100}$ in the second test.
    Wayne says he did better in the first test.
    Is he correct? Show why you think this.   **(2 marks)**

3   Write these in order of size. Start with the smallest.
    $0.88, \frac{4}{5}, 0.82, \frac{9}{10}, \frac{18}{25}$ **(2 marks)**

4   **a** Increase £75 by 5%.   **(2 marks)**
    b Decrease £150 by 12%.   **(2 marks)**

5   The rate of simple interest is 3% per year.
    Work out the simple interest paid on £8000 in one year.   **(2 marks)**

6   Work out
    a $2\frac{2}{3} \times 3$ **(2 marks)**
    b $\frac{1}{2} \div \frac{3}{5}$ **(2 marks)**

7   Work out $\frac{2}{3} + \frac{1}{5}$ **(2 marks)**

8   A tiling bill is £360 plus VAT at 20%.
    Work out the total amount.   **(3 marks)**

9   Here are two fractions: $\frac{3}{4}$ and $\frac{4}{5}$
    Which is the larger fraction?   **(2 marks)**

10 A company spends $\frac{1}{4}$ of its profit on new machinery.
    The company makes a profit of £$\frac{3}{4}$ million.
    How much money does the company spend on new machinery?   **(2 marks)**

11 Work out $5\frac{2}{3} - 2\frac{3}{4}$ **(3 marks)**

12 Work out $\frac{3}{8} + \frac{5}{12} - \frac{1}{4}$ **(3 marks)**

13 A gift shop buys ribbon in rolls.
    Each roll has 30 m of ribbon.
    Eloise uses a machine to cut ribbon into lengths of $\frac{2}{5}$ m.
    She needs 400 lengths.
    How many rolls of ribbon does she need?   **(4 marks)**

14 Rosemary has £18 200 in a bank account that pays simple interest of 1.2%.
    How much interest does she receive each year?   **(2 marks)**

15 William needs a new mobile phone contract for 1 year.
He can get 25% off the £150 contract cost or pay £24.61 for 1 month and
£7.99 for the next 11 months.
Show that whatever method of payment he chooses, William pays the same amount.

(4 marks)

16 Karim is mixing some ingredients for a cake.
He needs $1\frac{1}{3}$ cups of sugar, $1\frac{1}{3}$ cups of flour and $2\frac{1}{2}$ cups of raisins.
His mixing bowl holds 5 cups.
Will the bowl be large enough?

(3 marks)

17 Olive thinks of a number.
$\frac{3}{4}$ of Olive's number is 21.
What is $\frac{4}{7}$ of Olive's number?

(2 marks)

18 Last year, Abdul spent $\frac{2}{5}$ of his salary on rent, $\frac{1}{4}$ of his salary on entertainment and 30% of
his salary on living expenses.
He saved the rest of his salary.
Abdul spent £10 500 on entertainment.
How much money did he save?
You must show all your working.

(4 marks)

(TOTAL: 50 marks)

19 Challenge

a Write these in order. Start with the smallest.

31% $\frac{1}{3}$ $\frac{3}{10}$ 0.3%   0.303

b Write a fraction at the beginning and a 2-digit percentage starting with the digit 3 at the
end of your list from **a**. The list must still be in order, starting with the smallest.

20 Challenge

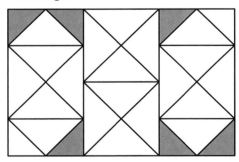

How many of the larger triangles must be shaded so that 75% of the shape is shaded?
What percentage of the shape is shaded?

21 Reflect Antony says,

'Fractions, decimals and percentages are all the same.'

Wendy says,
'They are different.'
Write one sentence explaining why Antony could be correct.
Write another sentence explaining why Wendy could be correct.

Mixed exercise 1

1   Ava has 25 sweets.
    She gives 9 of the sweets to Ben.
    What fraction of the 25 sweets does Ava have now?   **(2 marks)**

2   **Problem-solving** Which of these values is nearest to 25%?

    $\frac{14}{50}$ $\frac{1}{5}$ 0.27 $\frac{6}{25}$ 0.235

3   Work out the difference, in minutes, between 2 hours 25 minutes and $2\frac{1}{4}$ hours.   **(2 marks)**

4   **Problem-solving** Work out $\frac{3}{4} + 0.2$.

5   Find the number that is exactly halfway between $\frac{1}{12}$ and $\frac{5}{6}$.   **(2 marks)**

6   **Reasoning** The table shows information about the test results for eight students for two different maths papers.

Student	A	B	C	D	E	F	G	H
Paper 1	28	45	37	33	19	39	43	36
Paper 2	36	42	43	40	27	39	44	40

Fayaz draws this scatter graph and line of best fit to display these results.

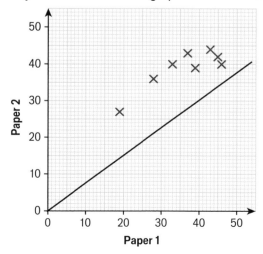

Write two mistakes Fayaz has made.

7 Here is the charge of a phone call with a network in France.

Network
0.036 euros per minute

Louisa uses this network to make a phone call.
Louisa started her call at 09:45.
The call cost 2.34 euros.
At what time did Louisa end her call? **(3 marks)**

8 Mr Davis buys 17 calculators for £8.50 each.
He pays with eight £20 notes.

a How much change should Mr Davis get? **(3 marks)**

The normal price of a laptop is £395.

In a sale there is $\frac{1}{5}$ off the normal price of the laptop.

b Work out the price of the laptop in the sale. **(2 marks)**

9 Strawberries are sold in two sizes of punnet.

Size of punnet	Weight of strawberries
small	0.4 kg
large	0.75 kg

Sarah buys 3 small punnets of strawberries and some large punnets of strawberries.
In total she buys 4.95 kg of strawberries.
Work out the number of large punnets of strawberries Sarah buys. **(3 marks)**

10 **Problem-solving** 800 students are asked if they have a sibling.

$\frac{4}{5}$ say yes.

15% of the students who say yes have at least two siblings.

a How many students have only one sibling?

b What fraction of the 800 students have at least two siblings?

11 There are 1200 students at a school.
Each student studies either French or Spanish.
65% of the students study French.
55% of the students are girls.
60% of the girls study French.
How many boys study Spanish? You must show all your working. **(4 marks)**

12 **Problem-solving/Reasoning** x is a common factor of 36 and 48 and y is a common multiple of 6 and 8.
What is the highest possible value of $\frac{x}{y}$?

Exam-style question

13 Jordan is x years old.
Kyle is twice as old as Jordan.
Lexi is three years younger than Jordan.
The total of Jordan's age, Kyle's age and Lexi's age is Y years.
Write a formula for Y in terms of x. **(3 marks)**

Exam-style question

14 The number of hours, h, that it will take to paint a house is given by
$$h = \frac{24}{n}$$
where n is the number of decorators used each day.
Dave's company has 5 decorators.
Emma's company has 8 decorators.
Emma's company will take less time to paint the house. How much less?
Give your answer in hours and minutes. **(3 marks)**

15 **Reasoning** $x + y = 7$
Work out the value of
  a $2(x + y)$
  b $(x + y)^2$
  c $5x + 5y$

Exam-style question

16 The pie charts show information about the favourite science of each student at school A
and of each student at school B.

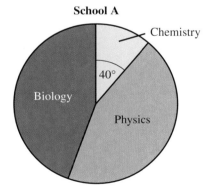

School A

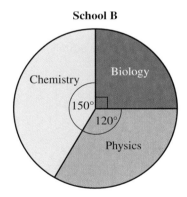

School B

There are 900 students at school A. There are 1140 students at school B.
The same number of students chose biology as physics at school A.
Taylor says, 'The same number of students at each school have physics as their
favourite science.'
Is Taylor correct? You must show how you get your answer. **(4 marks)**

17 Reasoning The table and frequency polygon show some information about the heights of some trees.

Height, h (metres)	Frequency
$8 < h \leqslant 12$	14
$12 < h \leqslant 16$	7
$16 < h \leqslant 20$	1

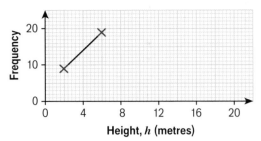

a Copy and complete the table.

b Copy and complete the frequency polygon.

18 a Reasoning Emma is asked to expand the brackets $3(2x + 7)$.
Emma writes:

$$3(2x + 7) = 5x + 7$$

Explain why Emma is incorrect.

b Emma and Felix are asked to factorise $12x^2 + 16x$.
Emma writes:

$$12x^2 + 16x = 2x(6x + 8)$$

Felix writes:

$$12x^2 + 16x = 4(3x^2 + 4x)$$

Are either of them correct? Explain your answer.

19 Reasoning Copy and complete each equation.

a $\dfrac{8a^5 b^{\square}}{\square a^{\square} b^3} = 2a^3 b^4$

b $3a^{\square} b^4 \times \square a^2 b^{\square} = 12a^5 b^6$

20 Reasoning A website sells concert tickets.
The options for a concert are standing tickets, seated tickets or VIP tickets.
2400 tickets are sold in an hour.
658 tickets are for children.
82 of the 100 VIP tickets sold are for adults.
1324 of the tickets sold were for seated tickets.
None of the children have standing tickets.
How many adults have seated tickets?

5 Equations, inequalities and sequences

Prior knowledge

5.1 Solving equations 1

- Understand and use inverse operations.
- Solve simple linear equations.

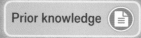
Active Learn
Homework

Warm up

1 Fluency a What is the inverse of each function machine?

$$\longrightarrow \boxed{\times 3} \longrightarrow \boxed{+6} \longrightarrow$$

$$\longrightarrow \boxed{-4} \longrightarrow \boxed{\div 5} \longrightarrow$$

b Work out each output when the input is 2.

2 Solve these equations.

a $a + 3 = 9$ **b** $b - 4 = 6$ **c** $7 + c = 10$

d $11 - d = 9$ **e** $8 = f - 5$ **f** $4h = 12$

> **Q2a hint**
>
>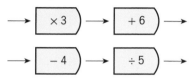
>
> $a \longrightarrow \boxed{+3} \longrightarrow 9$
>
> $\square \longleftarrow \boxed{-3} \longleftarrow 9$

Key point

An **equation** contains an unknown number (a letter) and an '=' sign.
When you **solve** an equation you work out the value of the unknown number.

3 Copy and complete to solve this equation.

$$\frac{y}{7} = 8 \qquad y \longrightarrow \boxed{\div 7} \longrightarrow 8$$

$$y = 8 \times \square \qquad \square \longleftarrow \boxed{\square} \longleftarrow 8$$

$$y = \square$$

4 Solve

a $\dfrac{a}{10} = 5$ **b** $\dfrac{b}{4} = 3$ **c** $\dfrac{c}{5} = 4$ **d** $\dfrac{d}{4} = 3$ **e** $\dfrac{e}{8} = 8$ **f** $\dfrac{f}{6} = 2$

5 Solve

a $\dfrac{p}{2} = 3.5$ **b** $\dfrac{h}{5} = 1.2$ **c** $\dfrac{w}{4} = 7.5$ **d** $\dfrac{x}{4} = \dfrac{1}{2}$

e $\dfrac{y}{5} = \dfrac{1}{4}$ **f** $\dfrac{z}{6} = \dfrac{1}{3}$ **g** $\dfrac{m}{3} = 1\dfrac{1}{2}$ **h** $\dfrac{n}{3} = 2\dfrac{1}{2}$

In an equation, the expressions on both sides of the $=$ sign have the same value. You can visualise them on balanced scales.

$$x + 3 \quad = \quad 5$$

The expressions stay equal if you use the same operations on both sides. You can use this **balancing method** to solve equations.

Example

Solve the equation $x - 3 = 7$

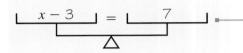

$$x - 3 \quad = \quad 7$$

Visualise the equation as balanced scales.

$$x - 3 + 3 \quad = \quad 7 + 3$$

The inverse of -3 is $+3$. Do this to both sides to keep the equation balanced.

$x = 7 + 3$

$x = 10$

Check: $x - 3 = 10 - 3 = 7$ ✓

6   Copy and complete to solve these equations using a balancing method.

a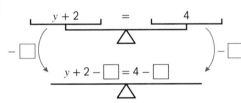

$y + 2$ $=$ 4

$y + 2 - \square = 4 - \square$

$y = \square$

b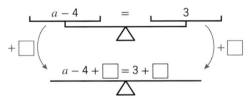

$a - 4$ $=$ 3

$a - 4 + \square = 3 + \square$

$a = \square$

c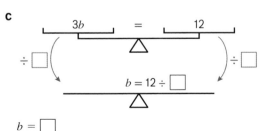

$3b$ $=$ 12

$b = 12 \div \square$

$b = \square$

d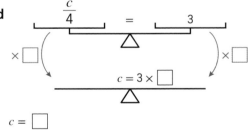

$\dfrac{c}{4}$ $=$ 3

$c = 3 \times \square$

$c = \square$

e **Reflect** Explain how you decide which operation to use on both sides.

7   Use a balancing method to solve
   a $5 + x = 7$ b $y - 7 = 11$ c $x - 4 = 1$ d $y + 2 = 6$

8   Simplify the left-hand side, then solve.
   a $2a + 5a = 28$ b $8p + 3p = 33$ c $4m - 2m = 6$ d $9q - q = 32$

9   **a**   Solve $n + n + n = 39$.   **(1 mark)**

       b   Solve $\frac{x}{5} = 2$.   **(1 mark)**

10   **Reasoning**   **a**   Copy and complete this equation for the sum of these angles.

       $\square + \square + \square = 180°$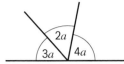

       b   Solve your equation to find the value of a.

11   **Reasoning**   The perimeter of this rectangle is 30 cm.

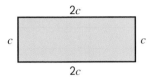

       a   Write an equation for the perimeter.

       b   Work out the value of c.

       c   Work out the length of the longer side.

12   **Problem-solving**   Calculate the size of the largest angle.

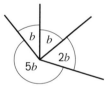

13   The perimeter of this rectangle is 60 cm. Work out the length and width.

       (3 marks)

Exam tip

Make sure you write down the length and the width.

14   A T-shirt costs £s. The cost of 4 T-shirts is £36.

       a   Write an equation involving s.

       b   Solve your equation to find the cost of a T-shirt.

15   Alex has sold 11 games from his collection of n games. He now has 18 games.

       a   Write an equation involving n.

       b   Solve your equation to find the original number of games in his collection.

5.2 Solving equations 2

- Solve two-step equations.

Warm up

1 Fluency Simplify

a $8d - 4d$ **b** $5f + 3f$ **c** $2p + 3 + p + 9$ **d** $3 - 2b - 4 + 3b$

2 Copy and complete the table for each function machine.

a
Input → ×4 → +2 → Output

b
Input → ×3 → −4 → Output

Input	Output
1	
2	
3	
4	
5	

3 Copy and complete these inverse function machines.

a
2 → ×3 → +4 → 10

2 ← ⬡ ← ⬡ ← 10

b
30 → ÷5 → −2 → 4

30 ← ⬡ ← ⬡ ← 4

4 a Solve the equation $2x + 5 = 9$ using this function machine.

x → ×2 → +5 → 9

□ ← ⬡ ← ⬡ ← 9

b Solve these equations using function machines.

  i $5k - 7 = 8$

  ii $4w + 8 = 28$

  iii $6w - 3 = 24$

Key point

To solve an equation, identify the operations used in the equation and then use inverse operations in reverse order.

Solve the equation $3a + 7 = 13$.

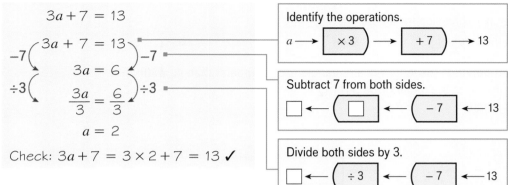

$$3a + 7 = 13$$

Identify the operations.

$$a \longrightarrow \boxed{\times 3} \longrightarrow \boxed{+ 7} \longrightarrow 13$$

Subtract 7 from both sides.

$$\square \longleftarrow \boxed{\square} \longleftarrow \boxed{-7} \longleftarrow 13$$

Divide both sides by 3.

$$\square \longleftarrow \boxed{\div 3} \longleftarrow \boxed{-7} \longleftarrow 13$$

$$3a + 7 = 13$$
$$3a = 6$$
$$\frac{3a}{3} = \frac{6}{3}$$
$$a = 2$$

Check: $3a + 7 = 3 \times 2 + 7 = 13$ ✓

5 Copy and complete.

a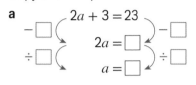

$$2a + 3 = 23$$
$$2a = \square$$
$$a = \square$$

b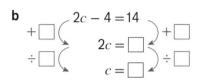

$$2c - 4 = 14$$
$$2c = \square$$
$$c = \square$$

6 Solve these equations.

a $2a + 1 = 5$ **b** $2a - 1 = 5$ **c** $3a + 2 = 8$

d $7f + 12 = -2$ **e** $-5c + 12 = 2$ **f** $-4d - 7 = 1$

7 Solve these equations.

a $3a - 5 = 4$ **b** $2p - 4 = -5$ **c** $5x + 4 = 3$

d $3y + 4 = 6$ **e** $8t + 2 = -3$ **f** $3a + 1 = 8$

Exam-style question

8 Solve $2w + 5 = 16$. **(1 mark)**

Exam tip

Write your solution clearly.

$w = \square$

9 **Reasoning**

a Write an equation for the diagram.

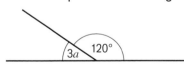

b Solve your equation to find the value of a.

10 **Reasoning** The sizes of the angles on a straight line are $a + 25°$, $2a + 40°$ and $55° - a$.

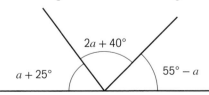

Find the value of a.

11 **Problem-solving** I think of a number. I multiply it by 6 and subtract 8.
The result is 46.
Find the number.

12 **Reasoning** The length of each side of a square is $3y - 3$ centimetres.
The perimeter of the square is $36\,\text{cm}$.

  a Draw a diagram.

  b Write an expression for the perimeter.

  c Find the value of y.

13 **Problem-solving** The length of a rectangle is $3\,\text{cm}$ greater than its width.
The perimeter of the rectangle is $54\,\text{cm}$.
Find its length.

14 Copy and complete to solve the equation.

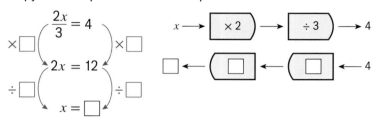

15 Solve these equations.

  a $\dfrac{4c}{5} = 4$ **b** $\dfrac{5d}{12} = 10$ **c** $\dfrac{7e}{2} = 21$ **d** $-\dfrac{2f}{3} = 2$

16 Copy and complete to solve the equation.

$$-\square\left(\dfrac{x}{2} + 3 = 7\right) - \square$$
$$\dfrac{x}{2} = 4$$
$$\times\square\left(\phantom{\dfrac{x}{2}}\right)\times\square$$
$$x = \square$$

$$x \longrightarrow \boxed{\div 2} \longrightarrow \boxed{+3} \longrightarrow 7$$
$$\square \longleftarrow \boxed{\square} \longleftarrow \boxed{\square} \longleftarrow 7$$

17 Solve these equations.

  a $\dfrac{a}{6} + 2 = 4$ **b** $\dfrac{b}{3} - 8 = 2$

  c $\dfrac{c}{4} + 6 = 9$ **d** $\dfrac{d}{2} - 4 = 2$

  e $\dfrac{e}{4} - 8 = -4$ **f** $\dfrac{f}{9} + 2 = 5$

  g **Reflect** Explain how you can check your solutions to equations.

18 Solve

  a $\dfrac{1}{2}x = 8$ **b** $\dfrac{x}{6} = 20$

  c $\dfrac{3}{4}x = 9$ **d** $\dfrac{2}{5}x = 12$

  e $\dfrac{2x}{3} = 3$ **f** $\dfrac{x}{2} = \dfrac{5}{7}$

5.3 Solving equations with brackets

Active Learn
Homework

- Solve linear equations with brackets.
- Solve equations with unknowns on both sides.

Warm up

1 **Fluency** Expand **a** $3(a+7)$ **b** $5(b-6)$

2 Expand

a $2(2a+3)$ **b** $5(6p-4)$ **c** $3(6-7c)$ **d** $-4(3-2x)$

3 Solve

a $4e+5=29$ **b** $4a+4=0$ **c** $-3b+5=-4$

d $-7g-4=-1$ **e** $\dfrac{3x}{2}=12$ **f** $\dfrac{x}{3}+5=2$

> **Q3 hint**
> Use inverse operations.

Key point

In an equation with **brackets**, you can expand the brackets first.

4 Expand and solve

a $3(d-5)=15$ **b** $6(b+5)=30$ **c** $3(2d-5)=27$

d $4(m-4)=12$ **e** $7(4-c)=35$ **f** $-2(e+2)=-10$

g $-3(7-f)=-3$ **h** $2(m-5)=-7$ **i** $5(n+3)=19$

5 **Reasoning** The diagram shows the plan of a room. Lengths are in metres. The area of the room is $45\,\text{m}^2$. Write an equation and solve it to find b.

$b+5$

5

6 **Reasoning** Steve is 30 years older than his son Jenson. He is also 11 times as old as Jenson. Write an equation and solve it. Use x for Jenson's age. How old is

a Steve **b** Jenson?

7 Copy and complete to solve the equation.

$$\frac{5x+3}{2}=6$$
$$5x+3=12$$
$$5x=9$$
$$x=\frac{9}{5}$$

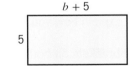

$x \to \boxed{\times 5} \to \boxed{+3} \to \boxed{\div 2} \to 6$

$\Box \leftarrow \boxed{\Box} \leftarrow \boxed{\Box} \leftarrow \boxed{\Box} \leftarrow 6$

8 Solve

a $\dfrac{4g-5}{5}=3$ **b** $\dfrac{5a-2}{3}=1$ **c** $\dfrac{3c+4}{3}=2$

d $\dfrac{2b-2}{5}=1$ **e** $\dfrac{5h+7}{4}=6$ **f** $\dfrac{2t+3}{5}=-4$

F
H

Key point

Whatever you do to one side of an equation, you must do to the other side.

Example

Solve $4d + 17 = 8d - 3$.

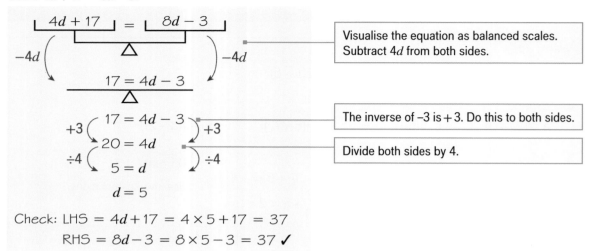

Visualise the equation as balanced scales. Subtract $4d$ from both sides.

The inverse of -3 is $+3$. Do this to both sides.

Divide both sides by 4.

Check: LHS $= 4d + 17 = 4 \times 5 + 17 = 37$

RHS $= 8d - 3 = 8 \times 5 - 3 = 37$ ✓

9   Solve these equations.

   a $2a = a + 14$ **b** $3c - 1 = c + 9$ **c** $5p - 7 = 2p + 11$

   d $2a + 9 = a + 5$ **e** $6v - 7 = 3v + 7$ **f** $3e = 7e - 18$

   g $2h + 7 = 8h - 1$ **h** $40 - 3x = 1$ **i** $9 - 5x = 3x + 1$

   j $1 - 6x = 9 - 7x$ **k** $8 + 3x = 1 - 4x$ **l** $3 - 9x = 5 - 6x$

   m $3(x - 6) = 2x + 4$ **n** $4(m + 2) = 6m - 3$ **o** $2(n + 1) = 8n - 1$

   p **Reflect** Explain how you decide which term to subtract first.

10   Expand the brackets on both sides, then solve these equations.

   a $3(s + 4) = 4(s - 4)$ **b** $4(f + 5) = 6(f + 4)$ **c** $3(x - 2) = 6(x + 3)$

   d $8(m - 2) = 4(m + 9)$ **e** $6(2y - 7) = 3(5y + 6)$ **f** $4(7t - 5) = 2(9t + 5)$

11   **Problem-solving** Work out the length and width of the rectangle.

                  $9 - 7x$ cm

   $1 - 6y$ cm $3y + 2$ cm

                 $10 - x$ cm

12   Copy and complete to solve these equations.

   a $\times 2 \left(\dfrac{x + 3}{2} = x - 1 \right) \times 2$

                $x + 3 = 2(x - 1)$

   b $\times \square \left(\dfrac{3 - x}{2} = 3x - 1 \right) \times \square$

                $3 - x = 2(3x - 1)$

Exam-style question

13   Solve $\dfrac{1 - w}{2} = 2w - 5$.   **(3 marks)**

5.4 Introducing inequalities

- Use correct notation to show inclusive and exclusive inequalities.
- Show inequalities on a number line.
- Write down whole numbers which satisfy an inequality.
- Solve simple linear inequalities.

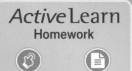
*Active*Learn
Homework

Warm up

1 **Fluency** Write the correct sign ($<$, $>$ or $=$) between each pair of numbers to make a true statement.

   a 4 ... 6   **b** 5 ... 2   **c** 15 ... 8   **d** 4.8 ... 4.79   **e** 4.5 ... 4.5

Key point

You can show solutions to inequalities on a number line.
An empty circle $\bigcirc$ shows the value is **not** included. A filled circle $\bullet$ shows the value is included.
An arrow $\bigcirc\!\!\longrightarrow$ shows that the solution continues in that direction towards infinity.

Example

Use a number line to show the values that satisfy each inequality.

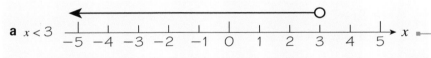

a $x < 3$

This includes all the numbers less than 3 (not including 3).

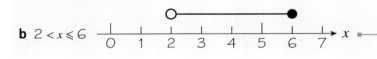

b $2 < x \leqslant 6$

This includes all the numbers greater than 2 (not including 2) and less than or equal to 6 (including 6).

2 Draw six number lines from −5 to +5. Show these inequalities.

   a $x > 1$ **b** $x \leqslant 3$ **c** $-2 < x < 5$ **d** $-4 \leqslant x < 1$

3 Write down the inequalities shown on these number lines.

   a

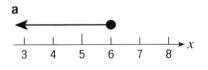

   b

   c

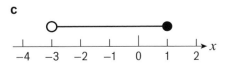

   d

4 Write down the integer values of x that satisfy each of these inequalities.

    a $4 < x < 6$ **b** $0 \leqslant x < 4$ **c** $-1 < x < 5$ **d** $-3 \leqslant x \leqslant 3$

5 **Reasoning** $n < 3$
Write an inequality for

    a $2n$ **b** $3n$ **c** $5n$ **d** $3n + 1$

6 Solve these inequalities. Show each solution on a number line.

    a $x + 3 < 7$ **b** $x + 5 \geqslant 1$ **c** $2x \leqslant 12$ **d** $\dfrac{x}{3} > 2$

    e $5x - 7 < 3$ **f** $3x + 7 \geqslant 1$ **g** $\dfrac{x}{4} + 5 \leqslant 3$ **h** $\dfrac{x}{3} - 2 > 0$

    i $\dfrac{x + 2}{3} \leqslant 4$ **j** $\dfrac{x - 5}{2} > 0$ **k** $3(x + 4) \leqslant 9$ **l** $2(2x + 3) \geqslant 22$

7 Solve

    a $2x + 6 < 5x$ **b** $7x > 3x + 12$ **c** $8x > 5x - 18$

    d $5x + 3 \geqslant 2x + 9$ **e** $9x - 7 < 5x + 3$ **f** $7 - 2x \geqslant 3x + 2$

    g $6 - 5x \leqslant 2 - 3x$ **h** $3 - 5x \geqslant 4 - 7x$ **i** $10 - 3x > 2x - 1$

Exam-style question

8 **a** On the number line, show the inequality $x > 3$.

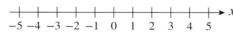
    (2 marks)

    b $-2 < y \leqslant 4$ where y is an integer.
    Write down all the possible values of y.   **(2 marks)**

    c Solve $5x + 2 < 3x + 10$.   **(3 marks)**

> **Exam tip**
> The solution to an inequality is also an inequality. Do not use = signs in your working.

Exam-style question

9 The sum of a number and 2 more than the number is less than 20.
What could the number be?   **(4 marks)**

> **Exam tip**
> Write an inequality. Use n for the number.

---

10 **Problem-solving / Reasoning** I think of a number, add 2 and multiply by 4.
My answer is greater than when I multiply the number by 3 and add 2.
Write an inequality.
Find three possible integer values for my number.

5.5 More inequalities

- Solve two-sided inequalities.

*Active*Learn
Homework

Warm up

1 **Fluency** What are the possible integer values of x?
$$-5 \leqslant x < 4$$

2 Solve

a $x+4 < 2$ **b** $x+2 < -3$ **c** $3x+4 < 16$ **d** $\dfrac{x}{4} > -3$

e $2(x-3) \geqslant -4$ **f** $5(x+2) > 15$ **g** $4(x-3) < 24$ **h** $\dfrac{x}{2}+4 \leqslant 6$

3 Show each pair of inequalities on a number line. Find the integer values of x which satisfy both inequalities. The first part has been done for you.

a $x \geqslant 2$ and $x \leqslant 5$ **b** $x > -3$ and $x < 2$

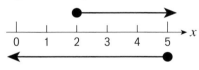

c $x \leqslant 4$ and $x < -1$ **d** $x > -1$ and $x > 3$

Exam-style question

4 Find the integer value of x that satisfies both of the inequalities $x+4 > 8$ and $2x-7 < 5$. **(2 marks)**

Exam tip

Solve each inequality. Show your solutions on a number line to get the final answer.

5 Solve

a $4 \leqslant 2x \leqslant 6$ **b** $-9 \leqslant 3x \leqslant 9$ **c** $-7 < 5x \leqslant 20$

$$\div 2 \left(\begin{array}{c} 4 \leqslant 2x \leqslant 6 \\ \square \leqslant x \leqslant \square \end{array} \right) \div 2 \qquad \div 3 \left(\begin{array}{c} -9 \leqslant 3x \leqslant 9 \\ \square \leqslant x \leqslant \square \end{array} \right) \div 3 \qquad \div \square \left(\begin{array}{c} -7 < 5x \leqslant 20 \\ \dfrac{-7}{5} < x \leqslant \square \end{array} \right) \div \square$$

d $-16 < 4x \leqslant 0$ **e** $-5 < 2x < 2$ **f** $-10 < 3x < 3$

Key point

You can solve **two-sided inequalities** using the balancing method.

Example

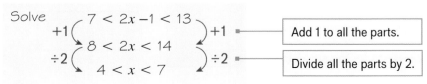

Solve

$$+1 \left(\begin{array}{c} 7 < 2x - 1 < 13 \\ 8 < 2x < 14 \end{array} \right) +1 \quad \text{—} \quad \boxed{\text{Add 1 to all the parts.}}$$

$$\div 2 \left(\begin{array}{c} 8 < 2x < 14 \\ 4 < x < 7 \end{array} \right) \div 2 \quad \text{—} \quad \boxed{\text{Divide all the parts by 2.}}$$

6 Solve these two-sided inequalities.

a $3 < x + 1 \leqslant 5$ b $0 \leqslant x + 1 \leqslant 5$ c $0 \leqslant x - 2 < 5$

d $-3 < x - 2 \leqslant 5$ e $-3 < x + 6 < 0$ f $-5 < x - 2 < -1$

7 Solve these two-sided inequalities.

a $5 < 3x - 1 < 14$ b $2 < 3x + 5 \leqslant 17$ c $1 \leqslant 4x - 4 < 11$

d $-1 < 2x + 1 < 5$ e $-7 \leqslant 3x - 1 < 2$ f $-6 \leqslant 4x - 6 \leqslant 15$

8 Solve each two-sided inequality and show its solution set on a number line.

a $-2 \leqslant 3x - 4 \leqslant 5$ b $5 \leqslant 4(x + 3) < 15$ c $10 < 2(3x + 4) \leqslant 24$

$\underset{0 \quad 1 \quad 2 \quad 3}{\longmapsto} x$ $\underset{-2 \quad -1 \quad 0 \quad 1}{\longmapsto} x$ $\underset{0 \quad 1 \quad 2 \quad 3}{\longmapsto} x$

9 On a number line you can see that $-3 < -2$

$\underset{-3 \quad -2 \quad -1 \quad 0 \quad 1 \quad 2 \quad 3}{\vdash\quad\vdash\quad\vdash\quad\vdash\quad\vdash\quad\vdash\quad\vdash}$

a Multiply both sides of the inequality by -1. Is the inequality still true?

b **Reflect** Explain what happens to the inequality sign when you multiply both sides of an inequality by -1.

c $-x < -2$. Write an inequality for x.

> **Key point**
>
> When you multiply or divide both sides of an inequality by a negative number, you reverse the inequality sign.

10 Solve these inequalities.

a $5 < -x$ b $5 < -x + 3$ c $5 < 2 - x$ d $4 \leqslant -2x$

e $6 < -2x$ f $6 \leqslant -3x$ g $1 \leqslant 1 - 2x$ h $6 \leqslant 1 - 2x$

11 Solve the inequality $11 - 2x < 2 - 5x$.
What is the smallest integer that satisfies it?

12 Solve these two-sided inequalities and show each solution set on a number line.

a $-5 < -x < 2$ b $-5 \leqslant -x < -2$ c $-6 < -2x < 4$

d $-6 \leqslant -3x < 6$ e $-6 \leqslant -3x < -3$ f $-4 < -2x + 6 < 4$

g $4 < 3(-2x + 1) \leqslant 7$ h $5 \leqslant 3(5 - 2x) < 12$

13 **Problem-solving** Harry has three parcels. The first parcel has a mass of x kg.
The second parcel has a mass twice that of the first parcel.
The third parcel has a mass 3 kg less than the first parcel.
The total mass of the parcels is less than or equal to 22 kg.
What is the largest possible mass of the smallest parcel?

14 **Problem-solving** The perimeter of the square P is greater than the perimeter of the rectangle Q.

a Explain why $x > 2$.

b Work out the range of values for the side length of the square.

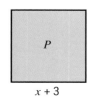
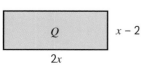

5.6 Using formulae

- Substitute values into formulae and solve equations.
- Change the subject of a formula.
- Know the difference between an expression, an equation and a formula.

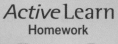
Active Learn
Homework

Warm up

1 **Fluency** What is the inverse of each operation? $+2$ -10 $\div 3$ $\times 8$

2 $T = pq + m$

Work out the value of T when

a $m = 5$, $p = 6$ and $q = 7$ **b** $m = -3$, $p = -2$ and $q = 4$ **c** $m = 7$, $p = -5$ and $q = -3$

3 $s = \frac{1}{2}at^2$

Work out the value of s when

a $a = 3$ and $t = 8$ **b** $a = 4$ and $t = -5$ **c** $a = -7$ and $t = -4$

Key point

You can substitute values into a formula and solve the resulting equation to work out the other value.

4 $A = bh$

a Copy and complete to find the value of h when $A = 36$ and $b = 4$.

$A = b \times h$

$36 = 4 \times h$

$\Box = h$

b Find the value of h when $A = 42$ and $b = 7$.

c Find the value of b when   **i** $A = 65$, $h = 5$ **ii** $A = 144$, $h = 6$

5 $y = 4x - 5$

Work out the value of x when

a $y = 3$ **b** $y = -31$ **c** $y = 75$ **d** $y = -6$

6 $V = lwh$

a Work out the value of h when $V = 100$, $l = 10$ and $w = 2$.

b Work out the value of l when $V = 40$, $h = \frac{1}{2}$ and $w = 4$.

c Work out the value of w when $V = 150$, $h = 5$ and $l = 3$.

7 $y = \frac{x}{7}$

Work out the value of x when

a $y = 3$ **b** $y = 8$ **c** $y = -5$

d $y = 4.3$ **e** $y = \frac{1}{7}$ **f** $y = \frac{3}{4}$

8 $P = a + 2b$

  a Work out the value of a when

    i $P = 11$ and $b = 4$ **ii** $P = 7$ and $b = 5.2$

  b Work out the value of b when

    i $P = 9$ and $a = 14$ **ii** $P = -23$ and $a = 6$

 9 **Future skills** The formula to work out the speed of an object is $s = \dfrac{d}{t}$, where d = distance in miles, t = time in hours and s = speed in mph.

  a Work out the distance when

    i $s = 50$ mph and $t = 4$ hours   **ii** $s = 65$ mph and $t = 5.5$ hours

  b Work out the time when

    i the distance is 220 miles and the speed is 55 mph

    ii a plane travels a distance of 1800 miles at 450 mph

 10 **Problem-solving** A doctor uses this formula to calculate the dose for children.

$$\text{Child's dose} = \text{adult dose} \times \left(\frac{\text{child's weight in pounds}}{150} \right)$$

The adult dose is 500 mg per day.
Estimate the age of each child.
Show your working.

Jon	120 mg
Fiona	165 mg

Average weight for girls and boys		
Age (years)	**Average weight (pounds)**	
	Girls	**Boys**
2	26.5	27.5
3	31.5	31.0
4	34.0	36.0
5	39.5	40.5
6	44.0	45.5
7	49.5	50.5

Exam-style question

11 $y = 3x + c$ $x = 4.5$ $c = 2.4$

  a Work out the value of y.   **(2 marks)**

  $y = 3x + c$ $y = 23.9$ $c = -1.6$

  b Work out the value of x.   **(2 marks)**

Exam tip

First substitute. Then solve the equation.

---

12 **Future skills** You can use this formula to work out the density of an object: $d = \dfrac{m}{V}$, where d = density, m = mass and V = volume.
Work out the mass when the density is 2.7 g/cm^3 and the volume is 24 cm^3.

13 **Future skills** Use the formula $v^2 = u^2 + 2as$ to work out s when $u = 10$, $a = 5$ and $v = 20$.

14 **Future skills** Use the formula $s = ut + \frac{1}{2}at^2$ to work out the value of a when $s = 200$, $u = 5$ and $t = 8$.

Key point

The **subject** of a formula is the letter on its own, on one side of the equals sign.

Example

Rearrange $y = 2x + 5$ to make x the subject of the formula.

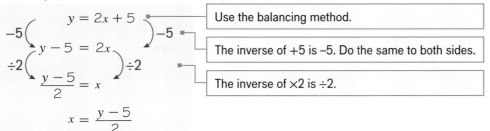

$$y = 2x + 5$$

Use the balancing method.

$-5 \quad \quad \quad -5$

$$y - 5 = 2x$$

The inverse of $+5$ is -5. Do the same to both sides.

$\div 2 \quad \quad \quad \div 2$

$$\frac{y - 5}{2} = x$$

The inverse of $\times 2$ is $\div 2$.

$$x = \frac{y - 5}{2}$$

15 Rearrange each formula to make the letter in the square brackets the subject.

a $y = x + 4$ $[x]$ b $y = x - 7$ $[x]$ c $x + 5 = y$ $[x]$

d $s = 8 + t$ $[t]$ e $y = -x + 2$ $[x]$ f $m = 6 - v$ $[v]$

g $v = a + t$ $[t]$ h $v = a + t$ $[a]$ i $x = u - s$ $[s]$

j $P = 5d$ $[d]$ k $y = 4x$ $[x]$ l $3x = y$ $[x]$

m $y = mx$ $[x]$ n $y = mx$ $[m]$ o $P = IV$ $[I]$

16 Rearrange each formula to make the letter in the square brackets the subject.

a $y + 1 = 2x$ $[x]$ b $y - 1 = 2x$ $[x]$ c $y - 3 = 5x$ $[x]$

d $y = 5x + 3$ $[x]$ e $y = 4x - 3$ $[x]$ f $4x - 2 = y$ $[x]$

g $M = 7N - 5$ $[N]$ h $M = 5N - 7$ $[N]$ i $s = 4t - 1$ $[t]$

j $V = \dfrac{W}{3}$ $[W]$ k $r = \dfrac{m}{4}$ $[m]$ l $k = \dfrac{n}{2}$ $[n]$

m $t = \frac{1}{2}x$ $[x]$ n $T = \dfrac{D}{V}$ $[D]$ o $d = \dfrac{m}{v}$ $[m]$

Exam-style question

17 $F = 3t + 7$

   a Work out the value of F when $t = 4$. **(2 marks)**

   b Make t the subject of the formula. **(2 marks)**

Exam tip

Show all your working out clearly.

Key point

A **formula** shows the relationship between two or more variables (letters).

An **equation** contains an unknown number (a letter) and an $=$ sign.

You can solve it to find the value of the letter.

18 a State whether each of these is an expression, an equation or a formula.

    i $x + 3 = 3(2x - 4)$ ii $V = IR$ iii $q(p - 8)$

    iv $y = 4x + 5$ v $10(x + 2) = 3x^2$ vi $s = \dfrac{d}{t}$

  b **Reflect**

    i What is the same about equations and formulae?

    ii What is the difference between equations and formulae?

5.7 Generating sequences

- Recognise and extend sequences.

Active Learn
Homework

Warm up

1 Fluency Find the missing numbers.

a 2, 4, 6, ☐, 10, 12, ☐   **b** 1, ☐, ☐, 7, 9, 11   **c** 10, 8, 6, ☐, 2, ☐

2 Write down the next two terms in each sequence.

a 41, 37, 33, 29, ☐, ☐   **b** 17, 14, 11, 8, ☐, ☐

Key point

A **sequence** is a pattern of numbers or shapes that follows a rule. The numbers in a sequence are called **terms**. The **term-to-term rule** describes how to get from one term to the next.

3 Write down the next two terms in each sequence.

a 1.5, 2, 2.5, 3, ☐, ☐   **b** $-\frac{2}{3}, -\frac{1}{3}, 0, \frac{1}{3}, $ ☐, ☐

c 3.5, 2.7, 1.9, 1.1, ☐, ☐   **d** −1.5, −2.5, −3.5, −4.5, ☐, ☐

e $\frac{3}{5}, -\frac{1}{5}, -1, -1\frac{4}{5}, $ ☐, ☐   **f** −10.6, −9.9, −9.2, −8.5, ☐, ☐

Key point

An **arithmetic sequence** goes up or down in equal steps.
You can describe an arithmetic sequence using the first term and the **common difference** (the difference between terms).

4 Use the first term and the common difference to generate the first five terms of each sequence.

a first term 3, common difference 0.4   **b** first term 10, common difference −0.2

c first term 7, common difference 3   **d** first term 7, common difference 2

e first term −3, common difference 2   **f** first term −7, common difference −5

Exam-style question

5 Here are the first four terms of a number sequence.

    3   5   9   17

The rule to continue this sequence is
   multiply the previous term by 2 and then subtract 1
Work out the 5th term of the sequence.   **(1 mark)**

6 Problem-solving The first term in a sequence is p.
The term-to-term rule is ×3 + 2.
The third term is 17.
Find the value of p.

Q6 hint

Write the 1st three terms in terms of p.

$p,$ ☐, ☐

7   Each sequence is made up of a pattern of sticks. For each sequence
    i   draw the next two patterns
    ii draw a table like this and complete it for each pattern

Pattern number	1	2	
Number of sticks			

    iii Is the sequence arithmetic? If so, find the common difference.
    iv Work out the number of sticks needed for the 10th pattern.

a

b

c

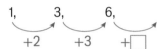

d

8   This sequence made from counters shows the first three triangular numbers.

    a   Draw the next two patterns in the sequence.
    b   Write down the number of counters in each pattern.
    c   Work out the differences between the numbers of counters.

        1,   3,   6,
         ⌣+2   ⌣+3   ⌣+☐

    d   Follow the sequence to work out the number of counters in the 10th pattern.
    e   Is this sequence arithmetic? Explain how you know.

9   This sequence made from counters shows the first three square numbers.
    a   Draw the next two patterns in the sequence.
    b   Work out the number of counters in the 8th pattern.

10 **Reasoning** The first three cube numbers are 1, 8, 27. What are the next two terms in the sequence?

Key point

In a **geometric sequence**, the term-to-term rule is 'multiply or divide by a number'.

11 Write down the next two terms in each geometric sequence.
    a   5, 50, 500, 5000, ☐, ☐   b   2, 1, $\frac{1}{2}$, $\frac{1}{4}$, ☐, ☐
    c   100, 10, 1, 0.1, ☐, ☐   d   2, 8, 32, 128, ☐, ☐
    e $\frac{1}{7}$, $\frac{2}{7}$, $\frac{4}{7}$, $\frac{8}{7}$, ☐, ☐   f   6.4, 3.2, 1.6, 0.8, ☐, ☐

12 Find the term-to-term rule for each sequence.
    a   2, 4, 8, 16, 32, 64, ...   b   100, 50, 25, 12.5, 6.25, ...
    c   1, 3, 9, 27, 81, 243, ...   d   5, –10, 20, –40, 80, –160, ...
    e   1, 10, 100, 1000, ...   f   18, 6, 2, $\frac{2}{3}$, ...

13 **Problem-solving** Write down the next three terms in

  a the arithmetic sequence that starts 1, 5, ☐, ☐, ☐

  b the geometric sequence that starts 1, 5, ☐, ☐, ☐

Key point

In a **Fibonacci** sequence, the term-to-term rule is 'add the two previous terms to get the next one'.

14 Write the next three terms in each Fibonacci sequence.

  a 1, 1, 2, 3, 5, ...

  b 3, 3, 6, 9, 15, ...

  c 5, 5, 10, 15, 25, ...

Exam-style question

15 The first three terms of a Fibonacci sequence are

   x x $2x$

   Find the 5th term of this sequence. **(2 marks)**

16 In these fraction sequences, the numerators form a sequence and the denominators form a sequence.

   Write down the next two terms in each sequence.

  a $\frac{1}{2}, -\frac{1}{3}, \frac{1}{4}, -\frac{1}{5}$, ☐, ☐

  b $\frac{1}{2}, \frac{3}{4}, \frac{5}{6}, \frac{7}{8}$, ☐, ☐

  c $\frac{2}{7}, \frac{5}{6}, \frac{8}{5}, \frac{11}{4}$, ☐, ☐

Exam-style question

17 Here are some patterns made from square tiles and triangle tiles.

 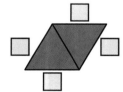 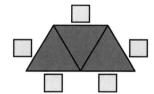

   Pattern 1   Pattern 2   Pattern 3

  a How many squares tiles are needed to make pattern number 6? **(2 marks)**

  b How many triangle tiles are needed to make pattern number 20? **(2 marks)**

Exam-style question

18 Here are the first three terms of a sequence.

   1   3   9

   Write down the two numbers that could be the 4th term and the 5th term of this sequence.
   Give the rule you have used to work out your numbers. **(2 marks)**

5.8 Using the *n*th term of a sequence

Active Learn
Homework

- Use the *n*th term to generate terms of a sequence.
- Find the *n*th term of an arithmetic sequence.

Warm up

1 Fluency Work out the value of $4n$ when

a $n = 1$ **b** $n = 2$ **c** $n = 3$ **d** $n = 4$

2 Solve   **a** $2x + 3 = 7$ **b** $2x - 7 = 15$ **c** $4x + 2 = 20$

Key point

The *n*th term of a sequence tells you how to work out the term at position n (any position). It is also called the **general term** of the sequence.

3 A sequence has *n*th term $2n + 1$. Copy and complete the table to work out the first five terms of the sequence.

n	1	2	3	4	5
Term	$2 \times ① + 1 = \square$	$2 \times ② + 1 = \square$			

4 a Write the first five terms of the sequences with these *n*th terms.

    i $3n$ **ii** $7n - 4$ **iii** $5n + 1$ **iv** $21 - 2n$

    v $14 - 3n$ **vi** $\frac{1}{2}n + 2$ **vii** $30 - 4n$ **viii** $-4n + 3$

b Look at the common difference for each sequence and its *n*th term. What do you notice?

Example

a Work out the *n*th term of the sequence 7, 11, 15, 19, 23, ...

b Is 33 a term of the sequence?

a

      $+4$ $+4$

    7,   11,   15,   19,   23, ...

$4n$ 4,   8,   12,   16,   20, ... $\Big)+3$

    *n*th term is $4n + 3$

> The common difference is 4.

> Write out the first five terms of the sequence for $4n$.
> Work out how to get from each term in $4n$ to the term in the sequence.

b

$-3 \Big($ $33 = 4n + 3$ $\Big) -3$

     $30 = 4n$

     $7.5 = 4n$

33 cannot be in the sequence because 7.5 is not an integer.

> Write an equation using the *n*th term and solve it.

> There is a 7th term and an 8th term, but not a 7.5th term.

5   Find the nth term for each sequence.

    a 2, 5, 8, 11, 14, 17, ...   **b** 2, 6, 10, 14, 18, 22, ...   **c** 2, 7, 12, 17, 22, 27, ...

    d 5, 7, 9, 11, 13, 15, ...   **e** 19, 17, 15, 13, 11, 9, ...   **f** 20, 18, 16, 14, 12, 10, ...

Exam-style question

6   The nth term of a sequence is $5n + 1$.

    Explain why 17 is not a term in this sequence.   **(2 marks)**

Exam-style question

7   Here are the first four terms of an arithmetic sequence.

    124, 122, 120, 118

    a Find an expression, in terms of n, for the nth term for this sequence.   **(2 marks)**

    b Can 9 be a term in this sequence? Give a reason for your answer.   **(2 marks)**

---

8   For each sequence, explain whether or not each number in the brackets is a term in the sequence.

    a 2, 5, 8, 11, 14, ... (50, 66)   **b** 5, 8, 11, 14, 17, ... (50, 62)

    c 1, 5, 9, 13, 17, ... (101, 150)   **d** 4, 9, 14, 19, 24, ... (168, 169)

    e 40, 35, 30, 25, 20, ... (85, 4)   **f** 5, 11, 17, 23, 29, ... (119, 72)

9   For each nth term, find the 20th term.

    a $2n$ **b** $3n + 1$ **c** $11 - 3n$

10   Find the nth term for each sequence. Use it to work out the 10th term.

    a 1, 3, 5, 7, ...   **b** 3, 6, 9, 12, ...   **c** 10, 8, 6, 4, ...   **d** 3, 7, 11, 15, ...

Exam-style question

11   Here are the first 4 terms of a sequence.

    2   8   14   20

    a i Write down the next term of the sequence.

       ii Explain how you worked out your answer. **(2 marks)**

    b Work out the 10th term of the sequence.   **(1 mark)**

Exam tip

'Explain' means show how you worked it out.

Exam-style question

12   Here is a pattern made from dots.

    a Draw the next pattern in the sequence.

                                    (1 mark)

    b Copy and complete this table for the numbers
       of dots used to make the patterns.   **(2 marks)**

4   7   10

Pattern number	1	2	3	4	5	6
Number of dots						

    c Write, in terms of n, the number of dots needed for pattern n.   **(1 mark)**

    d How many dots are needed for pattern 30?   **(1 mark)**

---

13 Solve nth term = 100 to find the first term that is bigger than 100 for each sequence.

    a 9, 18, 27, 36, 45, ...

    b 7, 10, 13, 16, 19, ...

    c 4, 9, 14, 19, 24, ...

    d 10, 15, 20, 25, 30, ...

Exam-style question

14 Here is a pattern sequence made with tiles.

    a Find an expression, in terms of n, for the number of tiles in pattern n.   **(2 marks)**

    b Sam has 35 tiles. Does he have enough to make the 20th pattern?   **(2 marks)**

    c Work out the pattern number of the biggest pattern he can make.   **(2 marks)**

---

15 Problem-solving Here is a pattern sequence of blue and white tiles.

    a Copy and complete the table for the numbers of white tiles and the numbers of blue tiles.

Pattern number	1	2	3	4	5
Number of white tiles	4	5			
Number of blue tiles	2	4			

    b Write down the nth term for the sequence of the numbers of blue tiles.

    c Write down the nth term for the sequence of the numbers of white tiles.

    d How many blue tiles are there in the 20th pattern?

    e How many white tiles are there in the 30th pattern?

    f Alex has 50 blue tiles and 45 white tiles.
       Which is the largest complete pattern she can make?

16 Write down the first five terms of the sequence with nth term

    a n^2 **b** $3n^2$ **c** n^2-1

    d $\frac{1}{4}n^2$ **e** n^2+4 **f** $65-n^2$

Exam-style question

17 The nth term of a sequence is $4n^2$.
Betty says that the 3rd term of this sequence is 144.
Is Betty correct?
Show how you work out your answer.   **(1 mark)**

---

5 Check up

Equations and formulae

1 Solve

a $c - 2 = -1$ **b** $3f + 7 = 1$ **c** $\dfrac{x}{2} + 5 = 12$ **d** $\dfrac{3a}{2} = 6$ **e** $\dfrac{d}{3} = \dfrac{1}{4}$

2 Solve

a $4(x + 3) = 24$ **b** $2(x - 2) = -14$ **c** $3a + 8 = 6a + 1$

3 This triangle has side lengths $3x$ cm, $3x$ cm and $(2x - 20)$ cm. The perimeter is 68 cm.

a Write an expression for the perimeter.

b Find the length of each side.

$3x$ $3x$

$2x - 20$

4 $V = IR$
Work out I when $V = 20$ and $R = 40$.

5 $P = 2w + 2l$
Work out l when $P = 25$ and $w = 3$.

6 Rearrange each formula to make the letter in the square brackets the subject.

a $P = \dfrac{M}{4}$ $[M]$ **b** $y = 3x - 5$ $[x]$

7 Solve   **a** $3(a + 5) = a + 21$ **b** $2(3t + 4) = 5(2t - 1)$

8 Solve   **a** $\dfrac{4x + 1}{3} = 7$ **b** $\dfrac{x + 5}{2} = x - 1$

Inequalities

9 Show each inequality on a number line.

a $x > 2$ **b** $x < -\frac{1}{2}$ **c** $-1 < x \leqslant 4$

10 Write down the inequalities shown on these number lines.

a   **b**

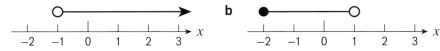

11 For each inequality, list the integers that satisfy it.

a $-3 \leqslant x < 4$ **b** $-6 < x \leqslant 1$

12 Solve each inequality.

a $x - 6 > 4$ **b** $6x \leqslant 36$ **c** $2x - 5 > 15$ **d** $2(x - 3) \geqslant 14$

13 Solve   **a** $-10 < 2x \leqslant 4$ **b** $-4 \leqslant x - 7 < 3$

Sequences

14 **a** Write down the next two terms in each sequence.

    i $\frac{1}{2}, 2, 3\frac{1}{2}, 5, \square, \square$

    ii $1, 2, 4, 8, 16, \square, \square$

    iii $100, -50, 25, -12.5, \square, \square$

  b State whether each sequence is arithmetic or geometric.

15 Here are the first four terms of a sequence.
    98, 96, 94, 92

  a Write down the next term in the sequence.

  b Write down the 7th term in the sequence.

  c Write a formula for the nth term of this sequence.

  d Can −3 be a term in this sequence? Explain how you know.

16 Here are the first five terms of a number sequence.
    2, 6, 10, 14, 18

  a Work out the nth term of this sequence.

  b Work out the 10th term.

17 Here is a sequence of patterns made from counters.

Pattern 1   Pattern 2   Pattern 3

  a Find an expression, in terms of n, for the number of counters in pattern n.

  b Lucy has 57 counters.
    Can Lucy make a pattern in this sequence using all 57 of her counters?
    Show how you work out your answer.

18 **Reflect** How sure are you of your answers? Were you mostly

Just guessing   Feeling doubtful 😐   Confident ☺

What next? Use your results to decide whether to strengthen or extend your learning.

Challenge

19 The solution to an equation is $x = 3$.
One possible question would be 'Expand and solve $2(3x + 1) = 20$'.
Write three more questions with this solution.

5 Strengthen

Active Learn
Homework

Equations and formulae

1 What is the inverse of each operation?

a $+7$ **b** -10 **c** $\times 2$

2 Use function machines to solve these equations.

a $2m + 7 = 13$ **b** $\dfrac{x}{3} - 1 = 2$

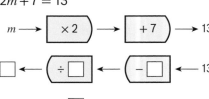

3 Use function machines to solve these equations.

a $3x + 4 = 19$ **b** $2y - 7 = 13$ **c** $5s - 12 = 18$

d $4t + 3 = 5$ **e** $27 = 3 + 4p$ **f** $50 = 9s + 14$

g $\dfrac{x}{2} + 1 = 8$ **h** $\dfrac{z}{3} - 2 = 0$ **i** $\dfrac{k}{5} + 7 = 9$

4 $F = ma$

Copy and complete to find the value of

a a when $F = 20$ and $m = 5$

$F = ma$

$20 = \square \times a$

$\dfrac{\square}{\square} = a$

b m when $F = 30$ and $a = 10$

$F = ma$

$\square = m \times \square$

$\dfrac{\square}{\square} = m$

5 $v = u + at$

Work out

a t, when $v = 20$, $u = 10$ and $a = 2$

b a, when $v = 15$, $u = 7$ and $t = 4$

6 Expand the brackets. Then solve these equations.

a $3(2x + 5) = 45$ **b** $6(x - 7) = 18$ **c** $5(4x - 4) = 100$

7 These scales are balanced.
Each box weighs x kg.

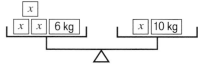

Use the balancing method to work out the weight of one box.

8 Use the balancing method to solve these equations.

a $2a + 5 = 3a$

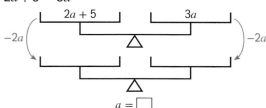

$a = \square$

b $5s = 3s + 4$

c $5x + 6 = 4x + 5$

d $4a + 1 = 5a - 3$

e $7p - 3 = 5p + 1$

f $5y + 2 = 3y + 12$

9 Expand the brackets. Then solve these equations.

a $2(5x + 7) = x - 13$　　　**b** $4(3a + 2) = 27a + 3$　　　**c** $3(4z + 6) = 4(2z + 5)$

10 a Copy and complete to make x the subject of $y = \dfrac{x}{3}$

b Copy and complete to make x the subject of $y = 4x - 1$.

11 Make x the subject of each formula.

a $n = x + 5$　　　**b** $m = x - 2$　　　**c** $y = 3x$　　　**d** $y = 3x - 2$　　　**e** $P = \dfrac{x}{2}$

Inequalities

1 Match each inequality to the number line that represents it.

a $x < 2$　　　　　　　　**b** $x > 2$　　　　　　　　**c** $x \geqslant 2$

d $-1 < x < 2$　　　　　　**e** $-1 \leqslant x < 2$　　　　**f** $-1 \leqslant x \leqslant 2$

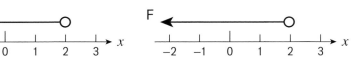

2 Follow these instructions to draw a number line that shows the inequality $n < -1$.

   a Draw a number line with –1 in the middle.

   b Draw a circle above –1.

   c Should your circle be empty or filled?

   d Draw an arrow from your circle.

$$\begin{array}{cccccccc} | & | & | & | & | & | & | \\ -4 & -3 & -2 & -1 & 0 & 1 & 2 \end{array}$$

3 Show each inequality on a number line.

   a $x \geqslant 3$ **b** $x < 8$ **c** $x \leqslant -5$

   d $x > \frac{1}{2}$ **e** $17 < x \leqslant 20$ **f** $-7 \leqslant x \leqslant -2$

4 List the integers that satisfy each inequality.

   a $2 < x < 7$ **b** $-8 < x < -3$ **c** $-1 \leqslant x < 4$ **d** $3 < x < 5$

5 Copy and complete to solve each inequality.

   a $x + 4 > 9$ **b** $4x < 12$

   c $3x - 10 \geqslant 8$ **d** $6(y + 3) < 36$

6 Copy and complete to solve each inequality.

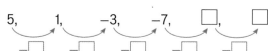

7 Copy and complete to solve each inequality.

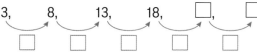

Sequences

1 Copy and complete to find the next two terms in each sequence.

   a 3,   7,   11,   15,   ☐,   ☐,

   $+☐$ $+☐$ $+☐$ $+☐$ $+☐$

   b 5,   1,   –3,   –7,   ☐,   ☐,

   $-☐$ $-☐$ $-☐$ $-☐$ $-☐$

   c 3,   8,   13,   18,   ☐,   ☐,

   ☐   ☐   ☐   ☐   ☐

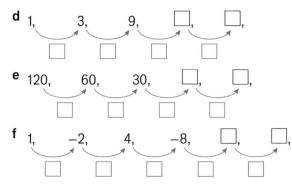

d 1, 3, 9, ☐, ☐,

e 120, 60, 30, ☐, ☐,

f 1, −2, 4, −8, ☐, ☐,

2 For each sequence in **Q1**, state whether it is arithmetic or geometric.

Arithmetic
term-to-term rule is
+ ☐ or − ☐

Geometric
term-to-term rule is
× ☐ or ÷ ☐

3 Match each sequence to its nth term.

a 2, 4, 6, 8, ...   **i** $-4n$

b 4, 8, 12, 16, ...   **ii** $2n$

c 5, 10, 15, 20, ...   **iii** $4n$

d −4, −8, −12, −16, ...   **iv** $5n$

4 For sequence **a** in **Q1**

a What is the term-to-term rule?

b Compare the terms of the sequence with the sequence $4n$.

Term	3	7	11	15
4n	4	8	12	16

What do you notice?

c Copy and complete the nth term of the sequence 3, 7, 11, 15, ... $4n - ☐$

d Check that your nth term works.

  i Substitute $n = 1$ into your answer for part **c**.
    What number do you get?

  ii Substitute $n = 2$ into your answer for part **c**.
    What number do you get?

e Substitute $n = 10$ into your answer for part **c** to find the 10th term.

5 The nth term of a sequence is $2n + 5$.

a Solve the equation $2n + 5 = 20$ to see if 20 is a term in the sequence.

b Is 33 a term in the sequence? Write and solve an equation to explain.

5 Extend

Exam-style question

1 Here is a number machine.

input ⟶ $\boxed{\times 4}$ ⟶ $\boxed{-7}$ ⟶ output

    a Work out the **output** when the input is 6. **(1 mark)**

    b Work out the **input** when the output is 13. **(2 marks)**

---

2 Solve these equations.

    a $1.7 + 2f = 8.4$ **b** $3.8 + 3g = 11.7$ **c** $4h - 0.22 = 1.1$

    d $\dfrac{a}{4} = 5$ **e** $\dfrac{b}{3} + 2 = \dfrac{7}{3}$ **f** $\dfrac{c}{8} - 2 = 0.1$

3 Solve

    a $\dfrac{9d - 4}{2} = 1$ **b** $\dfrac{18e + 12}{3} = -1$ **c** $\dfrac{x + 7}{3} = 2x + 1$

4 Here is a sequence: 120, 113, 106, ...

    a Work out the nth term.

    b Which term is the first negative term? State the term and its position in the sequence.

Exam-style question

5 Here is a rectangle. All measurements are in centimetres.

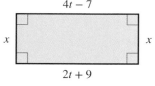

The area of the rectangle is $100\,\text{cm}^2$.
Show that $x = 4$. **(4 marks)**

---

6 **Problem-solving** Work out the size of the largest angle.

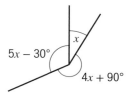

7 **Problem-solving** Jug A contains w litres of water. Jug B contains $w + 1.5$ litres of water.
Jim pours 0.5 litres out of jug A and into a glass. Jug B now has twice as much water as jug A.
Work out the value of w.

8 **Reasoning** The perimeter of rectangle A is equal to the perimeter of rectangle B.

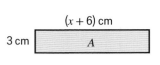
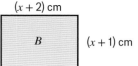

3 cm

$(x + 6)$ cm

A

$(x + 2)$ cm

B

$(x + 1)$ cm

a Write an equation using the perimeters.

b Solve the equation to find the perimeters.

Exam-style question

9 $10^x \times 100^y = 10^z$

Show that $z = x + 2y$. **(2 marks)**

---

10 Solve these inequalities.

a $\frac{1}{3}x \leqslant 2$ **b** $\frac{1}{5}x < 10$ **c** $\frac{2}{5}x > 2$ **d** $\frac{3}{4}x > 2$

11 Solve these inequalities.

a $3(x - 2) \geqslant 2x - 1$

b $3(x + 2) < 5x - 8$

c $2 - x < 3(1 - x)$

12 Find the integer values of x that satisfy each inequality.

a $\frac{7}{8} < \frac{1}{4}x \leqslant 2$ **b** $8 \leqslant 4.5x - 1 \leqslant 11$

c $-6 \leqslant \frac{2}{3}x + 4 < 0$ **d** $-1 < \frac{1}{3}(2x - 5) < 2$

e $-1 < -6x + 11 < 1$ **f** $0.12 \leqslant 4x - 1 \leqslant 1.8$

13 **Problem-solving** Ben chooses three different whole numbers.

The first number is a prime number.

The second number is 7 times the first number.

The third number is 10 more than the second number.

The sum of the three numbers is greater than 100 and less than 160.

a Write an expression for the sum of the three numbers.

b What are the three numbers?

> **Q13a hint**
>
> Use x for the first number.

> **Q13b hint**
>
> Write an inequality using your answer to part **a** and the information in the question.

14 Here are the first five terms of an arithmetic sequence.

10, 16, 22, 28, 34

Peter says that 555 is a term in this sequence.

Explain why Peter is wrong.

15 Change the subject of each formula to the letter in brackets.

a $M = n - 2x$ $[x]$

b $P = \dfrac{2k}{h}$ $[k]$

c $C = \dfrac{d}{3} + f$ $[d]$

5 Test ready

Summary of key points

To revise for the test:

- Read each key point, find a question on it in the mastery lesson, and check you can work out the answer.

- If you cannot, try some other questions from the mastery lesson or ask for help.

Key points

1 An **equation** contains an unknown number (a letter) and an '=' sign.
   When you **solve** an equation you work out the value of the unknown number. → **5.1**

2 In an equation, the expressions on both sides of the = sign have the same value.
   The expressions stay equal if you use the same operations on both sides.
   You can use this **balancing method** to solve equations. → **5.1**

3 To solve an equation, identify the operations used in the equation and
   then use inverse operations in reverse order. → **5.2**

4 In an equation with **brackets**, expand the brackets first. → **5.3**

5 Whatever you do to one side of an equation, you must do to the other. → **5.3**

6 You can show solutions to inequalities on a number line.

   - An empty circle ○ shows that the value is not included.

   - A filled circle ● shows that the value is included.

   - An arrow ○──→ shows that the solution continues in that direction
     towards infinity. → **5.4**

7 An **integer** is a positive or negative whole number or zero. → **5.4**

8 You can solve inequalities using the balancing method. → **5.5**

9 When you multiply or divide both sides of an inequality by a negative number,
   you reverse the inequality sign. → **5.5**

10 You can substitute values into a formula and solve the resulting equation to work out
   the other value. → **5.6**

11 The **subject** of a formula is the letter on its own, on one side of the equals sign. → **5.6**

12 A **formula** shows the relationship between two or more variables (letters). → **5.6**

13 A **sequence** is a pattern of numbers or shapes that follows a rule. → **5.7**

14 The numbers in a sequence are called **terms**. → **5.7**

15 The **term-to-term rule** describes how to get from one term to the next. → **5.7**

16 An **arithmetic sequence** goes up or down in equal steps. You can describe
   an arithmetic sequence using the first term and the **common difference**. → **5.7**

17 In a **geometric sequence**, the term-to-term rule is 'multiply or divide by a number'. → **5.7**

18 In a **Fibonacci sequence**, the term-to-term rule is 'add the two previous terms to get
   the next one'. → **5.7**

19 The nth term of a sequence tells you how to work out the term at position n (any position).
   It is also called the **general term** of the sequence. → **5.8**

Sample student answers

1 Here is a rectangle.

20   Diagram not accurately drawn

$2(x - 1)$ $x + 4$

20

All measurements are given in centimetres.
Find the perimeter of the rectangle.   **(4 marks)**

$$2(x - 1) = x + 4$$
$$2x - 2 = x + 4$$
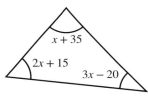
$$x - 2 = 4$$
$$x = 6$$

Is '$x = 6$' the correct final answer? Explain your answer.

2 Here is a triangle.

$x + 35$

$2x + 15$

$3x - 20$

Show that the size of the smallest angle in this triangle is $55°$.   **(3 marks)**

$$x = 25°$$
$$3x - 20 = 55°$$

Has the student clearly shown:
- how they found the value of x
- that the **smallest** angle in the triangle is $55°$?

Write a better answer, clearly showing all the working.

5 Unit test

Active Learn
Homework

1 Solve these equations.

    i $\dfrac{a}{5} = 3$ **ii** $b - 8 = 12$ **iii** $6p + 3 = 33$ **(3 marks)**

2 $a = \dfrac{b}{c}$

    Work out the value of b when

    i $a = 2$ and $c = 3$ **ii** $a = 8$ and $c = -4$ **iii** $a = \frac{1}{4}$ and $c = 12$ **(3 marks)**

3 Make the letter in square brackets the subject of each formula.

    i $d = 5t$ [t]   **ii** $y = x - 6$ [x]   **iii** $m = \dfrac{n}{4}$ [n]   **(3 marks)**

4 Make p the subject of $q = 3p + 8$.   **(2 marks)**

5 Solve

    i $\dfrac{5x}{6} = 10$ **ii** $\dfrac{x}{5} + 3 = 7$ **(2 marks)**

6

| expression   multiple   equation   factor   formula   identity   term   inequality |

    Choose words from the box to make this statement correct.

        at is a/an _____ in the _____ $v = u + at$ **(2 marks)**

7 The diagram shows a rectangle.

    The perimeter of the rectangle is 146 cm.
    Find the length of the shorter side.

$(3x - 6)$ cm
$(2x + 3)$ cm
     (3 marks)

8 a Solve $4(x - 4) = 14$.   **(2 marks)**

    b $-5 < n \leqslant 3$

       n is an integer. Write down all the possible values of n.   **(2 marks)**

9 The nth term of a sequence is $n^2 + 2$.
    Write down the first four terms of this sequence.   **(2 marks)**

10 Solve $\dfrac{x}{3} = 1\frac{1}{2}$.   **(1 mark)**

11 Solve $60 - 4x = 5x + 6$.   **(2 marks)**

12 Solve $4x - 3 = 2(x - 1)$.   **(3 marks)**

13 Solve $\dfrac{3 - x}{2} = 2x + 5$.   **(3 marks)**

14 Draw four number lines from −5 to +5. Show these inequalities.

    i $-4 \leqslant x \leqslant 4$ **ii** $x \leqslant 3$ **iii** $-3 < x < 0$ **iv** $-2 < x$ **(4 marks)**

15 Here is an isosceles triangle.

The angles at B and C are equal.
Show that $y = 40°$.

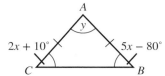

(4 marks)

16 Find the integer solutions that satisfy each inequality.

   a $5 < x + 6 < 11$ **(2 marks)**

   b $2 < 2x + 10 \leqslant 8$ **(2 marks)**

17 **a** Solve $13t = 9t + 8$.   **(2 marks)**

   b On the number line, show the set of values of x for which $-3 < x + 2 \leqslant 2$.   **(3 marks)**

```
  |   |   |   |   |   |   |   |   |   |   |   |  → x
 −6  −5  −4  −3  −2  −1   0   1   2   3   4   5
```

18 Penny chooses three different whole numbers.
The second number is 5 times the first number.
The third number is 20 less than the second number.
The sum of the three numbers is greater than 40 and less than 50.

   a Write an expression for the sum of the three numbers.   **(2 marks)**

   b What are the three numbers?   **(2 marks)**

19 Here are the first five terms of an arithmetic sequence.
    8, 13, 18, 23, 28

   a Write the next two terms in the sequence.   **(1 mark)**

   b Write an expression, in terms of n, for the nth term.   **(2 marks)**

20 Here are the first five terms of an arithmetic sequence.
    8, 15, 22, 29, 36
Peter says that 163 is a term in this sequence.
Explain why Peter is wrong.   **(2 marks)**

21 Here are the first five terms of a Fibonacci sequence.
    2   2   4   6   10
Write down the next two terms of the sequence.   **(1 mark)**

(TOTAL: 60 marks)

22 **Challenge** In this puzzle, the rows and columns
add to the totals given.

   Work out the value of ☐, △ and ✳.

✳	5	☐	11
△	△	3	11
✳	☐	☐	10
8	13	11	

23 **Reflect** Look back at this unit.
Which lesson made you think the hardest?
Write a sentence to explain why.

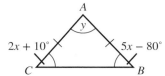

6 Angles

6.1 Properties of shapes

Prior knowledge

- Solve geometric problems using side and angle properties of quadrilaterals.
- Identify congruent shapes.

Active Learn
Homework

Warm up

1 Fluency How many lines of symmetry does each shape have? What is its order of symmetry?

2 Draw

  a an angle of 65°   **b** an angle of 128°   **c** a rectangle and its diagonals

3 **a** What do the four angles of each of these quadrilaterals add up to?

  b Copy and complete this rule.   Angles in a quadrilateral add up to ☐°.

4 Work out the missing angles. Copy and complete the reasons.

  a   **b**

 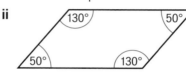

Angles on a straight line add up to ☐°.   Vertically opposite angles are _____ .

Key point

Shapes are **congruent** when they are exactly the same shape *and* size. Shapes are **similar** when they are the same shape. Similar shapes may be different sizes.

5 Write down the letters of three pairs of shapes that are congruent.

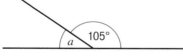

Q5 hint

Congruent shapes fit on top of each other exactly, if you cut them out.

6 These parts of shapes are drawn on a centimetre square grid.

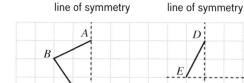

a Copy and complete the diagrams. **b** Name each shape.

c Use a protractor to measure the angles in each shape. Label each angle with its size.

d **Reflect** Write a sentence stating what you notice about the opposite angles in each of these shapes.

7 **a** Name each shape.

A    B   C    D   E

b Which of these shapes have

i two pairs of equal opposite angles

ii diagonals that bisect at $90°$

iii two pairs of parallel sides?

> **Q7b ii hint**
>
> '**Bisect**' means to cut in half.

8 **Reasoning** Name each quadrilateral being described.

a I have four equal sides and my opposite angles are equal.

b All my angles are $90°$ and my diagonals bisect at $90°$.

c My diagonals bisect at $90°$ but are not the same length.

d I have one pair of parallel sides and one pair of equal sides.

e **Reflect** Is there more than one answer for each part?
If so, name the other shape(s) that fit(s) the description.

9 **Problem-solving** Draw a coordinate grid with axes labelled from -5 to 5.
Plot these points.
$A(-5, 1)$, $B(-2, -1)$, $C(3, -2)$, $D(5, 2)$, $E(3, 3)$, $F(2, 1)$, $G(1, 2)$, $H(-2, 5)$, $I(-4, 3)$, $J(-5, 3)$
Which four points can you join to make

a a parallelogram   **b** a kite   **c** an isosceles trapezium?

10 Work out the sizes of the missing angles in each quadrilateral.

a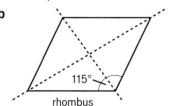

parallelogram

b

rhombus

c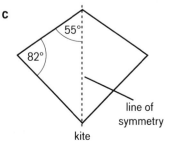

kite

11 **Reasoning** Bronwen is an engineer.
She draws a design for a bridge.
Bronwen uses these rules in her design.

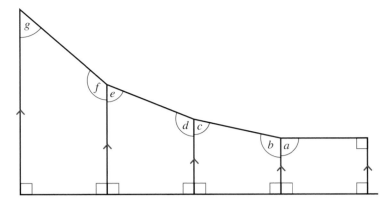

- Angle g is 10% less than angle e.
- Angle e is 10% less than angle c, rounded to the nearest degree.
- Angle c is 10% less than angle a, rounded to the nearest degree.

Work out the sizes of the angles a to g.

12 **Problem-solving** Draw a coordinate grid with axes labelled from −5 to 5.
Here are some coordinates for three shapes $ABCD$.

a $A(-1, 2)$, $B(1, 4)$, $C(5, 0)$ **b** $A(-2, 0)$, $B(-4, 1)$, $C(-2, 2)$
c $A(-1, -1)$, $B(-3, 2)$, $C(-1, 3)$

For each set of coordinates
i plot the points on your grid
ii join the points A and B, and B and C
iii draw a point D so that the shape $ABCD$ has rotation symmetry of order 2
iv name the shape

Exam-style question

13 The diagram shows quadrilateral $ABCD$ with each of its sides extended.

$AB = AD$
Show that $ABCD$ is a kite.
Give a reason for each stage of your working. **(4 marks)**

Exam tip

You must *show* (not assume) that $ABCD$ is a kite. Therefore, find all the angles in quadrilateral $ABCD$ and state why this shows the shape is a kite.

14 **Problem-solving** The diagram shows a pattern made from four identical parallelograms.
Show that angles a, b, c and d add up to 360°.

Q14 hint

Work out all the angles in one parallelogram first.

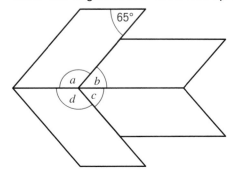

6.2 Angles in parallel lines

Active Learn
Homework

- Understand and use the angle properties of parallel lines.
- Find missing angles using corresponding and alternate angles.

Warm up

1 **Fluency** Find

a two parallel lines

b two perpendicular lines

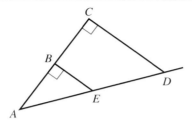

2 **a** Which of the angles marked with letters are obtuse?

b Which of them are acute?

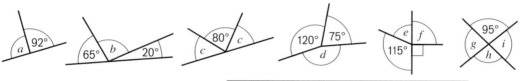

3 Work out each missing angle in **Q2**.
Give a reason for each answer.
Choose your reason from the box.

> Angles on a straight line add up to 180°.
> Angles around a point add up to 360°.
> Vertically opposite angles are equal.

Key point

Parallel lines are shown with arrows.
When a line crosses two parallel lines it creates a 'Z' shape.
Inside the Z shape are **alternate angles**.
Alternate angles are equal.
Alternate angles are on different or alternate sides of the line.

4 The diagram shows a line crossing two parallel lines
and angles labelled *p*, *q*, *r* and *s*.
Write down two pairs of alternate angles.

5 Find the sizes of the angles marked with letters.

a

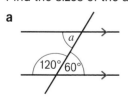

b

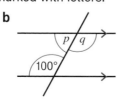

c

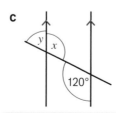

> **Q5b hint**
>
> Angles on a straight line add up to 180°.

6 Find the sizes of the angles marked with letters.

a
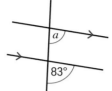

b
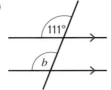

c
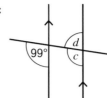

Exam-style question

Exam tip

Sometimes you must mark and find another angle to help you find the size of the angle asked for in the question. You must always give a reason for the angles you find.

7 Andy is asked to find the size of angle *s* in this diagram.

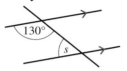

Here is Andy's working.

> Angle *x* = 130°
> (corresponding angles are equal)
> Angle *s* = 130°
> (vertically opposite angles are equal)

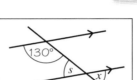

Explain what Andy has done wrong. **(2 marks)**

---

8 Copy each diagram.
Find the sizes of angles *t* to *w*.
Give reasons for each answer.
Choose your reasons from the box.

> Angles on a straight line add up to 180°
> Angles ☐ and ☐ are alternate. Alternate angles are equal.
> Angles ☐ and ☐ are corresponding. Corresponding angles are equal.

a
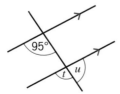

b
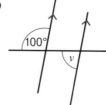

c
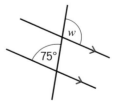

9 **Reasoning** An angle of 89° is shown on the diagram.

a Write down the letters of all other angles of size 89°.
Give reasons for your answers.

b Explain why *a* + *b* + *c* = 269°.

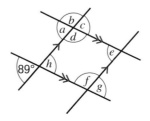

10 Reasoning Angles inside two parallel lines are called **co-interior angles**.
Use this diagram to explain why co-interior angles
add up to 180°.

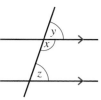

11 *ABCD* is a parallelogram with its sides extended.
AB is parallel to *CD*. *BC* is parallel to *AD*. Angle *BAD* is 105°.

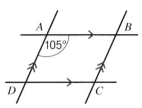

Copy and complete to find all angles in the parallelogram.

a Angle $ADC = 180° - \square° = \square°$
(*BAD* and *ADC* are co-interior angles. Co-interior angles add up to 180°.)

b Angle $BCD = 180° - \square° = \square°$
(____ and *BCD* are co-interior angles. Co-interior angles add up to 180°.)

c Angle $ABC = 180° - \square° = \square°$
(____ and *ABC* are co-interior angles. Co-interior angles add up to 180°.)

d Reflect What do you notice about opposite angles in a parallelogram?

12 *ABC* and *DEF* are straight lines. *ABED* is a parallelogram.

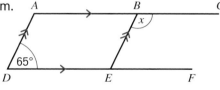

a Find the size of the angle marked *x*.
Give a reason for each stage of your working.

b Reflect There is more than one way to find angle *x*.
Answer part **a** again, finding a different angle and
using different reasons to find angle *x*.

13 Find the sizes of angles *a* to *j*. Give reasons for your answers.

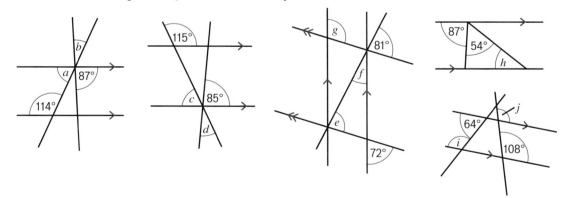

14 *AFB* and *CHD* are parallel lines.
EFD is a straight line.
Angle $AFE = 40°$
Angle $CHF = 108°$
Show that angle
$DFH = 68°$.
Give a reason for each
stage of your working. **(3 marks)**

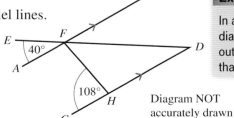

Diagram NOT
accurately drawn

Exam tip

In an exam, you can write on the
diagram any angles you work
out. Make sure you use the fact
that *AFB* is parallel to *CHD*.

6.3 Angles in triangles

Active Learn
Homework

- Solve angle problems in triangles.
- Understand angle proofs about triangles.

Warm up

1 Fluency Match each triangle to a name from the box.

a   b   c   d

> right–angled triangle
> scalene triangle
> equilateral triangle
> isosceles triangle

2 Copy and complete the sentence.
The three angles in a triangle add up to ☐.

3 Work out

a $180 - 100 - 50$ b $180 - 57 - 57$ c $\dfrac{180 - 35}{2}$

---

4 Calculate the sizes of the angles marked with letters.
Give a reason for each answer.
Choose your reason from the box.

> Angles around a point add up to 360°.
> Angles on a straight line add up to 180°.
> Angles in a triangle add up to 180°.

a 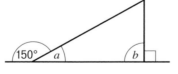   b    c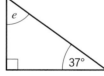

5 Reasoning The diagram shows an isosceles triangle
and an equilateral triangle.

a An isosceles triangle has a line of symmetry.
What does this tell you about the two angles a and b?

b An equilateral triangle has three lines of symmetry.
What does this tell you about angles d, e and f?

isosceles   equilateral

c **Reflect** Write a sentence explaining why you can work out the sizes of angles d, e and f,
but not angles a and b.

6 *ABCD* is a square.

a What type of triangles are *ABD* and *BDC*?

b Copy the diagram. Mark on all the angles in each triangle.

7 The diagram shows an equilateral triangle *PQR* and a square *QRST*.
The angle *PQS* is marked with an arc.
Work out the size of angle *PQS*.

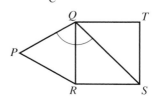

8 Ethel needs to work out the size of angle a in this diagram.

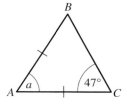

She writes

$BAC = 47°$ because base angles of an isosceles triangle are equal.

Ethel is wrong. Explain why. **(1 mark)**

9 Find the sizes of the missing angles.

a

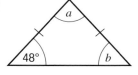

b

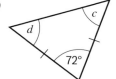

c

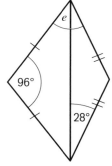

10 Problem-solving For safety, a ladder should make an angle of 15° with the vertical.

a How many degrees is this with the horizontal?

b Is each of these ladders safe? Explain your answers.

Q10 hint

vertical |
horizontal —

i

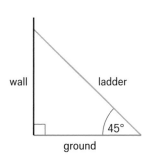

ii

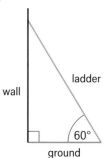

11 Reasoning Sketch a copy of this diagram.

Copy and complete these statements to prove that the sum of the angles in a triangle is 180°.

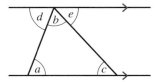

Statement	Reason
$d + b + e = \square$	Angles on a straight line add up to $\square°$
$d = \square$	Angles d and $\square$ are alternate. Alternate angles are _____ .
$e = \square$	Angles e and $\square$ are alternate. Alternate angles are _____ .

So $d + b + e = a + b + c$

$ = \square$

This proves that the angles in a triangle add up to $\square$.

12 Problem-solving Work out the sizes of the angles marked with letters.
Give a reason for each answer.

a

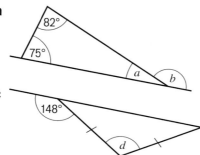

b

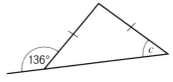

c

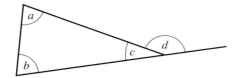

13 Reasoning Copy the diagram.

Copy and complete this proof to show that $w = x + y$.
$x + y = 180° - \square$ because the angles in a triangle add up to $\square°$.
$w = 180° - \square$ because the angles on a _____ line add up to $180°$.
So $x + y = w$.

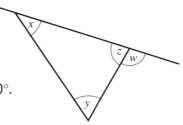

14 Reasoning Which statement about this triangle is true?

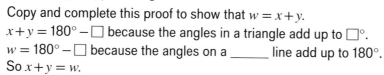

Q14 hint

Choose numbers for a, b and c, and work out d.
Which statement is true?

A Angle $d =$ angle b

B Angle $d =$ angle $a +$ angle c

C Angle $d =$ angle $a +$ angle b

D Angle $d =$ angle $b +$ angle c

Exam-style question

15 ABC is a straight line.
ABD is a triangle.
Show that triangle ABD
is an isosceles triangle.
Give a reason for each stage
of your working.

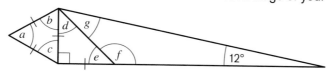

Exam tip

Work out the sizes of all the angles (giving reasons). Then state why the angles in triangle ABD show it is isosceles.

(4 marks)

16 Problem-solving Look at each triangle in this diagram, and work out the sizes of the angles marked with letters. Give a reason for each stage of your working.

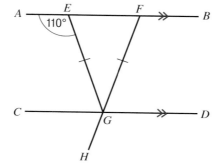

17 AB is parallel to CD.
$EG = FG$
Angle $AEG = 110°$
Sketch the diagram.
Then calculate the size
of angle DGH.

Q17 hint

Mark the angle DGH that you are trying to find. Work out the sizes of as many angles in the diagram as possible (giving reasons). There are parallel lines, so look for any alternate, corresponding or co-interior angles.

6.4 Exterior and interior angles

Active Learn
Homework

- Calculate the interior and exterior angles of regular polygons.

Warm up

1 Fluency Match the name of the shape with the number of sides.

triangle, pentagon, octagon, decagon, hexagon, quadrilateral, nonagon, heptagon

3 8 5 6
9 7 10 4

2 Work out

a $360 \div 10$ **b** $360 \div 12$ **c** $360 \div 18$

3 What size are the angles inside

a a square

b an equilateral triangle?

4 Work out

a $180 - 77$ **b** $180 - 137$ **c** $360 - 273$ **d** $180 - 37$ **e** $360 - 193$

Key point

You can draw an **exterior angle** of a shape by extending one of its sides.
The exterior angle is between the extended line and the next side of the shape.

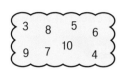

5 Copy the shapes. Draw one **interior angle** and one exterior angle for each shape.

a   **b** 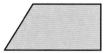   **c**

d   **e**    **f**

Key point

A **polygon** is a 2-dimensional shape bounded by straight sides.
A **regular polygon** has all equal side lengths and all equal angles.
An **irregular polygon** has unequal side lengths and unequal angles.

6 Look at the polygons in **Q5**.
Write down whether each polygon is regular or irregular.

7 **a** Work out the size of one exterior angle of each regular polygon.

i 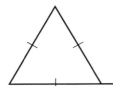   **ii** 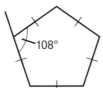   **iii**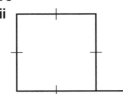

b Reflect In a regular polygon, all the exterior angles are equal.
Work out the sum of the exterior angles of each regular polygon in part **a**.
What do you notice?

c Reflect If you walked around the outside of a polygon, what size turn would you make?

> **Key point**
>
> The sum of the exterior angles of a regular polygon is 360°.

8 A regular polygon has 15 sides.
  a How many equal exterior angles does it have?
  b Work out the size of *one* exterior angle.

9 The exterior angles of some regular polygons are
  a 30°   **b** 45°   **c** 18°
  Work out the number of sides of each regular polygon.

> **Q9a hint**
>
> 30° × ☐ = 360°

10 The diagram shows an irregular pentagon.
  a What do the exterior angles add up to?
  b What is the sum of the exterior angles of any irregular polygon?

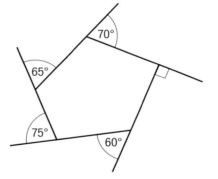

11 Work out the sizes of the missing exterior angles for each polygon.

a 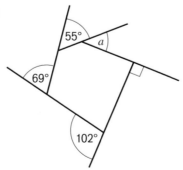   **b**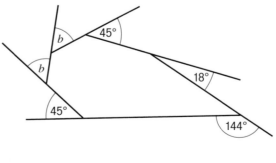

12 Reasoning Cerys says, 'I can work out the size of the exterior angles of an irregular hexagon by dividing 360° by 6.'
Is Cerys correct? Explain.

13 **Reasoning** The diagram shows an exterior and an interior angle of a regular polygon.

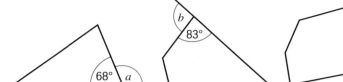

a Work out the number of sides of the polygon.

b What do the interior and exterior angles add up to?

c Copy and complete this rule.
The interior and exterior angles sum to □°.

d Explain why this rule is true.

14 Work out the sizes of the angles marked with letters.

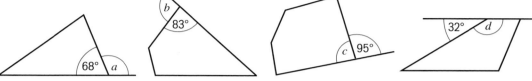

15 Copy this table.
Work out the exterior angles of each type of regular polygon.
Then work out the interior angles.

Regular polygon	Exterior angle	Interior angle
pentagon		
hexagon		
decagon		

16 A regular polygon has an interior angle of 144°.
Find the size of the exterior angle and work out how many sides the polygon has.

17 **Problem-solving** Point F lies on the midpoint of CD.
Find the size of angle x.
Give a reason for each stage of your working.

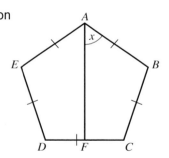

Q17 hint

AF is a line of symmetry so it bisects the interior angle EAB.

Exam-style question

18 $RSTUVXYZ$ is a regular octagon.
ARV is a straight line.
Angle $ARS = a°$

Work out the value of a.
Give a reason for each stage of your working.

(4 marks)

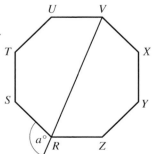

Exam tip

Start by working out the size of each exterior angle. Note that a is not an exterior angle.

6.5 More exterior and interior angles

- Calculate the interior and exterior angles of polygons.
- Explain why some polygons fit together and others do not.

Warm up

1 **Fluency** How many

    i vertices   **ii** interior angles

does each of these polygons have?

    a quadrilateral   **b** pentagon   **c** heptagon   **d** decagon

> **Q1 hint**
>
> 'Vertex' means corner.
> 'Vertices' means corners.

2 Work out the value of $180 \times (n-2)$ when

    a $n = 6$ **b** $n = 3$ **c** $n = 10$

3 **a** Is each lettered angle interior, exterior or neither?

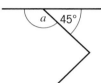 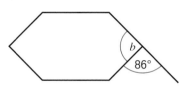 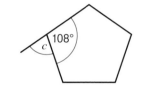

    b Work out the sizes of the lettered angles.

Key point

For shapes to fit together (**tessellate**), all the angles at the point where the shapes meet must add up to 360°.

4 Copy this parallelogram onto squared paper.

    a Continue your diagram to show four parallelograms meeting at point X.

    b What do the interior angles of the four parallelograms at point X add up to?

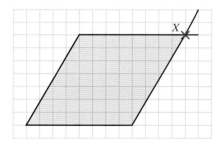

5 **a** Sketch three pentagons meeting at a point.

    b Do regular pentagons fit together? Explain your answer by considering the interior angles.

6 Copy this trapezium onto squared paper.
Show how trapeziums fit together by drawing
five more on your diagram.

7 **Reflect** 'Trapeziums will fit together no matter what their interior angles are.'
Explain whether this statement is always, sometimes or never true.

8 **Reasoning** Copy and complete these statements.
This polygon has ☐ sides. It is made of ☐ triangles.
The interior angles in a triangle add up to ☐°.
So the sum of the interior angles in this polygon is ☐ × ☐° = ☐°.

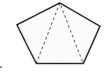

9 **Reasoning** **a** Copy this polygon. Then draw diagonals from
the top vertex to divide the polygon into triangles.

    b Copy and complete the calculation to work out
the sum of the interior angles of this polygon.

       ☐ × 180° = ☐°

    c **Reflect** Does it matter which vertex you choose to draw diagonals from?

10 **a** How many triangles can you divide an n-sided polygon into?

    b Show that your answer to part **a** works for

      i a 4-sided polygon   **ii** a 7-sided polygon

    c Write a formula to work out the sum of the interior angles, S, for a polygon with n sides.

11 For each irregular polygon, work out

    i its angle sum

    ii the size of the angle marked with a letter

Q11 i hint

Use your formula from **Q10c**.

a   **b**   **c**

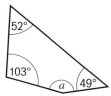 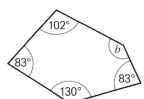 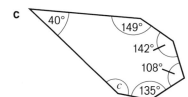

12 **Reflect** Does your formula in **Q10c** work for regular polygons? Explain.

13 For each shape, work out the size of   **i** the angle sum   **ii** the interior angle

    a regular hexagon   **b** regular pentagon   **c** 18-sided regular polygon

    d **Reflect** Write step-by-step instructions for finding the sum of the interior angles of any
regular polygon. Is there more than one method you could use? Which do you prefer?

14 For each polygon, work out the number of sides from the sum of its interior angles.

    a 1620°   **b** 2160°   **c** 2700°   **d** 3960°

15 The diagram shows a regular pentagon.
Calculate the sizes of the angles marked with letters.
Copy and complete a reason from the box for each answer.

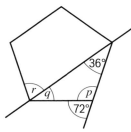

> An isosceles triangle has two equal sides and two equal angles.
> Vertically opposite angles are equal.
> Exterior and interior angles add up to 180°.
> The sum of the exterior angles of a polygon is 360°.
> The interior angles of a regular polygon are equal.

Example

AB, BC and CD are three sides of a regular octagon.

Find the size of angle BAC. Give reasons for your answer.

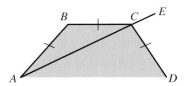

Angle *ABC* = interior angle

Exterior angle of a regular octagon
= 360° ÷ 8 = 45°

Interior angle = 180 − 45 = 135°

Angle *BAC* = angle *BCA* (triangle *ABC* is isosceles)

Angle *BAC* = (180 − 135) ÷ 2 = 22.5°
(angles in a triangle add up to 180°)

> Use the facts you know about interior and exterior angles.

> Exterior angle + interior angle = 180°

> Subtract the known angle from the sum of the interior angles of a triangle.
> Divide by 2 as angles *BAC* and *BCA* are identical.

16 Problem-solving *ABCDEF* is a regular hexagon.
Find the size of

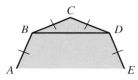

a angle *BAF*

b angle *AFB*

> **Q16 hint**
>
> Use the method in the worked example.

17 Problem-solving *AB*, *BC*, *CD* and *DE* are four sides of a 12-sided regular polygon.

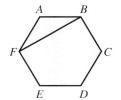

a Which other angles in the diagram are equal to angle *BCD*?

b Find the size of angle

  i *BCD*

  ii *CBD*

  iii *DBA*

Exam-style question

18 *PQ*, *QR* and *RS* are three sides of a regular decagon.
PRT is a straight line.
Angle *TRS* = *x*°

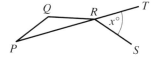

Diagram NOT accurately drawn

Work out the value of *x*. **(5 marks)**

> **Exam tip**
>
> Find as many angles as you can to help you find the size of *x*.
> Calculate the angles in the order shown here.
>
>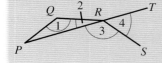

6.6 Geometrical problems

- Solve angle problems using equations.
- Solve geometrical problems showing reasoning.
- Use x for the unknown to help you to solve problems.

Active Learn
Homework

Warm up

1 Fluency Find the sizes of angles a, b, c, d and e. Give reasons for your answers.

 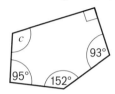 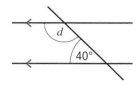 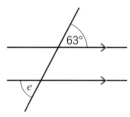

2 Solve these equations.

a $5x = 180$ **b** $3x + 30 = 180$ **c** $6x - 270 = 360$ **d** $60 + 8x = 10x - 105$

3 An angle is x. Another angle is $60°$ more than x.

a Write an expression for the larger angle in terms of x.

b Write an expression for the total of the two angles. Simplify your expression.

4 The diagram shows a straight line.

a Copy and complete this equation.

□ + □ = 180° (angles on a straight line add up to 180°)

b Solve your equation to find the value of x.

c Write down the sizes of the two angles.

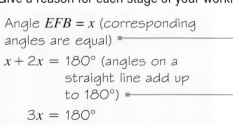

Key point

You can use angle facts to write an equation.
Then you can solve the equation to find the value of the unknown.

Example

Work out the size of angle x.
Give a reason for each stage of your working.

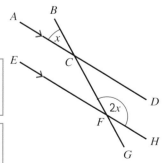

Angle **EFB** = x (corresponding angles are equal)

Use the angle facts you know to find the sizes of other angles in the diagram.

$x + 2x = 180°$ (angles on a straight line add up to 180°)

Write an equation and solve it to find x.

$3x = 180°$

$x = 60°$

5 For each diagram, write an equation in terms of x.
Then work out the value of x.
Give reasons for your answers.

a   **b**   **c**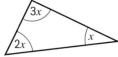

d 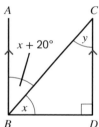   **e**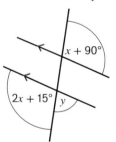

Q5d hint

Copy the diagram. Mark the angles on the diagram as you find them.

6 **Reasoning** Look at the diagram.
  a Explain why $x + 40° = 2x - 10°$.
  b Work out the value of x.
  c Substitute the value of x into $x + 40°$ to work out the sizes of the angles.

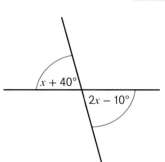

7 **Problem-solving** Work out the values of x and y. Give reasons for your answers.

a 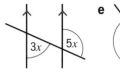   **b**   **c**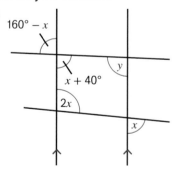

8 Look at the diagrams.

a 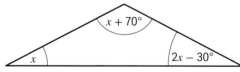   **b**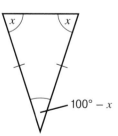

For each triangle
  i write an equation in terms of x
  ii solve this equation for x
  iii write down the sizes of the three angles of the triangle

9 **Reasoning** What type of triangle is this? Explain.

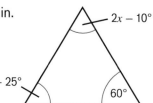

10 Reasoning In the box are sets of three angles, written as expressions in x.
For each set, which of these is true?
These three angles can always/sometimes/never make a triangle.

$x, 3x, 5x$
x, x, x
$x + 60°, x - 60°, 60° - 2x$
$x + 60°, x - 60°, x + 180°$
$60° - x, x + 90°, 30°$

Exam-style question

11

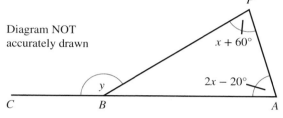

Diagram NOT accurately drawn

Exam tip

Do not assume $y = 3x + 40°$. You must show working that leads to the answer $y = 3x + 40°$. First find angle PBA. You can answer part **b** even if you cannot answer part **a**.

All angles are measured in degrees.
ABC is a straight line.
Angle $APB = x + 60°$
Angle $PAB = 2x - 20°$
Angle $PBC = y$

a Show that $y = 3x + 40°$.
Give reasons for each stage of your working. **(3 marks)**
b Given that $y = 151°$, work out the value of x. **(2 marks)**

12 Problem-solving ABD is a triangle. C is a point on the straight line DB.

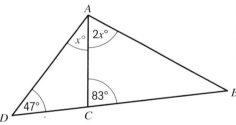

a Show that $x = 36°$.
b Find the size of angle ABD.

Q12 hint

Sketch the diagram. Find as many angles as you can. Then write an equation to work out x.

13 In this triangle, angle PQR is half the size of angle QPR. Angle QRP is $25°$ more than angle PQR.

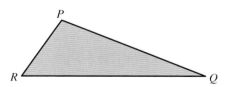

a Sketch the diagram. Label angle PQR as x. Read the question again and label the other two angles in terms of x.
b Write an equation in terms of x.
c What is the size of angle PQR?

Exam-style question

14 $ABCDE$ is a pentagon.

Angle $ABC = 2 \times$ angle AED
Work out the size of angle ABC.
You must show all your working.
(5 marks)

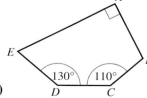

Exam tip

Label angle AED as x. Use what you know about the sum of angles in a pentagon to write and solve an equation.

6 Check up

*Active*Learn
Homework

Angles between parallel lines

1 Work out the size of angle x in each diagram. Give reasons for your answers.

a

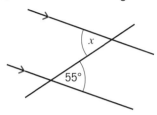

b

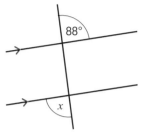

2 Work out the sizes of the angles marked with letters.
Give at least one reason for each answer.

a

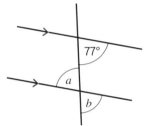

b

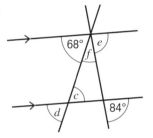

c

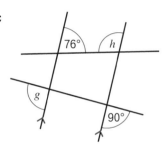

Triangles and quadrilaterals

3 Work out the size of angle x in each diagram. Give reasons for your answers.

a

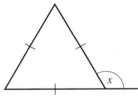

b

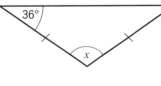

c

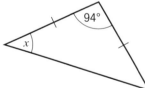

4 Find the sizes of angles a, b and c.
Choose at least one reason from the box to explain
each answer.

Reason 1	Alternate angles are equal.
Reason 2	Corresponding angles are equal.
Reason 3	Angles in a triangle add up to 180°.
Reason 4	Angles on a straight line add up to 180°.
Reason 5	Angles of a quadrilateral add up to 360°.
Reason 6	Co–interior angles add up to 180°.

5 Work out the sizes of the three angles in this triangle.

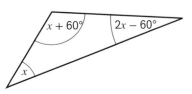

6 Work out the sizes of the angles marked with letters. State any angle facts that you use.

a

b
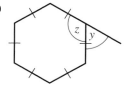

c
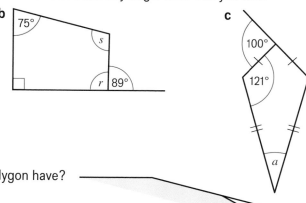

Interior and exterior angles

7 **a** How many sides does this regular polygon have?

b A regular polygon has 18 sides.
What is the size of its exterior angle?

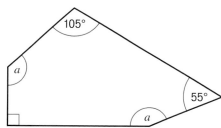

8 Work out the sizes of the angles marked with letters. Give reasons for your answers.

a
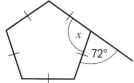

b

9 Find the sum of the interior angles of a polygon with 10 sides.

10 Find the size of angle a. Give reasons for your answers.

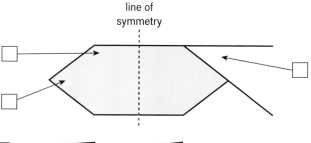

11 **Reflect** How sure are you of your answers? Were you mostly

Just guessing 😟   Feeling doubtful 😐   Confident 🙂

What next? Use your results to decide whether to strengthen or extend your learning.

Challenge

12 The diagram shows an irregular hexagon.
Here are some angle cards.

line of symmetry

| 29° | 31° | 56° | 58° | 60° | 62° | 64° | 116° | 118° | 120° | 122° | 124° | 149° | 151° |

Make some copies of the diagram.
Use the angle cards to complete the diagram in different ways.
How many ways can you find? You can use the same angle card more than once.

6 Strengthen

ActiveLearn
Homework

Angles between parallel lines

1 The diagrams show angles with straight lines.
Copy each diagram. Decide whether each pair of angles
are alternate, corresponding or vertically opposite.

a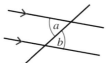

b

c

d

> **Q1 hint**
>
> Look for Z, X and F shapes and draw them on your diagrams in different colours.
>
> **Z N**
> alternate (Z shape)
>
> **X**
> vertically opposite (X shape)
>
> **F ⟩ ⟩⟩**
> corresponding (F shape)

2 a Write down the sizes of the angles marked with letters.
Part of each diagram is drawn in red to help you.

i
49°
p

ii 88°
q

iii s
37° r

iv u
76°
v t

b Beside each answer in part **a**, write the correct reason from the box.

> Vertically opposite angles are equal.
> Corresponding angles are equal.
> Alternate angles are equal.
> Angles on a straight line add up to 180°

3 Copy the diagrams. Find the sizes of the lettered angles in
each diagram.
Give at least one reason for each answer.

> **Q3b hint**
>
> You can find angle d using the fact the angles on a straight line add up to 180°.

a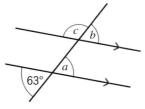
c b
a
63°

b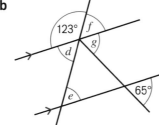
123° f
g
d
e
65°

c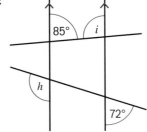
85° i
h
72°

Triangles and quadrilaterals

1 Copy and complete.

   a The three angles of any triangle add up to ☐°.

   b The four angles of any quadrilateral add up to ☐°.

   c Angles on a straight line add up to ☐°.

2 Use the rules in **Q1a** and **b** to help you to work out the sizes of the angles marked with letters. For each answer, write the rule (reason) that you used.

a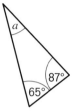

b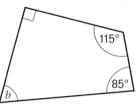

c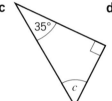

d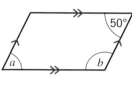

> **Q2d hint**
>
> Which other angle is the same as 50° in the parallelogram?

3 Copy and complete the sizes of angles in this kite $ABCD$, and the reasons.

   a Angle $a = $ ☐° (angles on a straight line add up to ☐°)

   b Angle $b = $ ☐° (the kite has a vertical line of symmetry ⟨ ⟩)

   c Angle $c = $ ☐° (angles in a quadrilateral add up to ☐°)

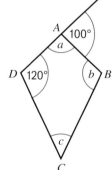

4 Copy and complete the equal sides and angles in each of these isosceles triangles.

a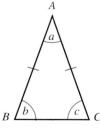

b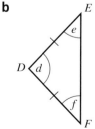

c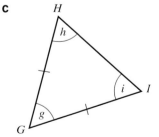

   Side $AB = $ side ☐   Side $DE = $ side ☐   Side $GH = $ side ☐

   Angle $b = $ angle ☐   Angle $e = $ angle ☐   Angle $h = $ angle ☐

5　**a** What type of triangle is PQR?

　　b What do its three angles add up to?

　　c Which two angles are equal?

　　d What do the angles at Q and R add up to?

　　e What is the size of the angle at Q?

Q5d hint

Use your answer to part **b**.

6　For each triangle, decide which angles are equal.
　　Then work out the sizes of the angles marked with letters.

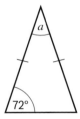 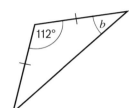 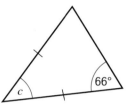

7　Find the sizes of all three angles in this triangle. Start with finding angle BAC.
　　Explain each stage of your working by choosing reasons from the box.

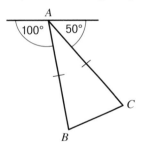

Reason 1　Angles in a triangle add up to $180°$.
Reason 2　Two angles in an isosceles triangle are equal.
Reason 3　Angles on a straight line add up to $180°$.

8　**a** What type of triangle is EFG?

　　b What size is angle EGF?

　　c What size is angle EGH?
　　　Give a reason for your answer.

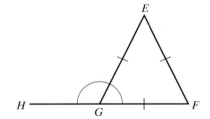

9　Here is a triangle.

　　a Write an expression for the sum of the three angles.

　　b Use your expression in part **a** to copy and complete this equation.
　　　　$\square x = 180°$

　　c Solve your equation in part **b** to find the value of x.

　　d Write down the sizes of the three angles of the triangle.

Q9d hint

When $x = \square°$
　$3x = \square°$
　$5x = \square°$

Interior and exterior angles

1 a Copy and complete.

The exterior angles of a polygon add up to ☐°.

b Here is the exterior angle of a regular polygon.

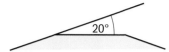

Use the information in part **a** to find the number of sides of this polygon.

c What is the size of the exterior angle of a regular polygon that has 30 sides?

Q1a hint

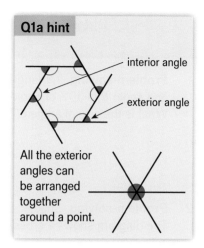

interior angle

exterior angle

All the exterior angles can be arranged together around a point.

2 The diagram shows a regular hexagon.

a Which angle is the interior angle?

b Which angle is the exterior angle?

c Work out the size of the exterior angle.

d The interior and exterior angles lie on a straight line. What do they add up to?

e Work out the size of the interior angle.

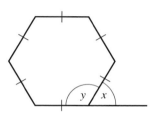

3 For each irregular polygon

i sketch the shape

ii divide the shape into triangles using diagonals from the same vertex

iii work out what the interior angles add up to

iv work out the size of angle x

a

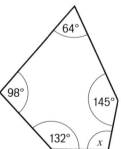

b

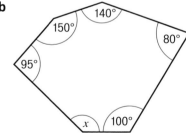

Q3a hint

$3 \times 180° = ☐°$

4 a Trace this regular octagon.

b Rotate your tracing to compare the other interior angles with x.

c Divide your octagon into triangles.

d Work out the angle sum for the octagon.

e Work out the size of angle x.

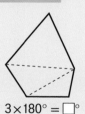

6 Extend

1 **Problem-solving** In triangle ABC, angle A is $80°$ and angle B is 60% of angle A.
What is the size of angle C?

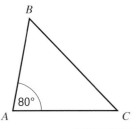

2 **Problem-solving** Shaojun wants to build a skate ramp.
The diagram shows a cross-section of the ramp.

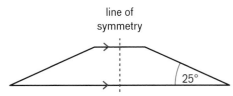

Q2 hint

The **angle of elevation** is the angle between the horizontal base and the ramp.

The angle of elevation must be $25°$.
Work out the sizes of the other interior angles.

3 **Problem-solving / Reasoning**
Find the size of angle x in this diagram.
Give reasons for each stage of your working.

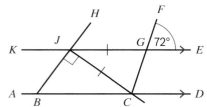

4 **Problem-solving / Reasoning**
Find the size of angle ABJ.

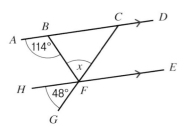

Exam-style question

5 BEG is a triangle.
ABC and DEF are parallel lines.
Work out the size of angle BGE.
Give a reason for each stage of your working. **(4 marks)**

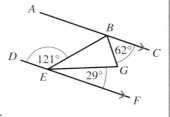

Exam tip

When a diagram includes parallel lines, look for alternate, corresponding, vertically opposite and co-interior angles.

6 **Problem-solving** Triangle BDC is an isosceles triangle.
Triangle ACE is a right-angled triangle.
Show that triangle ABC is an equilateral triangle.

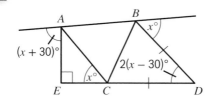

7 Problem-solving

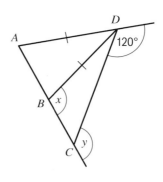

a What type of triangle is triangle ABD?

b Write an expression in terms of x for angle ADB.

c Follow these steps to write an expression in terms of y for angle ADB.

  i Write an expression in terms of x for angle CDB. Use angles on a straight line.

  ii Write an expression in terms of x for angle BCD. Use angles in a triangle.

  iii Write an equation for x in terms of y. Use angles on a straight line.

  iv Substitute your expression for x into your answer to part **b**.

8 **Reasoning** Can a polygon have an angle sum of $500°$? Explain how you know.

Q8 hint

$180° \times \square = 500°$

Exam-style question

9 The size of each interior angle of a regular polygon is 9 times the size of each exterior angle.

Work out how many sides the polygon has.

(3 marks)

Exam tip

When problem solving, sometimes it can be helpful to draw a diagram and/or label an unknown as x. For example

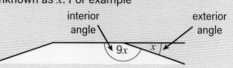

interior angle — exterior angle — $9x$ — x

Exam-style question

10 Four congruent regular octagons tessellate to make this pattern.

Show that quadrilateral $ABCD$ is a square.

You must show your working.

(4 marks)

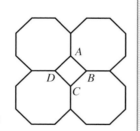

Exam tip

Write down what you know about the polygons in the question. For example

- the interior angles of an octagon
- angles around a point
- the interior angles of a square.

---

11 **Reasoning** $ABCDEF$ is a regular hexagon.

Angle HAF = angle HAB

G, F and B lie on a straight line.

Work out the sizes of angles x and y.

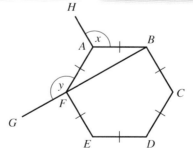

Exam-style question

12 $ABCDE$ is a regular pentagon.

ABP is an equilateral triangle.

Work out the size of angle x.

(4 marks)

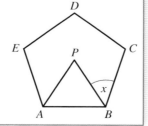

Exam tip

Write the sizes of all the angles you can on the diagram. For example, write the sizes of the angles of the equilateral triangle and the pentagon.

6 Test ready

Summary of key points

To revise for the test:

- Read each key point, find a question on it in the mastery lesson, and check you can work out the answer.

- If you cannot, try some other questions from the mastery lesson or ask for help.

Key points

1 Shapes are **congruent** when they are exactly the same shape *and* size. Shapes are **similar** when they are the same shape. Similar shapes may be different sizes. → **6.1**

2 Angles in a quadrilateral add up to 360°. → **6.1**

3 Angles on a straight line add up to 180°. → **6.1**

4 Vertically opposite angles are equal. → **6.1**

5 Opposite angles in a parallelogram are equal. → **6.1**

6 Angles around a point add up to 360°. → **6.1**

7 **Parallel lines** are shown with arrows. → **6.2**

8 **Alternate angles** are equal. → **6.2**

9 **Corresponding angles** are equal. → **6.2**

10 Angles inside two parallel lines are called **co-interior angles**. Co-interior angles add up to 180°. → **6.2**

11 Angles in a triangle add up to 180°. → **6.3**

12 An **isosceles triangle** has two equal sides and two equal angles. → **6.3**

13 An **equilateral triangle** has three equal sides and three equal angles. → **6.3**

14 An **exterior angle** of a triangle equals the sum of the two interior angles on the opposite side of the triangle.

$a = b + c$ → **6.3**

15 A **polygon** is a 2-dimensional shape bounded by straight sides. A **regular polygon** has all equal side lengths and all equal angles. An **irregular polygon** has unequal side lengths and unequal angles. → **6.4**

16 The sum of the exterior angles of a regular polygon is 360°. → **6.4**

Sample student answer

Exam-style question

CDEF is a parallelogram.
AFE is a straight line.
B is the point on CF so that ABD is a straight line.
Angle BDE = 30°
Angle AED = 72°

Show that angle BDC = 78°.
Give a reason for each stage of your working.

(4 marks)

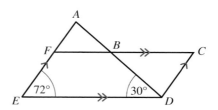

CBD = 30° (Z angles are equal)
BCD = 72° (opposite angles in a parallelogram)
BDC = 180° − 30° − 72° = 78° (angles in a triangle)

When giving reasons about angles you must be careful to state the correct reason and to use the correct mathematical language. Rewrite the reasons given by the student so they are fully correct.

6 Unit test

*Active*Learn
Homework

1 Work out the sizes of the angles marked with letters. Give reasons for your answers.

a
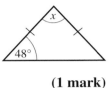

b
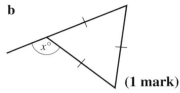

c
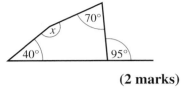

a (1 mark) **b** (1 mark) **c** (2 marks)

2 **a** The exterior angle of a regular polygon is 12°.
How many sides does the polygon have? **(1 mark)**

b A regular polygon has 36 sides.
Work out the sizes of the exterior and interior angles. **(2 marks)**

3 Work out the size of angle *x* in each diagram.
Give a reason for each answer.

a
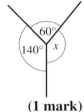

b

c

a (1 mark) **b** (1 mark) **c** (2 marks)

4 The diagram shows two triangles with an extended line.
The lines *AB* and *DC* are parallel.

Show that triangle *ABC* is a right-angled triangle.
Give a reason for each stage of your working. **(4 marks)**

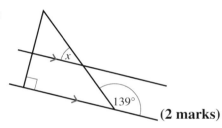

5 **a** The diagram shows a pentagon.
Work out the value of *x*. **(2 marks)**

b Work out the sizes of all the angles in the pentagon. **(3 marks)**

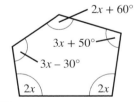

6 Work out the sizes of angles *a*, *b*, *c* and *d* in each polygon.

a
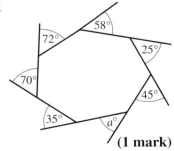

(1 mark)

b
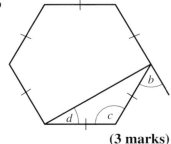

(3 marks)

7 The diagram shows angles with parallel lines.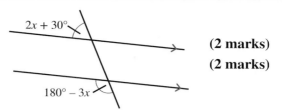

   a Work out the value of x. **(2 marks)**

   b Work out the sizes of the angles. **(2 marks)**

8 The diagram shows a hexagon.
The hexagon has one line of symmetry.

$AB = CD$

$AF = DE$

Angle $DEF = 105°$

Angle $BAF = 2 \times$ angle ABC

Work out the size of angle BAF.

You must show all your working. **(4 marks)**

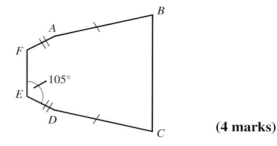

9 ABC is parallel to DEF.
EBG is a straight line.

$AB = EB$

Angle $GBC = 20°$

Angle $AED = x°$

Work out the value of x.

Give a reason for each stage
of your working.

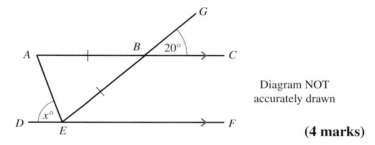

Diagram NOT
accurately drawn

(4 marks)

10 Write an expression for

   a angle BCD in terms of y **(1 mark)**

   b angle BDC in terms of x and y **(1 mark)**

   c angle BDA in terms of x and y **(1 mark)**

   d angle DBA in terms of x and y **(1 mark)**

   Give reasons for each answer.

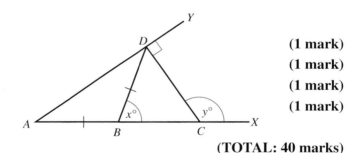

(TOTAL: 40 marks)

 11 Challenge

   a Find the size of one interior angle in a polygon with

      i 20 sides   **ii** 30 sides   **iii** 40 sides

      iv 50 sides   **v** 100 sides   **vi** 1000 sides

   b Write a sentence explaining why an interior angle of a regular polygon can never be equal
to or greater than $180°$.

12 Reflect 'Notation' means symbols. Mathematics uses a lot of notation.

   For example: $=$ means is equal to $°$ means degrees   ⌐ means a right angle.

   Look back at this unit. Write a list of all the maths notation used.

   Why do you think this notation is important?

   Could you have answered the questions in this unit without understanding the maths notation?

7 Averages and range

Prior knowledge

7.1 Mean and range

ActiveLearn
Homework

- Calculate the mean from a list and from a frequency table.
- Compare sets of data using the mean and range.

Warm up

1 Fluency The table shows the number of cars of different colours in a car park.

Colour	black	blue	green	red	white
Frequency	25	19	16	34	24

a Which colour was the mode?

b How many cars were in the car park altogether?

2 Find the range of each set of values.

a 5, 7, 8, 8, 10, 11   **b** 46, 50, 52, 58, 60   **c** 3, 9, 6, 4, 1, 5

Key point

The **mean** of a set of values is the total of the set of values divided by the number of values.

Example

Work out the mean of 0, 3, 6, 7 and 8.

$0 + 3 + 6 + 7 + 8 = 24$ ◦——— Add the values to find the total.

$\dfrac{24}{5} = 4.8$ ◦——— There are 5 values, so divide the total by 5.

The mean is 4.8

 3 Here are the numbers of books a teacher marked each day over a 10-day period.
0, 31, 45, 30, 58, 60, 0, 52, 30, 46

a Work out the total number of books he marked.

b Work out the mean number of books he marked each day.

Exam-style question

4 The table shows the midday temperature on each day for ten days.

Day	1	2	3	4	5	6	7	8	9	10
Temperature (°C)	18	19	17	15	18	21	19	18	23	21

Work out the mean temperature.   **(2 marks)**

Exam tip

Even if you are using a calculator, write down your calculations to gain any marks that might be available for working.

5 Lois is asked to find the range of these numbers.

2, 5, 4, 3, 4

Here is her working

$4 - 2 = 2$

This is wrong. Explain why. **(1 mark)**

6 **Problem-solving** Bill has 4 cards.
There is a number on each card.

The mean of the 4 numbers on Bill's cards is 6.
Work out the number on the 4th card.

7 Here are the times, in minutes, some patients waited at a doctors' surgery.
5, 7, 7, 6, 8, 5, 7, 6, 7, 8, 9, 6, 8, 7, 9, 7, 9, 6, 5, 8

   a Work out the mean waiting time.

   b Work out the range.

8 The waiting times in **Q7** are shown again in this table.

Waiting time, w (min)	Frequency, f	$w \times f$
5	3	15
6	4	24
7	6	
8		
9		
Total		

Q8a hint
3 people waited 5 minutes
$= 3 \times 5 = 15$ minutes in total.

Q8a hint
$7 \times 6 = \square$

Q8a hint
Add the numbers from the two columns.

   a Copy and complete the table.

   b Find the total waiting time of all the patients.

   c Find the total frequency, which is the total number of patients.

   d Use mean $= \dfrac{\text{total waiting time}}{\text{total number of patients}}$ to work out the mean waiting time.

   e **Reflect** Write down what is the same and what is different about the methods used in **Q7** and **Q8**. Explain which method you think is the most efficient.

9 100 children were asked how many portions of fruit and vegetables they eat each day. The results are shown in this table.

   a Work out the mean.

   b Work out the range of the number of portions.

Number of portions, p	Frequency, f	$p \times f$
4	12	48
5	33	
6	41	
7	14	
Total		

10 **Reasoning** The charts show the test scores achieved by students in 11W (left) and 11Y (right).

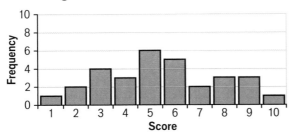
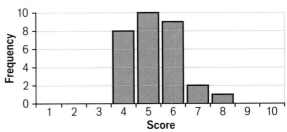

a Copy and complete the frequency table for 11W for all 10 scores.

b Work out the mean and range for 11W. Give your answer correct to 1 decimal place.

c Make a frequency table for 11Y.

d Work out the mean and range for 11Y.

Score, s	Frequency, f	$s \times f$
1	1	1
2	2	4
3	4	12

e Choose words from the box to fill in the gaps.

different	spread	11W	close	11Y	size

The mean score for 11W is _____ to the mean score for 11Y.

The range of the scores for _____ is larger than the range of the scores for _____.

A larger range shows the data is _____ out.

Key point

To compare two sets of data, **compare** an average (mode, median or mean) and the range of each set. A small range shows that the data values are all similar or consistent.

Exam-style question

11 Mia and Jade both took 6 times table tests. Their scores for each test are shown in the table.

Mia	4	12	5	13	7	9
Jade	8	1	7	9	9	8

Who had more consistent scores, Mia or Jade? Give a reason for your answer. **(2 marks)**

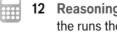
12 **Reasoning** Tess and Jo play cricket. Here are the runs they scored in the last 8 matches.

Tess	33	40	52	45	37	50	61	38
Jo	62	70	0	52	7	85	22	96

a Find the mean score and the range for each player.

b Which player would you like to have on your team? Give a reason.

c **Reflect** Sophie says, 'The mean is always between the lowest and highest data values'. Look back at all the means you have calculated. Is Sophie correct?

Exam-style question

13 The table gives information about the number of children in 25 families.

Work out the mean number of children per family. Give your answer correct to 1 decimal place.

Number	Frequency
0	4
1	8
2	9
3	3
4	1

Exam tip

You can draw an extra column on the table in your exam paper.

(3 marks)

7.2 Mode, median and range

- Find the mode, median and range from a stem and leaf diagram.
- Identify outliers.
- Estimate the range from a grouped frequency table.

Active Learn
Homework

Warm up

1 **Fluency** Find the mode, median and range of these data sets.

a 5, 3, 7, 11, 9, 6, 10, 4, 6 **b** 22, 17, 19, 26, 29, 25, 22, 28

2 Here are the pulse rates of 20 runners at the end of a half marathon.

| 164 | 172 | 158 | 186 | 160 | 165 | 152 | 177 | 149 | 155 |
| 176 | 180 | 163 | 157 | 148 | 167 | 173 | 181 | 175 | 162 |

Draw a stem and leaf diagram for this data.

3 This stem and leaf diagram shows the ages of people in a choir.

a How many people are in the choir?

b Write down the age of the oldest person in the choir.

c Work out the range of the ages.

1	6 8
2	3 7 7
3	7 8 8 9
4	0 1 4 4 7 8
5	0 0 3 4 6 6
6	1 1 2 3 7
7	6 9

Key: 1 | 6 means 16 years old

Key point

The **median** is the middle value when the data is written in order.

When n data values are written in order, the median is the $\frac{n+1}{2}$th value.

Example

This stem and leaf diagram shows the times, in seconds, for some swimmers to swim 100 m.

55	2 3 6
56	3 3 7 8
57	0 2 6 6 6 7
58	4 4 5
59	3

Count the number of values, 17.

Key: 55 | 2 means 55.2 seconds

$\frac{17+1}{2} = 9$

The median is the 9th value.

The median is the $\frac{n+1}{2}$th value.
There are 17 values, so $n = 17$.

The median is 57.2 seconds.

In a stem and leaf diagram the data is in order. So count up to the 9th value.

The mode is 57.6 seconds.

Look for repeated values in the rows.
57 | 0 2 **6 6 6** 7

4 **Reasoning** Here is a set of data.
4, 9, 12, 13, 16, 21, 22, 22, 27

a Find n, the number of data values.   **b** Work out $\dfrac{n+1}{2}$

c Show that the $\dfrac{n+1}{2}$ th value is the median.

5 This stem and leaf diagram shows the lengths, in cm, of some pencils.

5	7 7 9
6	5 6 8 8 9
7	0 0 0 2 5 5 7 8
8	4 4 5

Find

a the mode

b the median

c the range

Key: 5 | 7 means 5.7 cm

Exam-style question

6 Daisy measured 15 twigs.
The stem and leaf diagram shows information about their lengths.

0	5 7 8
1	1 3 4 7 9
2	0 1 5 5 5 7
3	2

Key: 1 | 1 represents 11 cm

Exam tip

Write 'Yes' or 'No' and then explain how you know.

Daisy says, 'The modal length is 5 cm because 5 occurs most often in the diagram.'
Is Daisy correct? You must explain your answer.   **(1 mark)**

---

7 **Reasoning** Look at the stem and leaf diagram in **Q3**.
The mode is not a very useful average for this data. Explain why.

8 Each act at a talent contest was given a score.
This stem and leaf diagram shows the results.

5	4 5 7 7 9
6	0 2 2 3 7 8
7	3 4 5 7 9 9
8	2 4 6 6 7 7 7 8
9	5 5 6

Find

a the mode   **b** the range

There are 28 values. The position of the median is $\dfrac{28+1}{2} = 14.5$

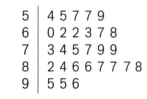
Key: 5 | 4 means 54 points

So the median is halfway between the 14th and 15th values.

c Find the median.

Key point

An **outlier** is an extreme data value that doesn't fit the overall pattern.

9 **Reasoning**

a Identify the outlier in each set of data.
The outliers may be large or small compared to the rest of the data.

i 43, 41, 52, 55, 42, 23, 46, 50   **ii** 128, 112, 133, 116, 195, 131, 120, 115

iii 6.43, 6.72, 6.57, 2.76, 7.01, 6.88, 6.39, 6.98, 8.85, 7.11

b For each set of data work out

i the range   **ii** the range ignoring the outlier

c Which range value do you think best describes the data?

10 In each set of data the outliers are likely to be errors.
Work out the range of each set of data, ignoring any outliers.

   a 50, 47, 56, 58, 59, 45, 86, 49, 53

   b 5.8, 6.2, 2.9, 5.7, 6.1, 6.3, 5.8, 5.9

11 **Problem-solving** The diagram shows the results of a physics experiment.

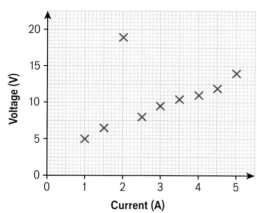

Find the range of the voltages ignoring any outliers.

12 A rugby team captain records the points scored by his team.

   a Estimate the range of points scored.

   b Explain why you can't work out the exact value of the range from this table.

Points scored	Frequency
0–10	1
11–20	8
21–30	7
31–40	5
41–50	3

13 Jill measured the heights of some seedlings.
The table shows her results.

   a From the table, what is the maximum possible height?

A seedling can't actually have a height of 0 cm but it makes sense to use this to give an estimate.

   b Estimate the range.

Height, h (cm)	Frequency
$0 < h \leqslant 5$	31
$5 < h \leqslant 10$	40
$10 < h \leqslant 15$	55
$15 < h \leqslant 20$	38
$20 < h \leqslant 25$	16

14 Jim runs a spin class at a sports centre.
The table shows the ages of the people in one session.
Estimate the range of the ages.

Age, a (years)	Frequency
$15 \leqslant a < 20$	2
$20 \leqslant a < 25$	8
$25 \leqslant a < 30$	5
$30 \leqslant a < 35$	3
$35 \leqslant a < 40$	4

7.3 Types of average

Active Learn
Homework

- Recognise the advantages and disadvantages of each type of average.
- Find the mode, modal class and median from a frequency table.

Warm up

1 Fluency This table shows information about Y11 students' test scores.

How many students scored 40 or more?

Score (%)	Number of students
0–19	6
20–39	24
40–59	22
60–79	8

2 a List the first 20 odd numbers in order.

 b Use $\dfrac{n+1}{2}$ to find the position of the median.

 c Find the median of the first 20 odd numbers.

Key point

The average (mode, median or mean) that best represents a set of data should be a typical value for the data.

3 Find the mode of each data set.

 a 21, 29, 25, 23, 25, 28, 25, 30, 22   **b** 108, 132, 110, 126, 132, 111, 129

 c 1.4, 1.4, 1.4, 1.6, 1.7, 3.9   **d** 5200, 5800, 6100, 6100, 6100, 6500

 e Is the mode a good representative value for each data set?

 4 Find the mean of each data set.

 a 26, 28, 27, 153, 24   **b** 43, 52, 37, 44, 50

 c Is the mean a good representative value for both data sets?

5 Find the median of each data set.

 a 11, 12, 14, 17, 18, 18, 20, 21, 23   **b** 1.7, 1.8, 1.8, 1.9, 5, 8, 14

 c Is the median a good representative value for both data sets?

  6 Reasoning For each data set:

 i which averages can you work out

 ii which averages best represent the data?

 a 39, 43, 37, 40, 46, 37, 125, 43   **b** 1.7, 1.8, 1.9, 3, 6

 c 46, 48, 47, 48, 49, 48, 48, 50   **d** blue, yellow, blue, green, yellow, blue, red

7 Find the mode, median and mean of:

 a 24, 26, 26, 28, 30

 b 24, 26, 26, 28, 50

 c 10, 26, 26, 28, 30

8 Copy and complete the table by putting one boxed statement in each blank position.

May not change if a data value changes.

Every value makes a difference.

Easy to find; not affected by extreme values, and can be used with non–numerical data.

There may not be one.

Not affected by extreme values.

Affected by extreme values.

Average	Advantages	Disadvantages
Mean		
Median		
Mode		

9 **Problem-solving** The tables below show data on the first and second series of a TV drama. Each series had 8 episodes. Write a summary of the data for the production company.

- What statistics (mean, median, mode, range) could you find?
- Are there any outliers? What will you do about any outliers?
- Would it be useful to draw a graph? If so, what kind of graph?
- Are there any trends or patterns?
- What is the same/similar, or different about the two series?

Series One								
Episode	1	2	3	4	5	6	7	8
Ratings (millions)	6.3	5.9	5.6	5.7	5.5	5.9	6	6.3
Appreciation figure (%)	74	76	74	74	73	74	76	75

Series Two								
Episode	1	2	3	4	5	6	7	8
Ratings (millions)	7.2	7.1	6.9	6.9	6.7	7.0	7.1	7.5
Appreciation figure (%)	74	76	76	75	76	74	75	72

Ratings means the number of people who watch a programme. The **appreciation figure** is the percentage of viewers who describe it as either 'good' or 'excellent'.

Key point

The **modal class** is the class with the highest frequency.

Exam-style question

10 The table shows information about the number of people in 20 different houses.

Find the modal number of people.
(1 mark)

Number of people	Frequency
1	6
2	9
3	2
4	2
5	1

Exam tip

Your answer should be a number of people, not the frequency.

11 Members of a karate club did as many press-ups as they could in one minute.
This table shows the results.

Write down the modal class as ☐–☐.

Press-ups	Number of students
20–29	4
30–39	5
40–49	8
50–59	7
60–69	5

12 The table shows the points scored by the top 25 athletes in a competition.

Points scored, p	Frequency
$7500 \leqslant p < 7750$	4
$7750 \leqslant p < 8000$	10
$8000 \leqslant p < 8250$	4
$8250 \leqslant p < 8500$	4
$8500 \leqslant p < 8750$	2
$8750 \leqslant p < 9000$	1

Write down the modal class as $\square \leqslant p < \square$.

Key point

In a frequency table, the median is in the class that contains the $\dfrac{n+1}{2}$ th piece of data.

Example

Tom rolled a dice 25 times. This table shows his scores.

Score	Frequency
1	4
2	4
3	6
4	4
5	2
6	5

Find the median score.

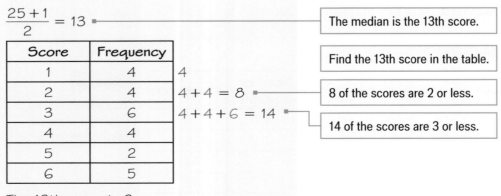

$\dfrac{25+1}{2} = 13$ — The median is the 13th score.

Score	Frequency	
1	4	4
2	4	$4+4 = 8$ — 8 of the scores are 2 or less.
3	6	$4+4+6 = 14$ — 14 of the scores are 3 or less.
4	4	
5	2	
6	5	

Find the 13th score in the table.

The 13th score is 3.

The median score is 3.

13 Jess spun a five-sided spinner 45 times.
The table shows her scores.

Score	Frequency
1	11
2	8
3	9
4	7
5	10

Find the median score.

14 Lily recorded the number of people in each car that passed her house in one hour.
This table shows her results.

Number of people	Frequency
1	23
2	16
3	6
4	3
5	1

   a How many cars passed Lily's house?

   b Find the median number of people in a car.

   c What is the mode?

   d Add an extra column to the table for number of people × frequency.
     Work out the mean number of people per car.

15 Visitors at a school fete guessed the number of marbles in a jar.
The table shows the results.

Guess	Frequency
100–125	9
126–150	12
151–175	15
176–200	8
201–225	3

Which class contains the median?

16 This table gives information about the distances, in metres, Ted hit a golf ball.

Distance, d (m)	Frequency
$200 < d \leqslant 220$	4
$220 < d \leqslant 240$	10
$240 < d \leqslant 260$	9
$260 < d \leqslant 280$	8

Exam tip

Write the class interval exactly as it is written in the table.

Find the class interval that contains the median. **(1 mark)**

7.4 Estimating the mean

*Active*Learn
Homework

• Estimate the mean of grouped data.

Warm up

1 **Fluency** How many minutes are there in 1 hour?

2 Find the value halfway between each pair of numbers.
   a 16, 20   **b** 20, 25   **c** 18, 20   **d** 4, 5

3 This six-sided spinner is spun 50 times.
   The results are shown in the table.

   Work out the mean score.

Score	Frequency
2	16
3	9
4	6
5	10
7	9

4 Write each time in hours and minutes to the nearest minute.
   a 3.74 hours = 3 hours + 0.74 × 60 minutes = 3 hours ☐ minutes to the nearest minute.
   b 2.831 hours   **c** 4.658 hours

Example

The table shows some test scores.

a Work out an estimate for the mean.

b Explain why the mean is only an estimate.

Score	Frequency
1–5	5
6–10	6
11–15	9
16–20	10

> Add a column to calculate the midpoint of each class. Use this as an estimate of the scores, because you don't know the exact values in each class.

a

Score	Frequency, f	Midpoint of class, m	$m \times f$
1–5	5	3	15
6–10	6	8	48
11–15	9	13	117
16–20	10	18	180
Total	30	Total	360

> Add a column, $m \times f$ to calculate an estimate of the total score for each class.

$$\text{Estimate of mean} = \frac{360}{30} = 12$$

> Divide the total of the $m \times f$ column by the total frequency.

b The mean is an estimate because we don't know the exact test scores.

5 In a survey, 30 small companies were asked how many employees they had.
This table shows the results.

Number of employees	Frequency, f	Midpoint of class, m	$m \times f$
1–5	12		
6–10	7		
11–15	6		
16–20	5		
Total	30	**Total**	

Calculate an estimate for the mean number of employees per company.

6 The table gives some information about cooking times, in minutes, of some ready-meals.

Copy the table and add the extra columns needed, as in **Q5**.

a Work out an estimate of the mean cooking time to the nearest minute.

b Explain why the mean is only an estimate.

Time (min)	Frequency, f
11–15	3
16–20	7
21–25	6
26–30	4

7 This table shows the distances that students in Class 11Y travel to school.

Distance, d (miles)	Frequency, f	Midpoint of class, m	$m \times f$
$0 < d \leqslant 2$	9	1	$1 \times 9 = 9$
$2 < d \leqslant 4$	7	3	$3 \times 7 = 21$
$4 < d \leqslant 6$	8		
$6 < d \leqslant 8$	6		
Total		**Total**	

a Copy and complete the table.

b How many students are in Class 11Y?

c Estimate the total distance travelled by all of the students in Class 11Y.

d Work out an estimate of the mean distance travelled.

8 **Problem-solving** 28 members of a running club took part in a marathon.
Information about their times is given in the table.

Finishing time, t (hrs)	Frequency, f
$2 \leqslant t < 3$	4
$3 \leqslant t < 4$	17
$4 \leqslant t < 5$	5
$5 \leqslant t < 6$	2

Work out an estimate of the mean time.
Give your answer in hours and minutes to the nearest minute.

9 Here are the temperatures, in °C, at 30 different locations in the UK one day in June.

16.1	17.2	18.3	18.7	18.8	19.1	19.2	20.0	20.4	21.2
21.7	21.8	22.3	22.5	22.5	22.8	22.8	23.0	23.1	23.4
23.4	23.7	24.2	24.2	24.3	24.9	25.3	25.3	25.8	25.9

Temperature, t (°C)	Frequency, f		
$16 \leqslant t < 18$			
$18 \leqslant t < 20$			
$20 \leqslant t < 22$			
$22 \leqslant t < 24$			
$24 \leqslant t < 26$			

a Work out an estimate of the mean temperature.

b What is the modal class?

c Which class contains the median?

d Estimate the range of the temperatures.

Exam-style question

10 The table shows information about the weights of 40 people.

Weight, w (kg)	Frequency
$30 < w \leqslant 40$	5
$40 < w \leqslant 50$	8
$50 < w \leqslant 60$	15
$60 < w \leqslant 70$	6
$70 < w \leqslant 80$	4
$80 < w \leqslant 90$	2

Show that an estimate for the mean weight is 55.5 kg.
You must show all your working. **(3 marks)**

Exam-style question

11 The table shows information about the lengths, in cm, of some eels.

Length, l (cm)	Frequency
$10 < l \leqslant 20$	1
$20 < l \leqslant 30$	0
$30 < l \leqslant 40$	25
$40 < l \leqslant 50$	14
$50 < l \leqslant 60$	10

Exam tip

Check that your mean is sensible – it should be inside the range of the original data.

a Work out an estimate of the mean length of the eels. **(3 marks)**

b Shula says, 'The mean may not be the best average to use to represent this information.'
Do you agree with Shula? You must justify your answer. **(1 mark)**

7.5 Sampling

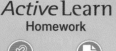

Active Learn
Homework

- Understand the need for sampling.
- Understand how to avoid bias.

Warm up

1 Fluency Work out

a $\frac{1}{3}$ of 60   **b** $\frac{1}{5}$ of 40   **c** $\frac{2}{5}$ of 40

2 Which way of picking one student from a class is most fair?
A Choosing the last name on the register
B Choosing the tallest student
C Putting all the names in a hat and choosing one without looking.

Key point

In a survey, a **sample** is taken to represent the **population**. A sample that is too small can **bias** the results. The population is the whole of the group that you are interested in.

3 Reasoning A TV company wants to find out what percentage of people living in the Midlands watch EastEnders.

a What is the population in this case?

b Explain why they will take a sample instead of interviewing every member of the population.

c The sample needs to be large enough to represent the population, but small enough to keep interviewing costs within a budget. Around 10 million people live in the Midlands.
How many people do you think they should choose for the sample?

100   500   1000   100 000

d Would it make sense for the entire sample to be taken from one Midlands town? Explain.

4 Reasoning Sean is writing an article for his local newspaper.
He wants to know what the local people think about a new supermarket.
He carries out a survey in his office.

a What is the population in this case?

b Think about the size of Sean's sample, and who is in the sample.
Write two reasons why the sample may be biased.

5 Reasoning Adam wants to find out if people in his town are satisfied with local refuse collection.
He asks all of the people in one street for their opinions.

a Explain why Adam's sample is not reliable.

b How could Adam improve his sample?

6 Reasoning Helen and John are conducting a survey to find out how far people drive in a typical week.
Helen chooses a sample of people at a London shopping centre.
John chooses a sample of people at a motorway service area on the M6.
Do you expect Helen and John to get similar results? Explain.

Unit 7 Averages and range   211

7 A teacher wants to know students' views on school dinners.
She plans to give questionnaires to her Y10 tutor group.

  a Would this give fair or biased results?   **b** How can she improve her survey?

> **Key point**
>
> In a **random sample**, every member of the population has an equal chance of being included.

8 Tristan and Sue are both selecting a sample of five students from Class 11W.

  a Tristan selects his sample by putting all 11W students' names into a hat and picking out five names without looking.
Will this give a random sample, where every student has the same chance of being picked?

  b Sue chooses her sample by picking the first five names on the 11W register.
Would this give a random sample? Explain your answer.

9 Here is a list of random numbers.

    103   883   041   033   381   839   122   109

  a Write down the last 2 digits from each number.

  b Explain how you could use this method to pick 6 people at random from a list numbered 1 to 99.

  c Alix has a list of 40 people who bought theatre tickets.
She wants to pick 5 at random to win a prize.
Explain how she could use the random numbers in part **a** to pick 5 people from her list.

Exam-style question

10 A hotel has 200 guests staying in single rooms.
The manager wants to carry out a survey using a random sample of 20 guests.
Describe two ways she could select a random sample.

                              (2 marks)

> **Exam tip**
>
> Explain both ways of picking the sample clearly.

> **Key point**
>
> An **unbiased** sample represents the population.

> **Example**

Seren asks a sample of 30 adults in her town where they do most of their shopping.
Each adult chooses one place.
The table shows information about her results.

There are 6000 adults in the town.

Place	Number of people
Retail Park	14
High Street shops	6
Online	10

  a Work out how many adults you think do most of their shopping online.

  b State any assumptions you have made.

  a Sample: $\dfrac{10}{30} = \dfrac{1}{3}$ shop online ——— Find the proportion of the sample who shop online.

     Sample: $\dfrac{1}{3}$ of 6000 = 2000 ——— Find the same proportion of the population.

     2000 people shop online.

  b I have assumed that the sample represents the population. ——— If the sample does not represent the population the answer to part **a** will be wrong.

11 Maisie is planning a primary school fun day.
She asks a sample of 40 pupils to choose one activity they would like.
The table shows the results.

Activity	Number of pupils
Face painting	15
Bouncy castle	10
Cake decorating	2
5-a-side football	13

Work out how many of the 220 school pupils you think would choose 'Bouncy castle'.

Exam-style question

12 A teacher is planning after-school clubs for 95 students.
She asks a sample of 20 students to choose one club.
The table shows information about her results.

Club	Number of students
Drama	4
Gym	3
Art	5
Music	8

 a Work out how many of the 95 students you think will choose Music. **(2 marks)**

 b State any assumptions you made and explain how this may affect your answer. **(1 mark)**

Key point

The population may be divided into groups, e.g. by age.
A **stratified sample** contains members of each group in the same proportion as the population.

13 There are 30 women and 20 men in a dance class.

 a What fraction of the dancers are women?

 b A stratified sample of 15 dancers is selected.
 Use your fraction from part **a** to work out the number of women in this sample.

14 There are 1200 students in a school.
There are 200 students in Year 11.
A stratified sample of 300 students is taken.
How many Year 11 students should there be in the sample?

15 The table shows the age groups of people in a tennis club.

 a Work out the total number of people in the tennis club.

 Megan wants to take a stratified sample of 10% of the people.

 b Calculate 10% of the number of people in each age group.
 Round your answers to the nearest whole number of people.

 c Work out the total number of people in her 10% sample.

 d **Reasoning** Show that your answer to part **c** is equal to 10% of the total number of people in the tennis club.

Age group	Number of people
5–16	32
17–21	28
22–40	43
Over-40	17

7 Check up

Active Learn
Homework

Averages and range

1 These are the numbers of people in the queue for each checkout at a supermarket.

5   4   7   3   6   3   4   2   4   5

Work out the mean number of people in a queue.

2 Mark recorded the duration, in minutes, of each of his favourite television programmes.

35   40   45   15   45   30   30   30   25   35   60   50   60   65

a Work out the mean duration.

b Work out the range.

3 Work out the range of this set of data, ignoring any outliers.

4.3, 4.9, 5.1, 5.3, 5.5, 5.5, 5.7, 60

4 This table shows the number of goals scored in 25 Premier League football matches.

Work out

a the mean

b the range

c the median

Goals scored	Number of matches
0	3
1	5
2	9
3	6
4	2

5 Ria and Jade both play for a cricket team.
Here are the numbers of runs that Ria made in her last 5 matches.

27   42   16   11   53

Jade scored a mean of 32.4 runs with a range of 28 in her last 5 matches.

a Who had the highest mean? Explain your answer.

b Who is the more consistent player? Explain your answer.

6 This stem and leaf diagram shows weights of parcels in a post office van.

Find

a the mode

b the range

c the median

```
0 | 5 7 7
1 | 2 3 5 5 6 9
2 | 0 1 6
3 | 4 4 4 5 7
4 | 2 3 6 8
```

Key: 1 | 2 means 1.2 kg

7 Ed plays snooker. Here are the scores from his last 7 breaks.

28   21   105   21   32   35   38

a Which is the most appropriate average to use for Ed's scores?

b Explain why the other two do not represent the scores very well.

Averages and range for grouped data

8 The table shows the number of emails Sophie received per day in one month.

a Work out an estimate for the mean number of emails Sophie received per day.

b Work out an estimate for the range.

c Which is the modal class?

d Which class contains the median?

Number of emails	Frequency
1–10	3
11–20	5
21–30	7
31–40	16

9 Here are 10 people's annual salaries.

| £16 200 | £18 900 | £20 750 | £22 000 | £23 500 |
| £25 800 | £27 250 | £28 500 | £32 000 | £74 000 |

Which average (mode, median or mean) best represents these salaries? Justify your answer.

Sampling

10 Dan wants to select three students at random from Class 8W. There are 30 children in the class. Describe how he can select a random sample.

11 Jo wants to find out people's views on whether VAT should be charged for hairdressing.
She asks a random sample of ten people at a shopping centre on a Monday morning for their views.
Give two reasons why her survey may be biased.

12 A school cook asks a sample of 30 students which type of lunch they want.
Each student chooses **one** type of lunch.

180 students have lunch.
Work out how many you think will have sandwiches.

Type of lunch	Number of students
sandwiches	20
hot meal	4
salad	6

13 **Reflect** How sure are you of your answers? Were you mostly

Just guessing 😞   Feeling doubtful 😐   Confident 🙂

What next? Use your results to decide whether to strengthen or extend your learning.

Challenge

14 This table gives information about the ages of members of a gym.

Age (years)	Frequency, f	Class midpoint, m	$f \times m$
18–24	32		672
25–☐		27	1107
☐–40	70		
41–59			2650
60–70	14		

a Copy and complete the table to fill in the missing values.

b Work out an estimate of the mean age.

c Which class contains the median?

d Which is the modal class? Explain why this may be misleading.

7 Strengthen

Active Learn
Homework

Averages and range

1 Here are 5 sets of counters

  a How many counters are there in total?

  b The counters are divided in 5 equal sets. How many counters are there in each set?

2 These are Lisa's times, in seconds, to run 200 m.
24, 25, 24, 24, 23, 24, 25, 23

  a Copy and complete this sentence to predict the mean.

    I think Lisa's mean time will be about ☐ seconds.

  b How many times are there?

  c Work out the total of her times.

  d Work out Lisa's mean time $= \dfrac{\text{total of times}}{\text{number of times}}$.

3 Here are the top ten midday temperatures, in °C, in Bridlington during October 2019.
16, 16, 16, 15, 14, 14, 14, 13, 12, 12
Work out the mean of these temperatures.

4 These are the amounts that Anya earned in the last 6 weeks.
£86   £94   £110   £80   £75   £88
Calculate Anya's mean weekly wage. Round your answer to the nearest penny.

5 Kyle and George both play for a cricket team.
Here are the numbers of runs that Kyle made in the last five matches.
18, 53, 19, 14, 62
In the last five matches, George scored a mean of 31.6 runs with a range of 21.

  a Calculate Kyle's mean number of runs.

  b Who scored more runs? Explain your answer.

  c Calculate the range for Kyle.

  d Who is the more consistent player? Explain your answer.

> **Q5d hint**
>
> Consistent means more or less the same every time. Results are more consistent when the range is small.

6 Here are the numbers of runs scored by two cricketers in three matches.
Ellie: 35, 40, 36   Sam: 62, 0, 34

  a Who had the highest score in a match?

  b Work out the range of scores for Ellie and for Sam.

  c Whose results are the most consistent?

  d Who had the highest mean score?

  e Who would you pick for your team? Explain why.

7 This table shows the number of emails a group of boys received.

Number of emails	Frequency	Number of emails × frequency
1	8	$1 \times 8 = \square$
2	5	$2 \times 5 = \square$
3	3	
4	2	
Total		

a i How many boys received 1 email?

ii How many emails did these boys receive in total?

b i How many boys received 2 emails?

ii How many emails did these boys receive in total?

c Copy and complete the table.

d What is the total number of boys?

e What is the total number of emails?

f Work out mean $= \dfrac{\text{total number of emails}}{\text{total number of boys}}$.

8 A company has an office building with 32 rooms.
This table shows the number of people working in each room

Number of people	Frequency
1	7
2	10
3	8
4	6
5	1

a Copy the table.
Draw an extra column for the number of people × frequency.
Complete this column and work out the total.

b How many people work in the building in total?

c Work out the total frequency.

d Work out the mean number of people per room.

e Work out the range = largest number of people in a room − smallest number of people in a room.

f The data could be written in a list like this.

1 1 1 1 1 1 1   2 2 2 2 2 2 2 2 2 2   3 3 3 3 3 3 3 3   4 4 4 4 4 4   5

7 ①s   10 ②s   8 ③s   6 ④s   1 ⑤

i Find n, the number of values in the list.

ii Use $\dfrac{n+1}{2}$ to find the position of the middle value.

iii Find the median value.

9 This stem and leaf diagram shows masses of fish caught in a competition.

```
0 | 3 4 5
1 | 1 2 2 4
2 | 4 6 6 8 8 8 9
3 | 2 3 5 5 9
4 | 0 2
```

Key: 0 | 3 means 0.3 kg

a Which number appears most in the diagram?

b Write down the mode.

c Work out the range = highest number − lowest number.

d Count the number of values.

The position of the median is $\dfrac{\square + 1}{2} = \square$

e Find the median.

Averages and range for grouped data

1 The table gives information about the number of puzzles completed by Y11 students in 30 minutes.

Puzzles completed	Frequency, f	Midpoint of class, m	$m \times f$
1–5	11	3	$3 \times 11 = 33$
6–10	15		
11–15	23		
16–20	16		
Total		**Total**	

a Copy and complete the table.

b What is the total number of students?

c Estimate the total number of puzzles solved.

d Do you know how many students actually completed 3 puzzles in the given time?
Perhaps all 11 students in the 1–5 class completed 5 puzzles.
Explain why your answer to part **c** is an estimate.

e Work out an estimate of the mean using

$\dfrac{\text{estimated total number of puzzles solved}}{\text{total number of students}}$.

f Which is the modal class?

g Work out $\dfrac{\text{total frequency} + 1}{2}$ to find the position of the median.

h Which class contains the median?

2 The table gives information about the number of points Steve scored in general knowledge quizzes.

Score	Frequency, f	Midpoint of class, m	$m \times f$
0–20	9	10	$10 \times 9 = 90$
21–40	6	30.5	
41–60	2		
61–80	2		
Total		**Total**	

a Copy and complete the table.

b What is the total number of quizzes?

c Estimate the total number of points Steve scored.

d Explain why Steve's points cannot be worked out exactly.

e Work out an estimate of Steve's mean score per quiz.

f Estimate the range of Steve's scores using range = highest possible score – lowest possible score.

Sampling

1 Cherry wants to select 4 students at random from her class.
Which of these methods give a random sample, where every student is equally likely to be picked?

A Put students' names in a hat and pick 4.

B Pick the first 4 students to arrive in the morning.

C Write students' names in a list.
Give each name a number.
Generate 4 random numbers and pick the students with these numbers.

D Pick the last 4 students on the register.

E Give each student a raffle ticket.
Pick the raffle tickets from a box.

2 Jenny wants to find out how many people do their weekly shopping online.
She asks people in her local town centre on a Saturday morning to complete her questionnaire.
Do you think that the people in a town centre on a Saturday morning include a fair proportion of people who do their weekly shopping online?
Explain why her results may be biased.

3 In a hobbies survey of 20 people, 6 people choose 'running'.

a What fraction choose 'running'?

b Out of 200 people, what fraction do you think will choose 'running'?
Work out this fraction of 200.

7 Extend

1 There are only red, white, yellow and green counters in a bag. The chart shows the number of red, white and yellow counters.

The total number of counters is 30.

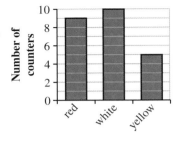

a Work out the number of green counters. **(1 mark)**

b Write down the mode. **(1 mark)**

2 Find the mean of the first 5 square numbers.

 3 The mean of 7 amounts of money is £4.30. Work out the total amount of money.

4 The mean of 6 numbers is 11. The mean of a different set of 9 numbers is 6. What is the mean of all 15 numbers?

5 Trey is x years old.
Will is $x + 7$ years old.
Ali is $3x$ years old.
Write an expression, in terms of x, for the mean of their ages. **(2 marks)**

Exam tip

Find the total by collecting like terms.

6 The table shows the results for two competitors in a javelin competition.

	1st throw	2nd throw	3rd throw	4th throw
Liz	28.45 m	29.20 m	27.38 m	28.35 m
Katie	30.24 m	29.76 m	29.84 m	20.47 m

a Work out the mean distance thrown by Liz and Katie.

b Work out the median distance for Liz and for Katie.

c Which value most affected Katie's mean distance?

d Why did the value you identified in **c** not affect Katie's median distance thrown?

e Which average best represents their performances?

f Who do you think is the better javelin thrower?

Q6c hint

Is there any data value that stands out as being very high or very low?

Q6d hint

Which type of average is not affected by an outlier?

7 **Problem-solving** This back-to-back stem and leaf diagram shows the heights, in cm, of Class 11W students.

Compare the distribution of heights of the boys with the distribution of heights of the girls.

Boys		Girls
	15	7 7
7	16	2 3 6 7 8 9 9
9 8 7 5 2 0	17	0 1 2 8
9 5 4 3 2 1 1 0	18	3
3	19	

Key: (Boys) 7 | 16 means a height of 167 cm
(Girls) 15 | 7 means a height of 157 cm

8 A netball team played six games.
  Here are the number of goals the team scored.
     3, 5, 4, 6, 2, 0
  a Work out the mean score for these six games. (2 marks)

  The netball team played one more game.
  The mean number of goals for all seven games is 4.
  b Work out the number of goals the team scored in the seventh game. (2 marks)

9 This table gives information about the prices of cars.
  a Work out an estimate of the mean price.
  b Which is the modal class?
  c When n is large use $\frac{n}{2}$ to give the position of
    the median.

    Which class contains the median?
  d Work out an estimate of the range.

Price, p (£)	Frequency
$7000 < p \leqslant 8000$	652
$8000 < p \leqslant 9000$	583
$9000 < p \leqslant 12000$	976
$12\,000 < p \leqslant 15\,000$	354
$15\,000 < p \leqslant 20\,000$	185

10 **Problem-solving** The histogram shows the
  distribution of times for an Austrian downhill
  ski race.
  a Make a frequency table for the times.
  b Work out an estimate of the mean time.
  c Work out an estimate of the range.

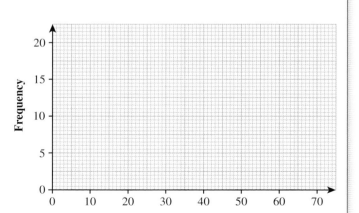

11 This table shows information about the length of 60 snakes.

Length, l (cm)	Frequency
$20 < l \leqslant 30$	5
$30 < l \leqslant 40$	16
$40 < l \leqslant 50$	18
$50 < l \leqslant 60$	14
$60 < l \leqslant 70$	7

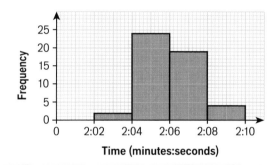

  a Find the class interval that
    contains the median.
    (1 mark)
  b Draw a frequency polygon for
    the information
    in the table. (2 marks)

12 1200 people attend a concert. The organisers would like some feedback from members of the
  audience to help them plan their next event. They decide to sample 15% of the audience.
  a How many people should they include in their sample?
  b How could they select a representative sample?

7 Test ready

Summary of key points

To revise for the test:

- Read each key point, find a question on it in the mastery lesson, and check you can work out the answer.

- If you cannot, try some other questions from the mastery lesson or ask for help.

Key points

1   The **mean** of a set of values is the total of the set of values divided by the number of values. → **7.1**

2   To calculate the mean from a frequency table, use: $\text{mean} = \dfrac{\sum f \times x}{\sum x}$

   where f is the frequency, x is the variable and Σ means 'the sum of'. → **7.1**

3   To compare two sets of data, compare an average (mean, mode or median) and the range. → **7.1**

4   The **median** is the middle value when the data is written in order.
   The median is in the $\dfrac{n+1}{2}$ th position. → **7.2**

5   An **outlier** is an extreme value that doesn't fit the overall pattern. → **7.2**

6   To estimate the range of grouped data, work out
   maximum possible value – minimum possible value. → **7.2**

7   The **modal class** is the class with the highest frequency. → **7.3**

8   When the data is grouped, you can calculate an **estimate** for the mean using the midpoints of the classes as estimates for data values. → **7.4**

9   In a survey, a **sample** is taken to represent the **population**.
   A sample that is too small can **bias** the results.
   The population is the whole of the group that you are interested in. → **7.5**

10   When considering sample size, you need to balance the need for accuracy with the costs and time involved in taking a large sample. → **7.5**

11   In a **random sample**, every member of the population has an equal chance of being included. → **7.5**

12   An **unbiased** sample represents the population. → **7.5**

13   The population may be divided into groups, e.g. by age.
   A **stratified sample** contains members of each group in the same proportion as the population. → **7.5**

Sample student answers

Exam-style question

1 14 students did a quiz.
Here are the results.

Girls	2	8	3	5	3	4	10	6
Boys	3	5	3	2	1	4		

Compare the distributions of the girls' scores and the boys' scores. **(2 marks)**

Girls' range is $10 - 2 = 8$
Boys' range is $5 - 1 = 4$
The range for the boys is smaller than the range for the girls, so the boys' scores are more consistent.

Why might the student not get full marks?

Exam-style question

2 A hockey team played eight matches.
Here are the number of goals the team scored.
2, 0, 3, 1, 4, 2, 0, 4
Tom is asked to calculate the mean score.
Here is his working

$$2+3+1+4+2+4=16$$
$$\frac{16}{6}=2.666\ldots$$

This is wrong. Explain why. **(2 marks)**

He missed out the zeros.

a Has the student explained how Tom's working gives the wrong answer?

b Write a better answer, explaining how 'missing out the zeros' affects the calculation, and calculating the correct mean.

7 Unit test

Active Learn
Homework

1 Here is a list of numbers
2, 5, 13, 8, 7, 6, 9
Kim says, 'The median is the middle number, so the median of these numbers is 8.'
Kim's answer is not correct.

   a What is wrong with Kim's method? **(1 mark)**

   b Work out the range of the numbers in the list. **(2 marks)**

   c Work out the mean of the numbers in the list. **(2 marks)**

2 Peter rings 20 people on their home phone numbers between 10 am and 12 pm on a Monday.
He asks their views on traffic in the town.
Give two reasons why his sample is likely to be biased. **(2 marks)**

3 The mean of the numbers shown on these cards is 6.

   Work out the value of the fifth card. **(2 marks)**

4 This stem and leaf diagram shows the prices of some bars of chocolate.

```
3 | 2 9
4 | 2 5 5 6
5 | 1 3 3 5 7 9
6 | 0 2 4 4 5 7 8 8
7 | 3 5 9
8 | 6
9 | 0
```

Key: 3 | 2 means £0.32

   a Find the median amount spent. **(1 mark)**

   b Work out the range. **(1 mark)**

5 This back-to-back stem and leaf diagram shows the lengths, in cm, of some grass snakes.

Male		Female
6 5 2 0	7	
9 8 7 7 5 3 3 1	8	6 8
5 5 4 3 3	9	0 2 2 3 5 6 8
	10	3 5 5 5 6 7
	11	0 1 1

Key: (Male) 1 | 8 means a length of 81 cm
     (Female) 8 | 6 means a length of 86 cm

Compare the distribution of lengths of the male grass snakes with the distribution of lengths of the female grass snakes. **(2 marks)**

6 This table shows the number of goals scored in each match of a knockout tournament.

Goals scored	Frequency
0	1
1	3
2	7
3	11
4	8
5	6

    a Work out the mean number of goals scored.   **(2 marks)**

    b Find the median.   **(1 mark)**

    c Find the modal number of goals scored.   **(1 mark)**

    d Work out the range.   **(1 mark)**

7 3 female kittens have a mean mass of x kg.
4 male kittens have a mean mass of y kg.
Write down an expression, in terms of x and y, for the mean weight of all the kittens. **(2 marks)**

8 This table gives information about the number of visitors to a National Trust property in June.

Number of visitors, n	Frequency
$200 \leqslant n < 300$	4
$300 \leqslant n < 400$	11
$400 \leqslant n < 500$	12
$500 \leqslant n < 600$	3

    a Work out an estimate of the mean number of visitors.   **(3 marks)**

    b Work out an estimate of the range.   **(1 mark)**

    c Which is the modal class?   **(1 mark)**

    d Find the class interval that contains the median.   **(1 mark)**

    e Draw a frequency polygon for this data.   **(2 marks)**

9 A school has 600 students in KS3 and KS4.
240 of the students are in KS4.
A stratified sample of 50 students is selected.
How many KS4 students should be in the sample?   **(2 marks)**

(TOTAL: 30 marks)

10 Challenge Which average should you use for each of these?
Choose mean, median or mode. Explain your choices.

    a One value appears a lot more often than any other and it is not at one end of the data set.

    b There are no extreme values and every value should be taken into account.

    c The data is not numerical.

    d The data is fairly evenly spread but there is one extreme value.

11 Reflect Look back at this unit.
Which lesson made you think the hardest?
Write a sentence to explain why.
Begin your sentence with: Lesson _____ made me think the hardest because _____ .

Mixed exercise 2

1 **Problem-solving** The comparative bar chart shows information about the number of students absent from a school last week.

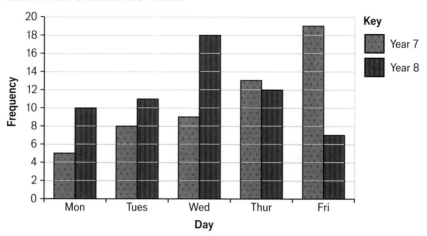

Work out the mean number of students absent in each year group.

2 **Reasoning** Work out the size of angle x.
You must give reasons for your answer.

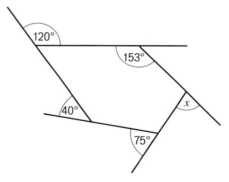

3 **Reasoning** Here are some numbers.
   12   14   16   21   23   34
12, 14, 16 is an arithmetic sequence.
Use three of the numbers to make a different arithmetic sequence.
Describe the rule for your sequence.

Exam-style question

4 Grace has 5 cards.
There is a number on each card.
Two of the numbers are hidden.

The mode of the 5 numbers is 5.
The mean of the 5 numbers is 6.
Work out the 2 numbers that are hidden.

(2 marks)

---

5   Here is a rectangle.

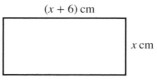

Emily solves a problem about this rectangle to find the value of x.
Her answer is
$x = -3$

Explain why Emily's answer must be wrong.   **(1 mark)**

6   **Reasoning** Ewan is asked to solve the equation $\dfrac{x+5}{4} = 7$.

Ewan writes:

$$-5\left(\dfrac{x+5}{4} = 7\right)-5$$

$$\times 4\left(\dfrac{x}{4} = 2\right)\times 4$$

$$x = 8$$

Is Ewan correct? Give reasons for your answer.

7   PQR is a straight line.
$PQ = QR = QS$
Angle $SPQ = 68°$
Work out the size of the angle marked x.
Give a reason for each stage of your working.   **(4 marks)**

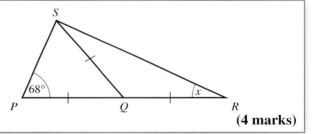

8   **a** **Reasoning** A sequence has 4 terms.
The term-to-term rule for the sequence is 'multiply by 5 and then add 8'.
The first term of the sequence is -1.
Work out the other 3 terms of the sequence.

   b The order of the 4 terms is reversed to make a different sequence.
Work out the term-to-term rule for this sequence.

   c A term-to-term rule for a sequence is subtract 3 and then multiply by 4.
What is the term-to-term rule for the reverse of this sequence?

9   **Problem-solving**

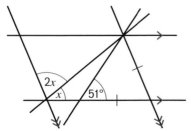

Work out the size of angle x.

Exam-style question

10 The diagram shows a hexagon.
   The hexagon has two lines of symmetry.
   $AB = 5x$
   $BC = 2x - 1$
   All these measurements are given in centimetres.
   The perimeter of the hexagon is 95 cm.

   a Show that $18x - 4 = 95$. **(3 marks)**

   b Find the value of x. **(2 marks)**

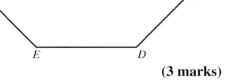

11 **Reasoning** Toby has a list of five numbers:
   6, 10, 12, 14, 16
   Toby works out the mean of four of the numbers and gets 11.
   Which number did Toby not use?

12 **Reasoning** Sequence A has nth term $3n + 4$.
   Sequence B has nth term $3 - 2n$.
   Do the sequences have any numbers that are the same?
   Give a reason for your answer.

13 **Reasoning** The diagram shows three straight lines.

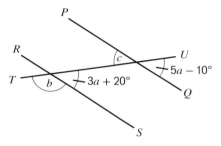

   a If PQ and RS are parallel lines, work out the value of angle b.

   b If PQ and RS are not parallel lines, work out angle b when angle $c = 50°$.
   You must show your working.

Exam-style question

14 6 salad tomatoes have a mean mass of 100 g.
   3 beef tomatoes have a mean mass of 250 g.
   1 cherry tomato has a mass of 25 g.
   Chris says, 'The mean mass of the 10 tomatoes is less than 175 g.'
   Is Chris correct? You must show how you get your answer. **(3 marks)**

15 **Problem-solving** The nth term of a sequence is $8n + 9$
   Find all the two-digit square numbers that are in this sequence.

Exam-style question

16 The diagram shows parallelogram $ABCD$.

   Angle ABC is an obtuse angle.
   Find the greatest whole number value of x.
   You must show your working. **(3 marks)**

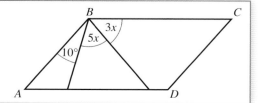

17 Reasoning Tom records how long the teachers have taught at his school.
He records his results in the table.

Number of years, y	Number of teachers
$0 < y \leqslant 10$	42
$10 < y \leqslant 20$	33
$20 < y \leqslant 30$	5

a Work out an estimate of the mean number of years.

b Tom decides to record his results more accurately.

Number of years, y	Number of teachers
$0 < y \leqslant 5$	39
$5 < y \leqslant 10$	3
$10 < y \leqslant 15$	30
$15 < y \leqslant 20$	3
$20 < y \leqslant 25$	5
$25 < y \leqslant 30$	0

Tom estimates the mean for his more accurate results.
Should his estimate be higher or lower than your estimate?
Give reasons for your answer.

c Write down the modal class.

d Which class contains the median?

18 The diagram shows a regular octagon and a rhombus.

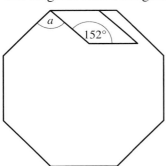

Work out the size of the angle marked a.
You must show all your working. **(4 marks)**

19 Reasoning Akram is asked to solve these two inequalities:
Inequality 1: $4 < 2x \leqslant 14$
Inequality 2: $4 \leqslant 2x < 14$
Akram says the two inequalities have the same integer solutions.
Is Akram correct? Explain why.

8 Perimeter, area and volume 1

Prior knowledge

8.1 Rectangles, parallelograms and triangles

- Calculate the perimeter and area of rectangles, parallelograms and triangles.
- Calculate a missing length, given the area.

Active Learn
Homework

Warm up

1 Fluency

a What does perpendicular mean?

b Which of these show perpendicular lines?

i ii iii iv

2 Work out the **perimeter** of each shape. Make sure you write the correct units.

a 2 cm 6 cm 6 cm 2 cm

b 3 m 10 m

c 5 m 4 m 8 m

d 120 mm 50 mm 80 mm

3 Work out the **area** and perimeter of each shape.

a 5 cm

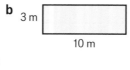

b 10 mm 37 mm

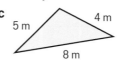

c 4 m 2 m

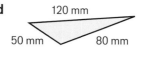

4 Reasoning

a Which of these expressions is for

i the area of this rectangle

ii the perimeter of this rectangle?

$2l + 2w$

$l \times w$

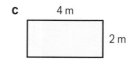

b Write a formula for the area of a rectangle. Start $A =$

c Write a formula for the perimeter of a rectangle. Start $P =$

Exam-style question

5 The length of a rectangle is three times as long as the width of the rectangle.
The area of the rectangle is 48 cm².
Draw the rectangle on a centimetre grid.   **(2 marks)**

6 Problem-solving The diagram shows a 3 cm by 5 cm rectangle.

What is the maximum number of 3 cm by 5 cm rectangles that can be cut from this piece of card?

3 cm
5 cm

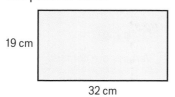
19 cm
32 cm

7 The diagram shows 7 identical squares inside a rectangle.

The length of the rectangle is 12 cm.
Work out the width of the rectangle. **(3 marks)**

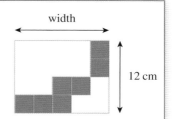
width
12 cm

Key point

The **dimensions** of a rectangle are its length and width.

8 Problem-solving Here is a square of side length 2 cm.

a Sketch a rectangle that can be made with 8 of these squares.

b Write the dimensions of the rectangle on your diagram.

2 cm
2 cm

Key point

The base of a parallelogram is b and its **perpendicular height** is h.
Cutting a triangle from one end of a parallelogram and putting it on the other end makes a rectangle.
Area of parallelogram = base length × perpendicular height

$$A = bh$$

 9 Calculate the area of each parallelogram.

a
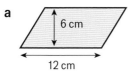
6 cm
12 cm

b
9 cm
7 cm

c
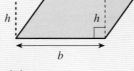
32 mm
76 mm

d
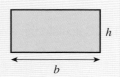
2.4 cm
6.1 cm

10 Calculate the perimeter and area of each shape.
All lengths are in centimetres.

a
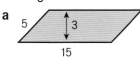
5
3
15

b
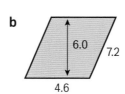
6.0
7.2
4.6

c Reflect How did you know which measurements to use in your area calculations?

11 Reasoning

    a Work out the area of this parallelogram.

    b Trace the parallelogram. Join 2 opposite vertices with a straight line to make two triangles.

    c What fraction of the parallelogram is each triangle?

    d Write down the area of one of the triangles.

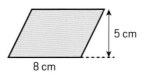

Q11b hint

Vertices are corners.

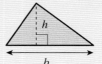

Key point

The diagonal splits a parallelogram into two identical triangles.

Area of 2 triangles $= b \times h$

Area of a triangle $= \frac{1}{2} \times b \times h$

Area of a triangle $= \frac{1}{2}bh$

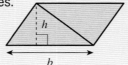

Example

Calculate the area of the triangle.

$b = 7, h = 4$ — Write down the values of b and h.

Area $= \frac{1}{2}bh$

$= \frac{1}{2} \times 7 \times 4$ — Substitute into the formula for area of a triangle.

$= 14\,cm^2$ — Write the units with your answer.

12 Calculate the area of each triangle.

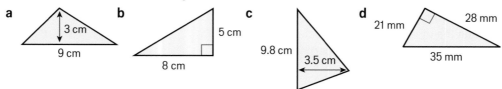

    a 3 cm, 9 cm
    b 5 cm, 8 cm
    c 9.8 cm, 3.5 cm
    d 21 mm, 28 mm, 35 mm

    e Reflect Is the base measurement you use always along the bottom of a triangle?

13 Problem-solving Calculate the perimeter and area of each shape.
All lengths are in centimetres.

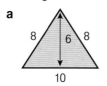
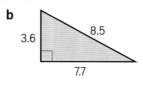
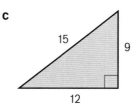

    a 8, 6, 8, 10
    b 3.6, 8.5, 7.7
    c 15, 9, 12

14 Problem-solving A triangle has base length 6.4 cm and height 5 cm.
Calculate its area.

15 Problem-solving Ali makes bunting from triangles of fabric like this.
Each has braid all around the three sides.

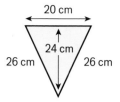

Ali makes 20 of these triangles. Work out

a the area of fabric needed

b the length of braid needed

16 Problem-solving Sketch and label three triangles that have area 18 cm².

17 These shapes are drawn on a centimetre grid.
Find the area of each shape. **(6 marks)**

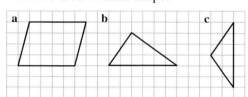

18 Problem-solving Each shape has an area of 25 cm². Write down the area formula for the shape.
Substitute the values you know into the formula. Solve the equation to find the missing length.

a
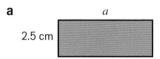
2.5 cm

b
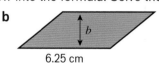
6.25 cm

c
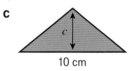
10 cm

19 Here are a right-angled triangle and a rectangle.

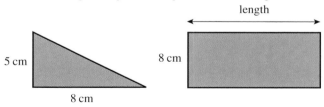

The area of the rectangle is 5 times the area of the triangle.
Work out the length of the rectangle. **(4 marks)**

20 Future skills A rectangular room measures 3.7 m by 2.9 m, and is 2.6 m high.

a Sketch four rectangles to represent the four walls. Label their lengths and widths.

b Calculate an estimate for the total area of the walls.

c A tin of paint costs £11.99 and covers 6 m². Estimate the cost of paint for the room.

d Write down any assumptions you have made.

8.2 Trapezia and changing units

- Calculate the area and perimeter of trapezia.
- Find the height of a trapezium given its area.
- Convert between area measures.

Active Learn
Homework

Warm up

1 **Fluency** Convert

    a 2.6 cm to mm   **b** 6.1 m to cm   **c** 54 cm to mm   **d** 127 cm to m

2 Work out   **a** $\frac{1}{2} \times 16$ **b** $\frac{1}{3}(3+9)$ **c** $\frac{1}{2} \times (2+4) \times 10$ **d** $\frac{1}{2} \times 5 \times 6$

3 Solve   **a** $5h = 20$ **b** $6.4x = 19.2$ **c** $7y = 21$ **d** $2p = 11$

4 Work out the area of this parallelogram.

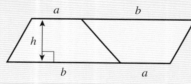

4 cm

5 cm   3 cm

Key point

This **trapezium** has parallel sides a and b and perpendicular height h.
Two trapezia put together make a parallelogram, with base $(a+b)$ and perpendicular height h.

    Area of 2 trapezia = base × perpendicular height = $(a+b) \times h$

Area of a trapezium = $\frac{1}{2}(a+b)h$

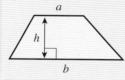

Hint

Trapezia is the plural of trapezium.

Example

Calculate the area of this trapezium.
All lengths are in centimetres.

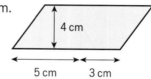

6.4

8

15.6

Area $= \dfrac{1}{2}(a+b)h$ •——— Write down the formula for the area of the trapezium.

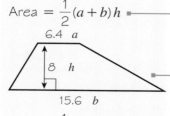

6.4 a

8   h

15.6 b

•——— Sketch the trapezium. Label a, b and h.

Area $= \dfrac{1}{2} \times (6.4 + 15.6) \times 8$ •——— Substitute the values in the formula.

$= \dfrac{1}{2} \times 22 \times 8 = 88\,\text{cm}^2$

5 Calculate the areas of these trapezia.
All lengths are in centimetres.
Round answers to 1 decimal place where necessary.

a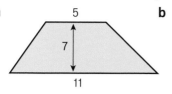

b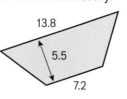

c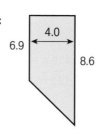

d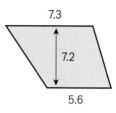

Exam-style question

6 Here is a trapezium drawn on a centimetre grid.

On a centimetre grid, draw a triangle equal in area to the trapezium. **(2 marks)**

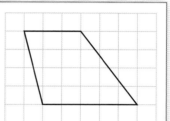

Exam tip

Use the area formula instead of trying to count parts of squares.

---

7 An **isosceles trapezium** has one line of symmetry.
Its two sloping sides are equal.
Calculate the area and perimeter of this isosceles trapezium.

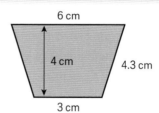

8 This trapezium has area 40 cm².

a Substitute the values for A, a and b into the formula
area $= \frac{1}{2}(a+b)h$

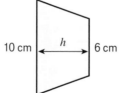

b Simplify your answer to part **a**.

c Solve your equation from part **b** to find h.
Give the units with your answer.

9 **Problem-solving** This trapezium has area 35 mm².
Work out its height.

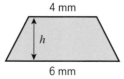

Key point

These two squares have the same area.
To convert from cm² to mm², multiply by 100.
To convert from mm² to cm², divide by 100.

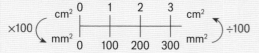

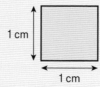

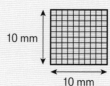

Area $= 1\,\text{cm} \times 1\,\text{cm}$
$= 1\,\text{cm}^2$

Area $= 10\,\text{mm} \times 10\,\text{mm}$
$= 100\,\text{mm}^2$

10 Convert

    a $2750\,\text{mm}^2$ to cm^2 **b** $275\,\text{mm}^2$ to cm^2 **c** $1.4\,\text{cm}^2$ to mm^2

    d $14\,\text{cm}^2$ to mm^2 **e** $140\,\text{cm}^2$ to mm^2 **f** $0.8\,\text{cm}^2$ to mm^2

    g $90\,\text{mm}^2$ to cm^2 **h** $9\,\text{mm}^2$ to cm^2 **i** $9\,\text{cm}^2$ to mm^2

11 **a** Calculate the area of this square in cm^2.

    b Convert your answer to mm^2.

2.1 cm

12 **a** Work out the area of each square to help you work out the number of cm^2 in $1\,\text{m}^2$.

    b Copy and complete the double number line.

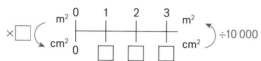

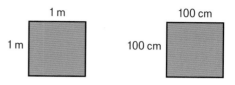

13 Convert

    a $7\,\text{m}^2$ to cm^2 **b** $50\,000\,\text{cm}^2$ to m^2 **c** $5000\,\text{cm}^2$ to m^2 **d** $3.2\,\text{m}^2$ to cm^2

14 **a** Choose four decimal measures between $7\,\text{m}^2$ and $8\,\text{m}^2$. Convert each measure to cm^2.

    b Choose four measures between $10\,000\,\text{cm}^2$ and $12\,000\,\text{cm}^2$. Convert each measure to m^2.

15 The diagram shows a yoga mat.

    a Work out the area of the mat in m^2.

    b Convert your answer to part **a** to cm^2.

    c Convert the length and the width of the mat to cm.
       Use these measurements to work out the area of the mat in cm^2.

    d **Reflect** Which do you think is the easier way to find the area in cm^2?

       • Find the area in m^2, then convert to cm^2.

       • Convert the lengths to cm, then find the area.

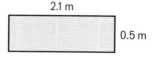
2.1 m
0.5 m

16 **Problem-solving** Work out these areas in cm^2.

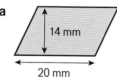
a 14 mm   20 mm

b 1.5 m   2.7 m

c 3.2 cm   20 mm   8 cm

17 Work out the area of this rectangle

    a in cm^2

    b in mm^2

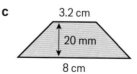
5.5 cm
2.4 cm

18 **Problem-solving** A biologist counts 170 bacteria on slide A in an area of $340\,\text{mm}^2$.
On slide B she counts 120 bacteria in an area of $2\,\text{cm}^2$.
Which sample has more bacteria per square millimetre?
Show your working to explain.

8.3 Area of compound shapes

- Calculate the perimeter and area of shapes made from triangles and rectangles.
- Calculate areas in hectares, and convert between ha and m².

Active Learn
Homework

Warm up

1 Fluency How many metres are there in 1 km?

2 Work out the area and perimeter of these shapes. All lengths are in cm.

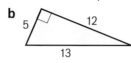

3 Convert these areas into the units given.

a 5 m² to cm²　　b 2.2 m² to cm²　　c 20 000 cm² to m²　　d 7200 cm² to m²

Key point

1 hectare (ha) is the area of a square 100 m by 100 m.
1 ha = 100 m × 100 m = 10 000 m².
Areas of land are measured in hectares.

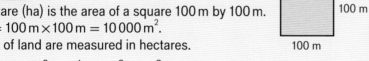

4 Convert these areas to the units given.

a 8 ha to m²　　b 40 ha to m²　　c 3.5 ha to m²　　d 0.5 ha to m²
e 40 000 m² to ha　　f 120 000 m² to ha　　g 225 000 m² to ha　　h 3000 m² to ha

5 a Use these diagrams to help you work out the number of m² in 1 km².
b Copy and complete the double number line.

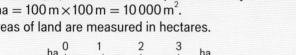
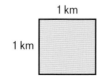
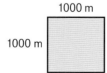

6 Reasoning The diagram shows 1 km² divided into 100 m squares.
a What is the area of each 100 m square?
b How many hectares are there in 1 km²?
c Copy and complete the double number line.

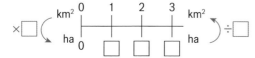
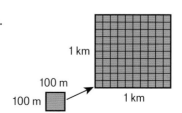

7 **Reasoning** Windermere and Coniston Water are two lakes in the Lake District.
The area of Windermere is 1473 ha. The area of Coniston Water is 4.9 km^2.

  a Which is larger, Windermere or Coniston Water?

  b What is the difference in area?

Example

Calculate the perimeter and area of this compound shape.

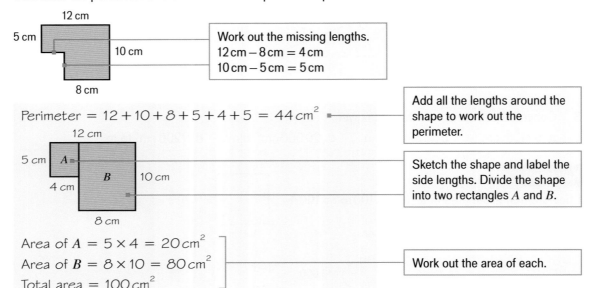

Work out the missing lengths.
12 cm − 8 cm = 4 cm
10 cm − 5 cm = 5 cm

Perimeter = 12 + 10 + 8 + 5 + 4 + 5 = 44 cm^2

Add all the lengths around the shape to work out the perimeter.

Sketch the shape and label the side lengths. Divide the shape into two rectangles A and B.

Area of A = 5 × 4 = 20 cm^2
Area of B = 8 × 10 = 80 cm^2
Total area = 100 cm^2

Work out the area of each.

8 Here is the shape from the example above, split into two different
rectangles C and D. Work out

  a the areas of C and D

  b the total area

  c the total perimeter

  d **Reflect** Compare the area from part **b** with the area in the example above.
  Does it matter how you divide up the shape into rectangles to find the area?

9 **a** Copy each shape. Work out the missing lengths.

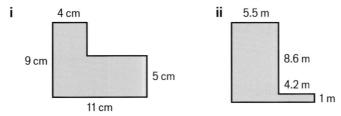

  b Find the perimeter of each shape.

  c Calculate the total area of each shape.

10 Here is a rectangle.

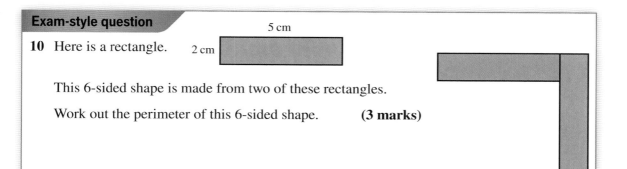

5 cm

2 cm

This 6-sided shape is made from two of these rectangles.

Work out the perimeter of this 6-sided shape.   **(3 marks)**

11 Calculate the area and perimeter of this compound shape.

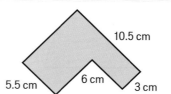

10.5 cm

5.5 cm   6 cm   3 cm

12 Here is a compound shape.

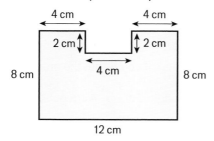

4 cm   4 cm

2 cm   2 cm

8 cm   4 cm   8 cm

12 cm

Here are two ways of working out its area.

E

F   G   H

Method 1
Work out the area of the whole rectangle
Subtract area E.

Method 2
Work out the areas of F, G and H
Add them together.

a Work out the area using both methods.

b **Reflect** Which do you prefer? Explain why.

13 Calculate the area and perimeter of each shape.

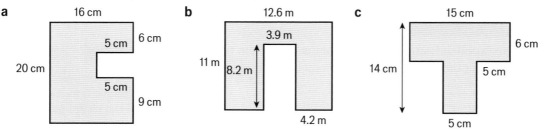

a   16 cm
          5 cm   6 cm
20 cm
          5 cm
                 9 cm

b   12.6 m
          3.9 m
11 m   8.2 m
                 4.2 m

c   15 cm
                      6 cm
14 cm   5 cm
          5 cm

14 **a** Sketch this shape. Draw a line to split it into a rectangle and a triangle.

b Work out the base length and height of the triangle.

c Calculate the area of the triangle.

d Work out the area of the whole shape.

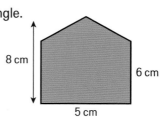

8 cm

6 cm

5 cm

15 Work out the area of each shape.

a

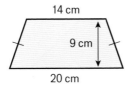

b

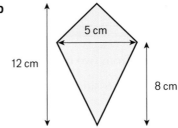

16 a Sketch this isosceles trapezium.

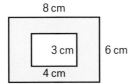

b Divide it into two triangles and a rectangle.

c Work out the area of each part, then add to find the total area.

d **Reflect** Explain how you could work out the area of a trapezium if you forget the area formula.

17 **Problem-solving** Tia makes this photo frame by cutting a rectangle 3 cm by 4 cm out of a larger rectangle of card.

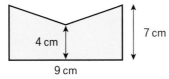

Work out the area of the larger rectangle.
Subtract the area of the cut-out rectangle to work out the area of card in the frame.

18 **Problem-solving** Calculate the area of this shape.

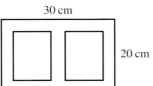

Exam-style question

19 This wooden frame holds two photos, each 8 cm by 15 cm.
Work out the area of wood in this picture frame.

30 cm

20 cm

Exam tip

Show clearly your calculations for the area of the two photos and how this leads to the area of the frame.

(3 marks)

8.4 Surface area of 3D solids

 Active Learn
Homework

- Calculate the surface area of a cuboid.
- Calculate the surface area of a prism.

Warm up

1 Fluency

a Name these 3D solids.

b For each solid, write down

   i the number of faces, edges and vertices

   ii the shapes of the faces

2 Work out the area of these shapes.

a 2 cm

6 cm

b

8 cm

6 cm

c 5 cm

3 cm

7 cm

3 Copy the net of this cuboid.
Label the lengths on the net.

3 cm

6 cm

4 cm

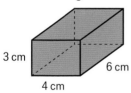

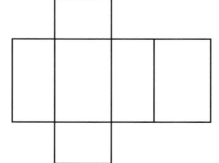

Key point

The **surface area** of a 3D solid is the total area of all its faces.
To find the surface area of a 3D solid, sketch the net and work out the areas of the faces.

4 Sketch the net of this cuboid.
Label the lengths on your net.

10 cm

8 cm

2 cm

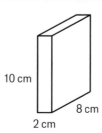

Example

Work out the surface area of this cuboid.

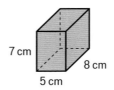

7 cm

8 cm

5 cm

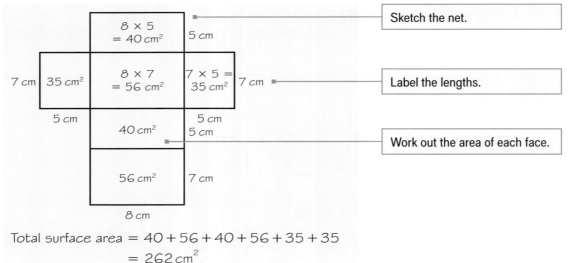

8 × 5 = 40 cm²	5 cm

Sketch the net.

7 cm | 35 cm²

8 × 7 = 56 cm²

7 × 5 = 35 cm² | 7 cm

Label the lengths.

5 cm

40 cm² | 5 cm / 5 cm

Work out the area of each face.

56 cm² | 7 cm

8 cm

Total surface area = 40 + 56 + 40 + 56 + 35 + 35

= 262 cm²

5 Work out the surface area of this cuboid.

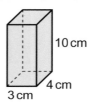

10 cm

4 cm

3 cm

6 Reasoning

  a i What is the area of the top of this cuboid?

    ii Which other face of the cuboid is identical to this one?

  b i What is the area of the front of this cuboid?

    ii Which other face of the cuboid is identical to this one?

  c i What is the area of the side of this cuboid?

    ii Which other face of the cuboid is identical to this one?

TOP

2 cm

FRONT

SIDE

4 cm

9 cm

  d Copy and complete this calculation to work out the total surface area of the cuboid.

$$2 \times \square + 2 \times \square + 2 \times \square = \square \text{ cm}^2$$

  e Reflect Explain how you can calculate the surface area of a cuboid without drawing its net.

7 Calculate the surface area of each cuboid.

  a

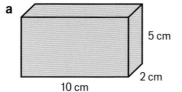

5 cm

2 cm

10 cm

  b

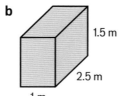

1.5 m

2.5 m

1 m

  c

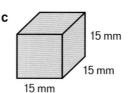

15 mm

15 mm

15 mm

  d Reflect Explain a quick way to work out the surface area of a cube.

Key point

The dimensions of a cuboid are its length, width and height.

8 The diagram shows a cube of side length 3 cm.

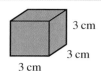
3 cm
3 cm
3 cm

   a Draw a cuboid that can be made with 6 of these cubes.
     Write the dimensions of the cuboid on your diagram. **(1 mark)**

   b Work out the surface area of your cuboid. **(2 marks)**

Key point

A **prism** is a 3D solid that has the same cross-section all through its length.

9 **a** Sketch the net of this triangular prism. Label the lengths.

   b Work out

     i the area of each face   **ii** the total surface area

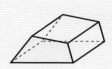
5 cm
3 cm
7 cm
4 cm

10 Calculate the surface area of each prism.
All the measurements are in centimetres.

a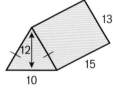
13
12
15
10

b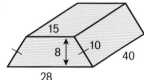
15
8
10
40
28

11 **Problem-solving** A central heating oil tank is in the shape of a cuboid with cross-section 120 cm tall and 100 cm wide. The tank is 150 cm long. Pete paints all the faces except the base.

   a Work out the total area he paints, in square metres.

   b One tin of paint covers 6 m². How many tins of paint does he need?

12 **Problem-solving** These soft play blocks are covered with plastic fabric.

   a Work out how many square metres of fabric you will need to cover them both.

The plastic fabric costs £12.99 per square metre.

   b Estimate the cost of the fabric needed to cover them both.

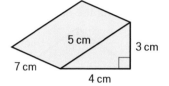
80 cm
110 cm
150 cm

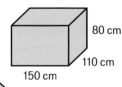
65 cm
97 cm
130 cm
72 cm

13 **Reasoning** This cube has surface area 384 cm².
What is the length of one side of the cube?

14 Children's building blocks are 4 cm wooden cubes, painted on all faces.

   a Work out the area to be painted on one building block.

   b One litre of paint covers an area of 5 m².
     How many cubes can be painted with one litre of paint? **(4 marks)**

Exam tip

In exam questions, the answer to the first part often helps you with the next part. Use your answer to part **a** to help you answer part **b**.

8.5 Volume of prisms

- Calculate the volume of a cuboid.
- Calculate the volume of a prism.
- Use a flow diagram to help solve problems.

*Active*Learn
Homework

Warm up

1 Fluency Sketch the shape of the cross-section of a

a cuboid **b** cube **c** triangular prism **d** pentagonal prism

2 Work out the area of each shape.

a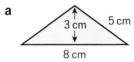
3 cm, 5 cm, 8 cm

b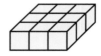
3 mm, 6 mm

c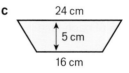
24 cm, 5 cm, 16 cm

3 Work out

a $8 \times 4 \times 2$ **b** $6 \times 5 \times 7$ **c** $9 \times 4 \times 2$ **d** $3 \times 2 \times 5$

4 Convert each length to the units given.

a 240 cm to m **b** 52 cm to m **c** 4.3 m to cm **d** 0.7 m to cm

Key point

The **volume** of a 3D solid is the amount of space inside it.
Volume is measured in cubic units:
millimetre cubed (mm^3), centimetre cubed (cm^3), metre cubed (m^3).

5 This cuboid is made of centimetre cubes.
Each cube has volume $1 \, cm^3$.
Work out the volume of the cuboid.

6 Here are some more cuboids made of $1 \, cm^3$ cubes.
To work out their volumes, count the cubes in the top layer and multiply by the number of layers.

a **b** **c** **d**

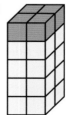

Exam-style question

7 A solid shape is made with 4 of these cubes.

3 cm, 3 cm, 3 cm

Show that the volume of the solid is $108 \, cm^3$. **(2 marks)**

Key point

Volume of a cuboid = length × width × height = lwh

8   Work out the volume of each cuboid. Give the correct units with your answer.

a   4 cm   2 cm   6 cm

b 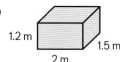   1.2 m   1.5 m   2 m

c   50 mm   15 mm   10 mm

d 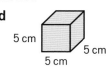   5 cm   5 cm   5 cm

e   Reflect Explain a quick way to work out the volume of a cube of side 4 cm.

9   a This prism is made from centimetre cubes. There are 3 'slices' with 4 cubes in each 'slice'. Work out the volume of the prism.

   cross–section   length 3 cm   1 cm   1 cm   1 cm

b What is the area of the cross-section of this prism?

c Work out area of cross-section × length. What do you notice?

d Work out area of cross-section × length for the cuboids in **Q8**.

Key point

Volume of a prism = area of cross-section × length

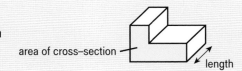
area of cross–section   length

Example

Work out the volume of this prism.

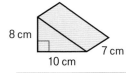
8 cm   7 cm   10 cm

Volume = area of cross-section × length

Area of ▷ = $\frac{1}{2}$ × 10 × 8 ■——— Write down the formula.

= 5 × 8 ■——— Work out the area of the cross-section.

= 40

Volume = 40 × 7 ■——— Substitute the area of the cross-section and the length into the formula.

= 280 cm³ ■———

Write the units.

10   Calculate the volume of each prism.

a 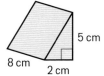   5 cm   8 cm   2 cm

b   area 18 cm²   4 cm

c 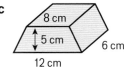   8 cm   5 cm   6 cm   12 cm

11 Problem-solving The cross-section of a prism is an equilateral triangle with area $6.4\,\text{cm}^2$. The prism is 9.2 cm long.

  a Sketch the prism.

  b Work out the volume of the prism.

12 Problem-solving A cube has volume $343\,\text{cm}^3$. How long is one side of the cube?

13 Problem-solving Sketch and label the dimensions of three cuboids with volume $24\,\text{cm}^3$.

Example

Leah packs soap into gift boxes.
Each soap is a cuboid, 60 mm by 40 mm by 20 mm.
Each box is a cuboid, 12 cm by 12 cm by 4 cm.
She charges £1 for each soap and £1.50 for the box.
How much does she charge for a full box of soaps?

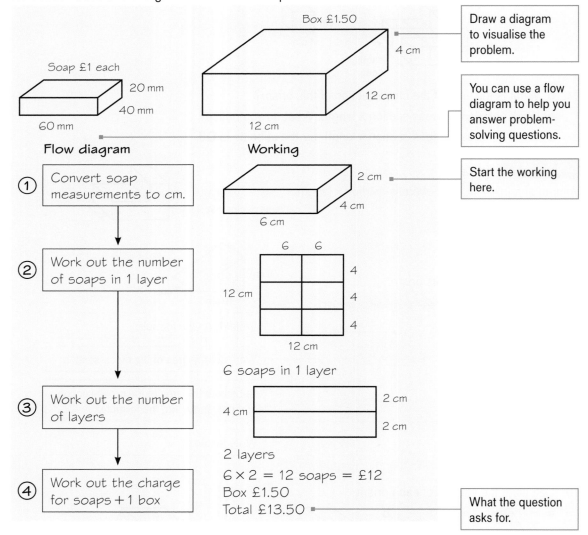

14 Problem-solving How many of these boxes will fit in the larger box?

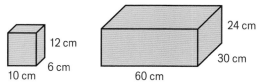

12 cm
6 cm
10 cm
24 cm
30 cm
60 cm

Exam-style question

15 Zara has a large cuboid-shaped box.
The dimensions of the box are:

    length   60 cm
    width   27 cm
    height   45 cm

Zara puts DVDs in the box.
Each DVD is a cuboid, 200 mm × 135 mm × 15 mm.

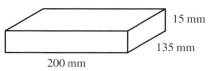

15 mm
135 mm
200 mm

Zara wants to put as many DVDs as possible into the box.
She can put 12 DVDs into the box in 1 minute.
Work out how many minutes it should take to put as
many DVDs as possible in the box. **(4 marks)**

Exam-style question

16 Tim builds a cuboid-shaped sandpit.
The sandpit is 40 cm tall, 1.6 m long and 2.3 m wide.

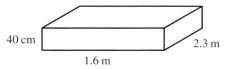

40 cm
1.6 m
2.3 m

Tim puts sand 35 cm deep in the sandpit.

a Work out how many cubic metres of sand he needs, to
1 decimal place. **(3 marks)**

b Sand costs £54 per cubic metre, plus £20 delivery.
How much will the sand cost for the sandpit?
(2 marks)

Exam tip

Mark the depth of the sand on
the diagram. Before you start
your working out, check what
units you need to use for your
answer. Do you need to change
any units before you do the
calculation?

8.6 More volume and surface area

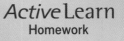
Active Learn
Homework

- Convert between measures of volume.
- Solve problems involving surface area and volume.

Warm up

1 Fluency Work out

a $100 \times 100 \times 100$ **b** 6×1000 **c** 5×1000000

d 7.2×1000 **e** 3.71×1000000 **f** $35000000 \div 1000000$

2 How many ml are there in 1 litre?

3 Match each object to the amount of liquid it can hold.

| teaspoon | drink can | bucket | juice carton |

| 330 ml | 5 ml | 5 litres | 1 litre |

4 Solve these equations.

a $54 = 6h$ **b** $222 = 74h$ **c** $360 = 18h$

5 For this prism, work out

    a the surface area

    b the volume

    Give your answers to 1 decimal place.

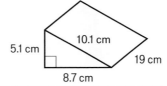

Key point

Volume is measured in mm^3, cm^3 or m^3. These two cubes have the same volume.

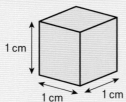

Volume $= 1\,cm \times 1\,cm \times 1\,cm$

$\qquad = 1\,cm^3$

$1\,cm^3 = 1000\,mm^3$

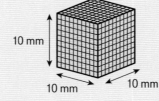

Volume $= 10\,mm \times 10\,mm \times 10\,mm$

$\qquad = 1000\,mm^3$

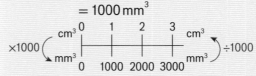

6 Convert

a $24\,cm^3$ to mm^3 **b** $2.4\,cm^3$ to mm^3 **c** $5.7\,cm^3$ to mm^3

d $0.57\,cm^3$ to mm^3 **e** $30000\,mm^3$ to cm^3 **f** $3000\,mm^3$ to cm^3

g $480\,mm^3$ to cm^3 **h** $450\,mm^3$ to cm^3 **i** $4.5\,cm^3$ to mm^3

7 **a** Work out the volume of this cuboid in cm^3, then convert it to mm^3.

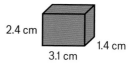

b Convert the measurements of the cuboid to mm.
Then work out the volume in mm^3.

2.4 cm   3.1 cm   1.4 cm

c **Reflect** Which method for finding the volume in mm^3 did you prefer? Explain why.

8 **a** Work out the volume of each cube.

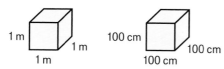

1 m   1 m   1 m   100 cm   100 cm   100 cm

b Copy and complete
1 m^3 = □ cm^3

c Copy and complete this double number line.   ×1 000 000

m^3 0   1   2   □ m^3
cm^3 0   cm^3
         1 000 000   □   3 000 000   ) ÷□

9 Convert

a 5 m^3 into cm^3 **b** 3.5 m^3 into cm^3

c 6.87 m^3 into cm^3 **d** 0.3 m^3 into cm^3

e 9 000 000 cm^3 into m^3 **f** 8 200 000 cm^3 into m^3

g 600 000 cm^3 into m^3 **h** 3 159 000 cm^3 into m^3

i 940 000 cm^3 into m^3 **j** 6 452 000 cm^3 into m^3

> **Key point**
>
> 1 cm^3 = 1 ml   1000 cm^3 = 1 litre

10 Copy and complete these conversions

a 35 cm^3 = □ ml   **b** 300 cm^3 = □ ml

c 3000 cm^3 = □ ml = □ litres   **d** 3500 cm^3 = □ ml = □ litres

e 40 ml = □ cm^3 **f** 400 ml = □ cm^3

g 4000 ml = □ cm^3 **h** 3000 ml = □ cm^3

11 Copy and complete these conversions

a 2 litres = □ cm^3 **b** 4 litres = □ cm^3

c 4.5 litres = □ cm^3 **d** 0.4 litres = □ cm^3

e 1 m^3 = □ cm^3 = □ ml = □ litres   **f** 2 m^3 = □ cm^3 = □ ml = □ litres

g 2.4 m^3 = □ cm^3 = □ ml = □ litres   **h** 0.5 m^3 = □ cm^3 = □ ml = □ litres

> **Key point**
>
> **Capacity** is the amount of liquid a 3D object can hold. It is measured in litres and ml.

12 **Future skills** The capacity of this beaker is 200 ml.
What is the capacity of this beaker in cm^3?

ml   200   160   120   80   40

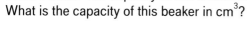

13 This cuboid is a fuel tank.

20 cm

80 cm

50 cm

a Work out its volume in cm^3.

b Convert the volume to ml and then litres to find the capacity of the fuel tank.

14 Problem-solving Work out the capacity of this vase in ml.

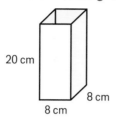

20 cm

8 cm

8 cm

Example

This cuboid has volume 2640 cm^3.

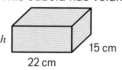

h

15 cm

22 cm

Work out its height.

Volume = $l \times w \times h$ ← Write down the formula.

$2640 = 22 \times 15 \times h$ ← Write in the values you know.

$2640 = 330 \times h$

$h = \dfrac{2640}{330} = 8$ cm ← Solve the equation to find h.

15 Work out the missing measurements in these cuboids.

a

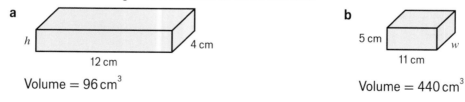

h

12 cm

4 cm

Volume = 96 cm^3

b

5 cm

11 cm

w

Volume = 440 cm^3

16 Problem-solving This cuboid has volume 600 cm^3.

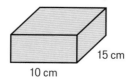

15 cm

10 cm

Work out its surface area.

17 Problem-solving / Reasoning Tracey pours melted chocolate into moulds like this.

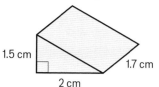

1.5 cm   1.7 cm   2 cm

a What volume of chocolate does she need for each mould?

She melts this block of chocolate.

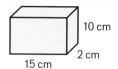

10 cm   2 cm   15 cm

b How many moulds can she fill with the chocolate?

c Is it sensible to round the number of moulds you calculated up or down?

18 Reasoning This steel bar is melted down to make ball bearings.

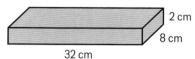

2 cm   8 cm   32 cm

Each ball bearing has volume $0.75\,\text{cm}^3$.
How many ball bearings can be made from the bar?

19 Problem-solving Here is the design for a swimming pool.

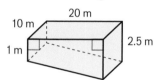

20 m   10 m   1 m   2.5 m

a How many litres of water will the swimming pool hold?

All the vertical sides of the pool will be tiled.

b How many square metres of tiles are needed?

20 A container is in the shape of a cuboid.

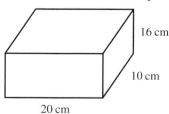

16 cm   10 cm   20 cm

The container is $\frac{3}{4}$ full of water.
A jug holds 250 ml of water.
What is the greatest number of jugs that can be completely filled with water
from the container?

(4 marks)

8 Check up

Active Learn
Homework

2D shapes

1 The diagram shows two shapes drawn on a centimetre grid.

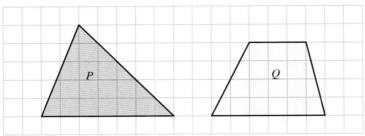

a Find the area of shape P.

b Write down the mathematical name of quadrilateral Q.

 2 For each shape, work out
   i the perimeter   ii the area
Give your answers to the nearest whole number.

a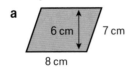

b 5.1 cm

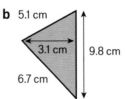

3 The area of this parallelogram is $27\,\text{cm}^2$.
Work out the length marked x.

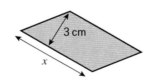

4 The area of this triangle is $15\,\text{cm}^2$.
Work out its height.

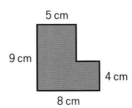

5 For this shape, work out
   a the perimeter
   b the area

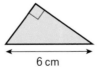

6 Work out the area of this arrow shape.

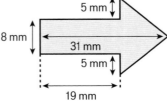

7 Work out the area of this trapezium.

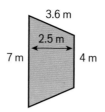

3.6 m

2.5 m

7 m   4 m

8 This wooden mirror frame was made by cutting out a right-angled triangle from the centre of a rectangle.
Work out the area of wood in the frame.

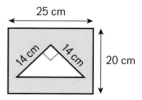

25 cm

14 cm   14 cm

20 cm

3D solids

9 For this cuboid, calculate

a the surface area

b the volume

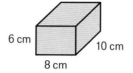

6 cm

10 cm

8 cm

10 Work out the volume of this 3D solid.

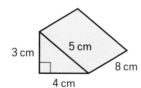

3 cm   5 cm

8 cm

4 cm

Measures

11 Convert these areas to the units given.

a $15\,cm^2 = \square\,mm^2$
b $54\,700\,cm^2 = \square\,m^2$
c $1.2\,m^2 = \square\,cm^2$
d $980\,mm^2 = \square\,cm^2$
e $18\,600\,mm^2 = \square\,m^2$
f $0.3\,m^2 = \square\,mm^2$

12 Convert these volumes to the units given.

a $6\,m^3 = \square\,litres$
b $450\,cm^3 = \square\,ml$
c $0.2\,litres = \square\,cm^3$
d $8400\,mm^3 = \square\,cm^3$
e $549\,000\,cm^3 = \square\,m^3$
f $1.6\,m^3 = \square\,cm^3$

13 Reflect How sure are you of your answers? Were you mostly

Just guessing 😞   Feeling doubtful 😐   Confident 🙂

What next? Use your results to decide whether to strengthen or extend your learning.

Challenge

14 Ria cuts a square of side 1 cm from each corner of a 10 cm by 8 cm piece of card.
Then she folds along the dashed lines to make an open cuboid like this.

a Work out the length, width and volume of the cuboid.

b Predict the volume of the cuboid made by cutting 3 cm squares from a piece of card, 10 cm by 8 cm.
Check by calculating the volume.

c Can you make a box by cutting 4 cm squares from each corner? Draw diagrams to explain.

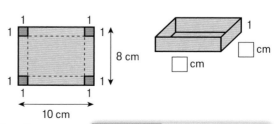

1   1

1   1

8 cm

1   1

1   1

10 cm

1 cm

cm

cm

Q14b hint

Sketch and label a diagram to work out the length, width and height.

8 Strengthen

Active Learn
Homework

2D shapes

1 Work out the perimeter of these shapes.
Matching dashes show equal lengths.
Start at one corner and trace your finger along the sides of the shape.
Add up the lengths as you go.

a
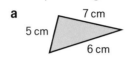
7 cm
5 cm
6 cm

b
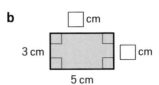
☐ cm
3 cm
☐ cm
5 cm

c
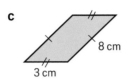
8 cm
3 cm

d

3.5 cm

2 Reasoning Work out the missing lengths marked with ☐ in each diagram.

a
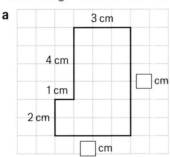
3 cm
4 cm
☐ cm
1 cm
2 cm
☐ cm

b
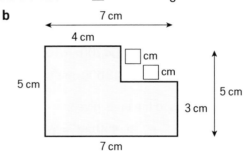
7 cm
4 cm
☐ cm
☐ cm
5 cm
3 cm
5 cm
7 cm

3 i Find the missing lengths.   **ii** Work out the perimeter of each shape.

a
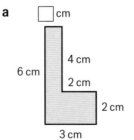
☐ cm
4 cm
6 cm
2 cm
2 cm
3 cm

b
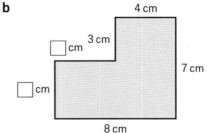
4 cm
3 cm
☐ cm
7 cm
☐ cm
8 cm

4 a Copy the diagram. Draw a vertical line to split the shape into two rectangles.
b Work out the area of each rectangle.
c Work out the total area of the shape.

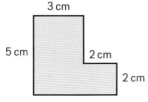
3 cm
5 cm
2 cm
2 cm

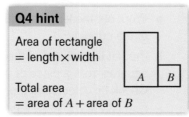
Q4 hint

Area of rectangle
= length × width

Total area
= area of *A* + area of *B*

5 Copy each diagram.

  i Work out the missing lengths.

  ii Work out the area of each rectangle.

  iii Work out the area of the whole shape.

a

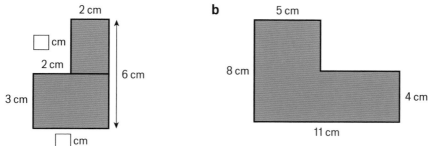

b

6 Copy the diagrams. In each one, mark two lines that are perpendicular to each other.

a   **b** 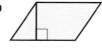   **c**

7 **a** In this triangle the base is labelled b and the height is labelled h.

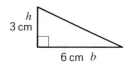

Use the formula $A = \frac{1}{2}bh$ to work out the area, A.

  b This is the same triangle turned on its side.

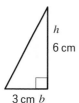

Use the formula $A = \frac{1}{2}bh$ to work out the area, A.

Check that you get the same answer as in part **a**.

  c The base and height are perpendicular.
  Does it matter which you label b and which you label h?

8 **i** Copy the diagram for each parallelogram.

  ii Find two perpendicular lengths. Label them b and h.

  iii Work out the area of the parallelogram using $A = bh$.

a 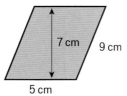   **b** 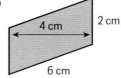   **c**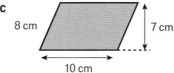

9 Copy the diagrams.

   i Draw a line to split the shape into a rectangle and a triangle.

   ii Work out the area of the rectangle and the triangle.

   iii Work out the total area = area of rectangle + area of triangle.

a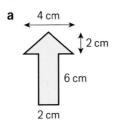

b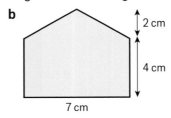

10 a Work out the area of this green rectangle.

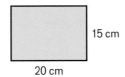

   b Pat cuts a white rectangle out of the centre of the green rectangle.
Work out the area of the white rectangle.

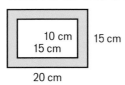

   c Work out area remaining = area of green rectangle − area of white rectangle

11 For each trapezium

   i copy the diagram

   ii label the two parallel sides a and b

   iii label the height h

   iv use the formula $A = \frac{1}{2}(a+b)h$ to work out the area

a

b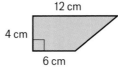

12 This parallelogram has area $12\,\text{cm}^2$.

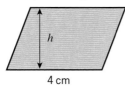

Copy and complete to find its height.

$A = bh$

$12 = \square \times h$

$h = \square\,\text{cm}$

13 This triangle has area 20 cm².

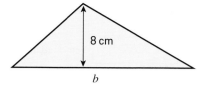

Copy and complete to find its base.

$A = \frac{1}{2}bh$

$\square = \frac{1}{2} \times b \times 8$

$\square = \frac{1}{2} \times 8 \times b$

$\square = \square \times b$

$b = \square$ cm

3D solids

1 a Copy and complete these sketches of the front, side and top faces of this cuboid.

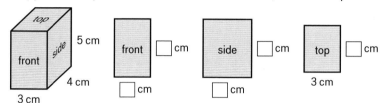

b Make labelled sketches of the back, other side, and bottom faces.

c Which faces are identical?

d Work out the areas of the faces.

e Add the areas of all the faces to work out the total surface area.

2 This triangular prism has 3 rectangular faces and 2 triangular faces.

a Sketch all its faces.

b Work out their areas.

c Work out the total surface area.

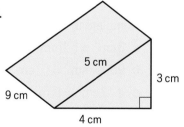

3 Follow these steps to work out the volume of these prisms.

i Sketch the front face.

ii Work out the area of the front face.

iii Multiply the area by the length of the prism.

a

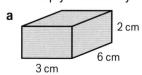

b

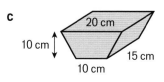

c

Measures

1 **a** Work out the area of each rectangle in cm^2.

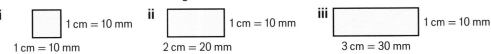

i ☐ 1 cm = 10 mm   **ii** ☐ 1 cm = 10 mm   **iii** ☐ 1 cm = 10 mm

1 cm = 10 mm   2 cm = 20 mm   3 cm = 30 mm

b Work out the area of each rectangle in mm^2.

c Use the areas of the rectangles from parts **a** and **b** to help copy and complete this double number line for cm^2 and mm^2.

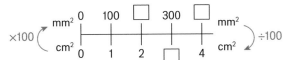

2 **a** Work out the area of these rectangles in cm^2 and m^2.

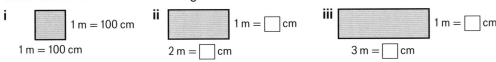

i ☐ 1 m = 100 cm   **ii** ☐ 1 m = ☐ cm   **iii** ☐ 1 m = ☐ cm

1 m = 100 cm   2 m = ☐ cm   3 m = ☐ cm

b Copy and complete this double number line for cm^2 and m^2.

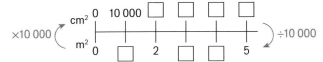

3 Use your double number lines to copy and complete these conversions.

a $5\,\text{cm}^2 = \square\,\text{mm}^2$

b $300\,\text{mm}^2 = \square\,\text{cm}^2$

c $6\,\text{m}^2 = \square\,\text{cm}^2$

d $80\,000\,\text{cm}^2 = \square\,\text{m}^2$

4 Complete these conversions. Use the double number lines.

a $3\,\text{cm}^3 = \square\,\text{mm}^3$

b $9000\,\text{mm}^3 = \square\,\text{cm}^3$

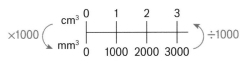

c $4.72\,\text{cm}^3 = \square\,\text{mm}^3$

d $2\,\text{m}^3 = \square\,\text{cm}^3$

e $1.2\,\text{m}^3 = \square\,\text{cm}^3$

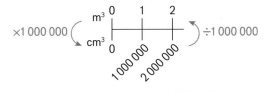

f $9\,000\,000\,\text{cm}^3 = \square\,\text{m}^3$

g $3\,200\,000\,\text{cm}^3 = \square\,\text{m}^3$

h $5000\,\text{cm}^3 = \square\,\text{ml} = \square\,\text{litres}$

i $875\,\text{cm}^3 = \square\,\text{ml} = \square\,\text{litres}$

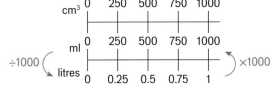

j $2.3\,\text{litres} = \square\,\text{ml} = \square\,\text{cm}^3$

k $1.345\,\text{litres} = \square\,\text{ml} = \square\,\text{cm}^3$

8 Extend

Exam-style question

1 A garden is in the shape of a rectangle 80 m by 50 m.
Vegetables are grown in 30% of the garden.
Work out the area of the garden that is not used for vegetables.

(4 marks)

80 m

50 m

Exam-style question

2 Here are two rectangles.
$XY = 12$ cm
$BC = WX$
The perimeter of $ABCD$ is 28 cm.
The area of $WXYZ$ is 90 cm^2.
Find the length of AB.

(4 marks)

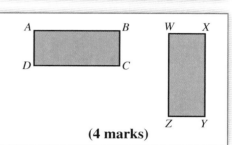

---

3 **Problem-solving** A water tank is in the shape of a cuboid.

All the faces except the base are to be painted.
Work out the total area to be painted, in square metres.
Give your answer to 1 decimal place.

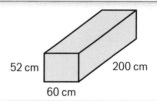

52 cm   200 cm

60 cm

---

4 **Problem-solving** A botanist counts 4 dandelion plants in a
square of lawn measuring 50 cm by 50 cm.
The whole lawn has area 15 m^2.
Estimate the number of dandelion plants on the lawn.

> **Q4 hint**
>
> Work out the area of the square in m^2. How many of these would fit in the lawn?

---

5 **Problem-solving** Jake is tiling this wall in a swimming pool.
He has 170 m^2 of tiles. Is this enough?

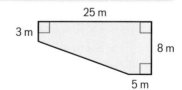

25 m

3 m

8 m

5 m

Exam-style question

6 The diagram shows a floor in the shape of a trapezium.

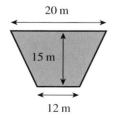

20 m

15 m

12 m

5 litres

£11.99

> **Exam tip**
>
> Show all the different steps in your calculation clearly. Underline the answer in each part.

Mark is going to paint the floor. Each 5 litre tin of paint costs £11.99.
1 litre of paint covers an area of 3 m^2. Mark has £200 to spend on paint.
Does Mark have enough money to buy all the paint he needs?
You must show how you get your answer.

(5 marks)

7 **Problem-solving a** Write an expression for the perimeter of this shape.

    b The perimeter of the shape is 28 cm.
      Use your expression from part **a** to write an equation for the perimeter
      of the shape.
      Solve the equation to find x.

    c Work out the area of the shape.

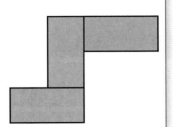

8 Here is a rectangle.

    The length of the rectangle is 5 cm greater than the width of
    the rectangle.
    3 of the rectangles are used to make this 8-sided shape.
    The perimeter of the 8-sided shape is 54 cm.
    Work out the area of the 8-sided shape.   **(5 marks)**

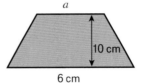

9 Work out the side length of each square. Round any decimal answers to 1 decimal place.

    a   **b**   **c**

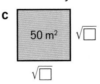

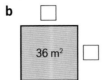

$$\square \times \square = 4$$

10 The area of this trapezium is $50\,\text{cm}^2$.

    a Substitute $A = 50$ and the values of b
      and h into the formula $A = \frac{1}{2}(a+b)\,h$

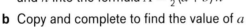

    b Copy and complete to find the value of a

$$\square = \tfrac{1}{2}(a + \square) \times 10$$
$$\square = \tfrac{1}{2} \times 10 \times (a + \square)$$
$$\square = 5 \times (a + \square)$$
$$\square = a + \square \quad \blacktriangleleft\!\!-\!\!- \text{Divide both sides by 5}$$
$$a = \square \text{ cm}$$

> **Q10 hint**
>
> You can do multiplication in any order.

11 A cube of side 6 cm has a cuboid-shaped hole cut through it from
    front to back.
    Work out the volume of the remaining shape.

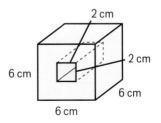

12 Sketch and label the measurements on different prisms that have volume $36\,\text{cm}^3$.

8 Test ready

Summary of key points

To revise for the test:

- Read each key point, find a question on it in the mastery lesson, and check you can work out the answer.

- If you cannot, try some other questions from the mastery lesson or ask for help.

Key points

1 The **dimensions** of a rectangle are its length and width. → **8.1**

2 The base of a parallelogram is b and its **perpendicular height** is h.
Cutting a triangle from one end of a parallelogram and putting it on the other end makes a rectangle.

Area of parallelogram = base length × perpendicular height

$$A = bh$$

→ **8.1**

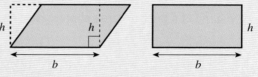

3 The diagonal splits a parallelogram into two identical triangles.

Area of 2 triangles $= b \times h$

Area of a triangle $= \frac{1}{2} \times b \times h$

Area of a triangle $= \frac{1}{2}bh$

→ **8.1**

4 This **trapezium** has parallel sides a and b and perpendicular height h.
Two trapezia put together make a parallelogram, with base $(a+b)$ and perpendicular height h.

Area of 2 trapezia = base × perpendicular height = $(a+b) \times h$

Area of a trapezium $= \frac{1}{2}(a+b)h$

→ **8.2**

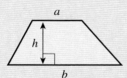 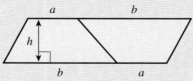

5 To convert from cm^2 to mm^2, multiply by 100.
To convert from mm^2 to cm^2, divide by 100.

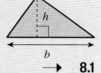

→ **8.2**

6 1 hectare (ha) is the area of a square 100 m by 100 m.

1 ha = 100 m × 100 m = 10 000 m^2

Areas of land are measured in hectares.

→ **8.3**

7 A **compound** shape is made up of simple shapes.
To find the area of a compound shape, split it into simple shapes like rectangles and triangles.
Find the area of each shape and then add them all together.

→ **8.3**

Key points

8 The **surface area** of a 3D solid is the total area of all its faces.
To find the surface area of a 3D solid, sketch the net and work out the areas of the faces. → **8.4**

9 The dimensions of a cuboid are its length, width and height. → **8.4**

10 A **prism** is a 3D solid that has the
same cross-section all through its length.

→ **8.4**

11 The **volume** of a 3D solid is the amount of space inside it.
Volume is measured in cubic units:
millimetre cubed (mm^3), centimetre cubed (cm^3), metre cubed (m^3). → **8.5**

12 Volume of a cuboid = length × width × height = lwh → **8.5**

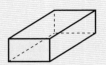

13 Volume of a prism = area of cross-section × length → **8.5**

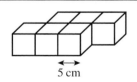

area of cross–section

length

14 **Volume** is measured in mm^3, cm^3 or m^3.
$1 cm^3 = 1000 mm^3$

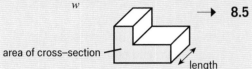

→ **8.6**

15 $1 cm^3 = 1 ml$ $1000 cm^3 = 1 litre$ → **8.6**

16 **Capacity** is the amount of liquid a 3D object can hold. It is measured in litres and ml. → **8.6**

Sample student answer

Exam-style question

6 identical cubes of side 5 cm are joined together like this.
Daisy says, 'The volume of this solid is $750 cm^3$.'

a Is Daisy correct?
You must show your working. **(2 marks)**

b Draw a cuboid that could be made with these cubes.
Write the dimensions of the cuboid on your diagram. **(1 mark)**

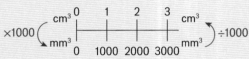

5 cm

---

a Yes

b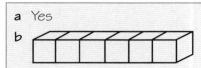

---

a In part **a**, 'yes' is correct but the student is unlikely to receive full marks. Explain why.

b What does the student need to add to their diagram in part **b**?

8 Unit test

*Active*Learn
Homework

1 Work out the area of this trapezium.
All measurements are in centimetres.

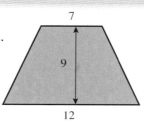

(2 marks)

2 A square has an area of $49\,\text{cm}^2$.

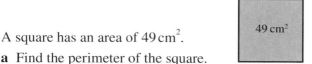

a Find the perimeter of the square. **(1 mark)**

The diagram shows a right-angled triangle and a parallelogram.

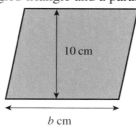

5 cm

8 cm

10 cm

b cm

The area of the parallelogram is 3 times the area of the triangle.
The base length of the parallelogram is b cm.

b Find the value of b. **(3 marks)**

3 The perimeter of this triangle is 30 cm.
Work out

A

12 cm

C

13 cm

B

a the length of the third side **(1 mark)**

b the area **(2 marks)**

4 a Convert $3700\,\text{cm}^2$ to m^2. **(1 mark)**

b A freezer has a volume of $1.5\,\text{m}^3$. Write this volume in cm^3. **(1 mark)**

5 The diagram shows a prism.
All measurements are in cm.

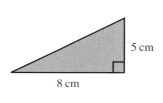

18

15

20

10

Calculate

a the volume **(2 marks)**

b the surface area **(3 marks)**

6 Jody makes a solid from 8 of these 2 cm cubes.
Jody says, 'The volume of my solid is $64\,\text{cm}^3$.'

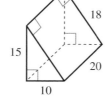

2 cm

2 cm

2 cm

a Is Jody correct? You must show your working. **(2 marks)**

b Draw a cuboid that can be made with 8 of these cubes.
Write the dimensions of the cuboid on your diagram. **(1 mark)**

c Work out the surface area of your cuboid. **(2 marks)**

7 A box of teabags is a cuboid 20 cm by 15 cm by 5 cm.

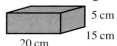

5 cm
15 cm
20 cm

24 boxes of teabags are put into a larger box.
The larger box is a cuboid 40 cm by 50 cm by 20 cm.
Work out the volume of empty space in the larger box. **(3 marks)**

8 This trapezium has area 21 cm^2.
Work out the value of x.

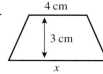

4 cm
3 cm
x

(3 marks)

9 A park is a rectangle, 350 m by 420 m.
1 hectare = 10 000 m^2
Work out the area of the park in hectares. **(2 marks)**

10 a How many litres of water will this tank hold? **(2 marks)**

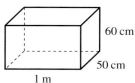

60 cm
50 cm
1 m

b The top and sides of the tank need painting.
Work out the surface area to be painted. **(3 marks)**

c 1 litre of paint covers 10 m^2.
Dan has $\frac{1}{2}$ litre of paint.
Is this enough to paint the tank with two coats of paint? **(1 mark)**

(TOTAL: 35 marks)

11 Challenge In a barn, each sheep and its lambs need 2.2 m^2 of floor space.

a What area of floor space do you need for 40 sheep
and their lambs?

b Sketch and label a rectangle with this area.
Work out the length of fence needed to make
this rectangle.

> **Q11b hint**
>
> Use 2 numbers that multiply
> to give the area as length and
> width.

c Sketch and label another rectangle with this area.
Work out the length of fence needed.
Does this take more or less fence than your rectangle in part **b**?

d What length of fence is needed to make a square with this area?

e Which shape takes least fence to make?

12 Reflect How can you make sure you remember the formulae for areas, volumes and
converting units from this unit? Would it help to write each formula in a diagram of the
shape, so you can visualise them?
How are the formulae for the area of a parallelogram and triangle the same as or different
from the formula for the area of a rectangle?
Can you make a phrase using the letters in each formula to help you remember it?

9 Graphs

9.1 Coordinates

- Find the midpoint of a line segment.
- Recognise, name and plot straight-line graphs parallel to the axes.
- Recognise, name and plot the graphs of $y = x$ and $y = -x$.

Prior knowledge

Active Learn
Homework

Warm up

1 Fluency Work out half of
   a 8   **b** 14   **c** 9   **d** 17

2 Add each pair of numbers and then halve the answer.
   a 6 and 4   **b** 2 and 5   **c** −5 and 7   **d** −2 and 8

3 $y = -x$. Work out the value of y when
   a $x = 2$ **b** $x = 5$ **c** $x = -1$ **d** $x = -4$

4 This is a coordinate grid from −5 to +5 on both axes. Lines A and B are drawn on the grid.
Which line is parallel to
   a the x-axis   **b** the y-axis?

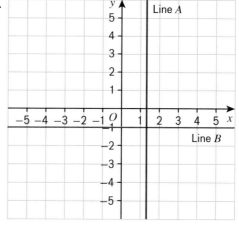

5 Reasoning a Copy the coordinate grid in **Q4**.

   b This is a rule to generate coordinates:
   The x-coordinate is always 1, no matter what the y-coordinate is.
   Which of these coordinate pairs satisfy the rule?
   (1, 5), (5, 1), (1, 1), (−1, 3), (1, 0), (1, 4), (3, 1), (1, 2)

   c On your grid, plot the points from part **b** that satisfy the rule.

   d Reflect Write a sentence describing what you notice about the points you have plotted.

   e This is another rule to generate coordinates:
   The x-coordinate is always 3, for any y-coordinate.
   These coordinates are generated by the rule: (3, 0), (3, −2), (3, 4) and (3, 2).
   Write a sentence describing where you expect these points to be on the grid.

   f Plot the points on the same grid. Were you correct?

> **Q5a hint**
>
> Only copy the grid. Do not copy lines A and B.

6 **a** Write down the integer coordinates of all the points you can see on line A.

   b Write a sentence describing what you notice about the coordinates on line A?

   c Copy and complete.

     i The equation of line A is $x =$ _____

     ii The equation of line B is $x =$ _____

     iii The equation of line C is $y =$ _____

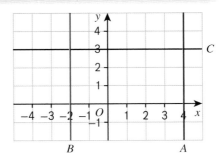

7 Write the equations of the lines labelled P, Q, R and S.

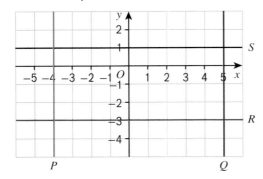

8 **Reasoning**

   a Draw a coordinate grid from -5 to $+5$ on both axes.

   b Draw and label these lines. When you draw a graph, extend it to the edge of the grid. Make sure you have drawn your graphs for parts **i** to **iii** correctly.

> **Q8 hint**
>
> Your coordinate grid should look like the one in **Q4**.

     i $y = 5$ **ii** $x = -3$ **iii** $y = -4$

9 **Reasoning** What is the equation of   **a** the x-axis   **b** the y-axis?

10 **Reasoning**

   a Write down the integer coordinates of the points you can see on this line.

   b What do you notice about the coordinates?

   c Complete the missing coordinates of these points on the line. $(5, \square)$, $(\square, -2)$

   d Describe the y-coordinate of every point on the x-axis.

   e The equation of the line is $y =$ _____

11 These coordinates are on the line $y = -x$
   $(3, -3)$ $(2, -2)$ $(0, 0)$ $(-1, 1)$ $(-5, 5)$

   a Draw a coordinate grid from -5 to $+5$ on both axes.

   b Plot the points and join them with a straight line.

   c Write the coordinates of three more points on the line $y = -x$.

12 Here are three straight lines.

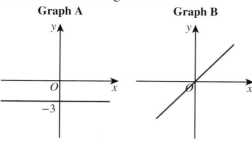

Graph A   Graph B   Graph C

Choose the correct equation for each graph.

| $x = 3$ | $x = -3$ | $y = x$ | $y = 3$ | $y = -3$ |

(2 marks)

13 **Problem-solving** Judith starts to draw a square on this grid.

a What is the equation of the line segment for the side of the square Judith has drawn?

b Judith says, 'One diagonal of my square has equation $y = -x$.' What is the equation of the other diagonal?

c Write down the equations of the other three line segments that could complete the sides of Judith's square.

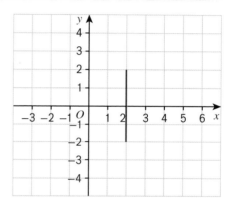

A **line segment** has a start and end point. The **midpoint** of a line segment is the point exactly in the middle of the line.

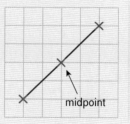

midpoint

14 Find the midpoints of each of these line segments.
Make a table to help you.

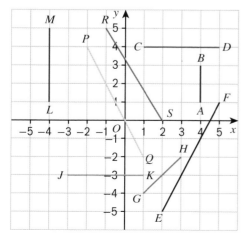

Line segment	Start point	End point	Midpoint
AB	(4, 1)	(4, 3)	
CD	(1, 4)	(5, 4)	
EF			
GH			
JK			
LM			
PQ			
RS			

Example

Find the midpoint of a line segment with start point (3, 2) and end point (7, 9).

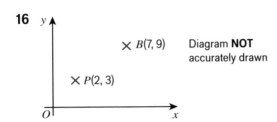

$\frac{(3+7)}{2} = 5$ ──────── Add the *x*-coordinates and divide by 2.

$\frac{(2+9)}{2} = 5.5$ ──────── Add the *y*-coordinates and divide by 2.

Midpoint = = (5, 5.5)

15 Look at the method in the Example. Use this method to check your answers to **Q14**.

16

× B(7, 9)   Diagram **NOT** accurately drawn

× P(2, 3)

The point *P* has coordinates (2, 3).
The point *Q* has coordinates (7, 9).
M is the midpoint of the line *PQ*.
Find the coordinates of *M*.

17 Work out the midpoints of the line segments with these start and end points.

  a (3, 5) and (7, 9)

  b (2, 7) and (5, 10)

  c (−3, 4) and (1, 6)

  d (−2, −5) and (0, 3)

Exam-style question

18

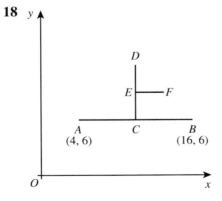
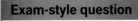

Exam tip

Read one sentence at a time. Write down the coordinates of any point you can find using the information in that sentence.

Horizontal line segment *AB* is from point *A*(4, 6) to point *B*(16, 6).
C is marked at the midpoint of *AB*.
A vertical line segment *CD* is drawn half the length of line segment *AB*.
E is marked at the midpoint of *CD*.
A horizontal line segment *EF* is drawn half the length of line segment *CD*.
Work out the coordinates of point *F*.   **(5 marks)**

9.2 Linear graphs

- Generate and plot coordinates from a rule.
- Plot straight-line graphs from tables of values.
- Draw graphs to represent relationships.

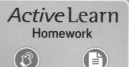

Active Learn
Homework

Warm up

1 Fluency Give three inputs and functions for this function machine.

2 Work out

   a $-2+3$ **b** $-0.8+1$ **c** 4×5 **d** 0.8×2

   e -3×6 **f** $3-5$ **g** $-3-5$ **h** 0.5×4

3 Work out the value of each expression when

   i $x = 2$ **ii** $x = -1$

   a $3x+2$ **b** $-2x+1$ **c** $\frac{1}{2}x-1$ **d** $\frac{1}{4}x+3$

4 Write the number that each arrow is pointing to.

5 This function machine generates coordinates.

   a Work out the missing coordinates.

   b Draw a coordinate grid from -3 to $+6$ on both axes.

   c Plot the coordinates from part **a** on the grid.

   d Join the points and extend the line to the edges of the grid.

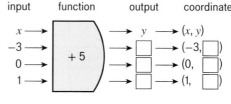

   e Write the coordinates of two more points that lie on the line.

   f What do you do to the x-value of each coordinate to get the y-value?

   g Copy and complete the rule for this function machine.

      $y = \square + \square$. This is the equation of the graph.

   h Label your graph by writing its equation next to the line.

6 Repeat **Q5** for this two-step function machine.

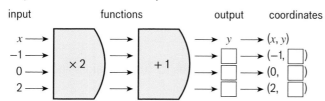

Key point

A table of values shows x- and y-values for the equation of a line.

Example

a Complete this table of values for the equation $y = 3x + 2$.

b Draw the line for $y = 3x + 2$

a

x	−2	−1	0	1
$y = 3x + 2$	−4	−1	2	5

When $x = -2$, $y = 3 \times -2 + 2$
$$= -4$$
When $x = -1$, $y = 3 \times -1 + 2$
$$= -1$$
and so on...

b

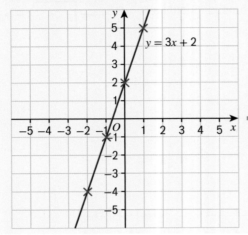

Plot the coordinate pairs from the table with crosses $(-2, -4)$, $(-1, -1)$, $(0, 2)$, $(1, 5)$.
Join them with a straight line and extend it to the edge of the grid.

---

7 **a** Copy and complete the tables of values for these straight-line graphs.

i

x	−3	−2	−1	0	1	2	3
$y = x + 1$			0	1			

ii

x	−3	−2	−1	0	1	2	3
$y = 2x - 3$			−5	−3			

Q7b hint

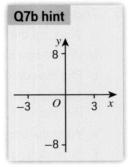

b Draw a coordinate grid with −3 to +3 on the x-axis and −8 to +8 on the y-axis. Draw and label the graphs of $y = x + 1$ and $y = 2x - 3$, using your tables of values from part **a**.

8 Make tables of values like the ones in **Q7** to find coordinates of these straight-line graphs.

a $y = 3x - 2$ **b** $y = 2x + 4$ **c** $y = 4x - 6$ **d** $y = 0.5x + 1$

9 **a** Copy the coordinate grid from **Q7**.

b Draw and label the four lines in **Q8** for $x = -3$ to $+3$ on the same grid.

c **Reflect** Your points should be in a straight line.
If they are not, how can you work out your mistake?

10 **a** Make tables of values like the ones in **Q7** to find coordinates for these straight-line graphs

i $y = -x + 1$ **ii** $-2x - 3$ **iii** $-4x + 2$

b Draw a coordinate grid with −3 to +3 on the x-axis and −10 to +10 on the y-axis. Draw the lines for $y = -x + 1$, $y = -2x - 3$ and $y = -4x + 2$ using your tables of values from part **a**.

11 Draw and label these straight lines on the same grid, for values of x from -4 to $+4$.

a $y = -3x + 2$ **b** $y = -2x - 1$

c $y = -4x + 6$ **d** $y = \frac{1}{4}x + 1$

e Reflect Would you expect these graphs to slope upwards or downwards from left to right? Explain why.

   i $y = 5x - 2$ **ii** $y = -4x + 3$

Q11 hint

Make a table of values for each graph, with x-values from -4 to $+4$. Find the lowest and highest values of y in your tables. Draw the y-axis between these values.

12 a Complete the table of values for $y = \frac{1}{2}x + 1$.

x	-2	-1	0	1	2	3	4
y	0		1		2		

b On graph paper, draw and label the line of $y = \frac{1}{2}x + 1$ for values of x from -2 to 4.

c Find the point on the graph where $x = 0.6$.
Read across the y-axis to find the value of y when $x = 0.6$.

d Use the graph to find the value of y when **i** $x = 1.2$ **ii** $x = -0.4$ **iii** $x = 2.8$

e Use the graph to find the value of x when $y = 2.8$.

f Reflect Write a sentence explaining what was different about your method in parts **d iii** and **e**.

13 The diagram shows the line for $y = 1 - \frac{1}{2}x$.

a Use the graph to find an estimate for the value of y when

   i $x = 1.6$ **ii** $x = 2.2$

   iii $x = -0.8$

b Use the graph to find an estimate for the value of x when

   i $y = 1.8$ **ii** $y = 0.6$

   iii $y = -0.4$

c Reflect Why are your readings estimates for the values of x and y?

d Use the graph to find approximate solutions to these equations.

   i $1 - \frac{1}{2}x = -1$

   ii $1 - \frac{1}{2}x = 2$

   iii $1 - \frac{1}{2}x = 1.2$

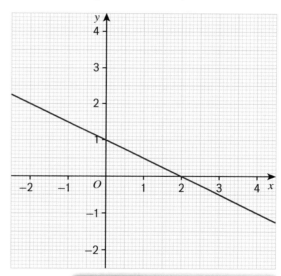

Q13d hint

Compare $-1 = 1 - \frac{1}{2}x$ and $y = 1 - \frac{1}{2}x$
Read the value when $y = -1$.

Exam-style question

14 a Copy and complete the table of values for $y = 0.8x + 1$.

x	-2	-1	0	1	2	3
y	-0.6			1.8		

(2 marks)

b On graph paper, draw and label the line $y = 0.8x + 1$ for values of x from -2 to 4.

(2 marks)

c Use the graph to

   i find the value of y when $x = \frac{1}{2}$

   ii estimate the value of x when $y = 0.7$

(2 marks)

Exam tip

Extend your graph to the edge of the grid.

9.3 Gradient

- Find the gradient of a line.
- Identify and interpret the gradient from an equation.
- Understand that parallel lines have the same gradient.

ActiveLearn
Homework

Warm up

1 Fluency What types of lines will never intersect? Why?

2 Work out

a $10 \div 2$	**b** $-6 \div 3$	**c** $8 \div 6$	**d** $2 \div 4$
e $0.3 \div 5$	**f** $0.03 \div 5$	**g** $0.3 \div 0.5$	**h** $0.03 \div 0.5$

Q2e hint

$5)\overline{0.3}$

3 a How many squares does line C go up for every

  i 1 square across to the right

  ii 2 squares across to the right?

b How many squares does line D go down for every

  i 1 square across to the right

  ii 2 squares across to the right

  iii 3 squares across to the right?

c Divide the number of squares up by the number of squares right for each of your answer in **a i** and **a ii**.

d Divide the number of squares down (shown as a negative integer) by the number of squares right for each of your answers in **b i, ii** and **iii**.

e Reflect Write a sentence describing what you notice when you divide the total distance up or down by the total distance across to the right.

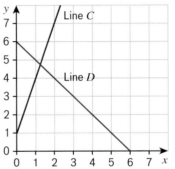

Q3a i hint

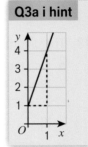

Q3b i hint

Write your answer as a negative number to represent the downward direction: $-\square$

Key point

The steepness of a graph is called the **gradient**.
To find the gradient work out how many units the graph goes up or down for each unit it goes across to the right.

4 Write down the gradient of

a line C in **Q3**

b line D in **Q3**

Key point

positive gradient   negative gradient

5 **a** Work out the gradient of each line segment by calculating

$$\frac{\text{total distance up or down}}{\text{total distance across to the right}}$$

b Write these lines in order of most steep to least steep.

 i A, E, F **ii** B, C, D

c **Reflect** Write a sentence explaining how to compare the steepness of lines by using their gradients.

Q5c hint

A gradient can be a fraction.

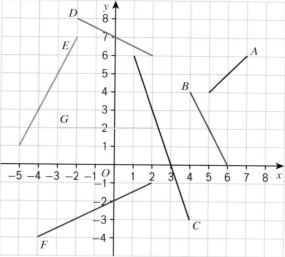

6 Here are the gradients of five different line segments.

 $A\ 1$ $B\ 3$ $C\ -2$ $D\ \frac{1}{2}$ $E\ -\frac{2}{3}$

a Which is the steepest? **i** line A, B or D **ii** line C or E

b On squared paper draw line segments with these gradients.

7 Here are the graphs of $y = 2x - 1$ and $y = 2x + 3$.

a Fill in the missing word.
The lines $y = 2x + 3$ and $y = 2x - 1$ are _____

b What are the gradients of the lines?

c Fill in the missing word.
Parallel lines have the same _____

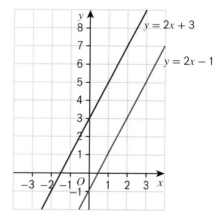

Exam-style question

8 Alfie finds the gradient of this line.
He writes the answer $\frac{3}{2}$.
Explain the mistake that Alfie has made. **(1 mark)**

Exam tip

Look carefully at the scales on the axes. Write a sentence explaining Alfie's mistake. Do not simply write the correct gradient.

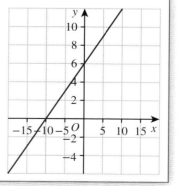

9 The sketch shows a ramp for wheelchair access.
Work out the gradient of the ramp.

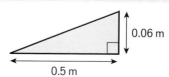

0.06 m

0.5 m

10 a Copy and complete these tables of values.

i

x	−3	−2	−1	0	1	2	3
$y = x + 1$	−2			1			

ii

x	−3	−2	−1	0	1	2	3
$y = -2x + 3$	9			3			

b Draw the lines

   i $y = x + 1$

   ii $y = -2x + 3$

c Work out the gradient of each line.

d The **coefficient** is the number in front of the x.
What is the coefficient in each of these equations?

   i $y = x + 1$

   ii $y = -2x + 3$

Q10d i hint

In algebra, $1x = x$

e **Reflect** Write a sentence explaining what you notice about the gradient of the line for an equation and the coefficient of x in the equation.

11 Copy and complete the equation of this line

   $y = \Box x - 1$

so that it is parallel to the line with equation $y = -4x + 1$

12 **Reasoning** Choose from these equations

a the line with the steepest gradient

b a pair of parallel lines

c a line that slopes down from left to right

$y = 3x + 2$	$y = 2x + 8$	$y = 3x$	$y = 5x + 2$
$y = -3x + 4$	$y = x + 5$	$y = 4$	$y = \frac{1}{2}x + 1$

Example

Find the gradient of the line joining the points $A(-2, 1)$ and $B(4, 3)$.

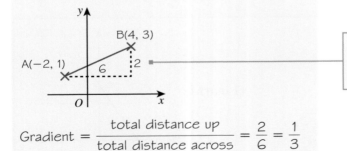

Sketch a diagram. Draw in lines across and up.
Work out the distances across and up.

$$\text{Gradient} = \frac{\text{total distance up}}{\text{total distance across}} = \frac{2}{6} = \frac{1}{3}$$

13 Find the gradient of the line joining the points

   a $E(1, 2)$ and $F(7, -4)$

   b $G(-1, 2)$ and $H(7, 6)$

   c $I(-1, 2)$ and $J(7, -6)$

9.4 $y = mx + c$

- Understand what m and c represent in $y = mx + c$.
- Find the equations of straight-line graphs.
- Sketch graphs given the values of m and c.

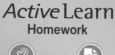

Active Learn
Homework

Warm up

1 Fluency For point $A(3, -2)$, which is the
   a x-coordinate   **b** y-coordinate?

2 For each equation, work out the value of
   i y when $x = 0$ **ii** x when $y = 0$
   a $y = 2x - 10$ **b** $y = -\frac{1}{2}x + 10$ **c** $x + y = 10$
   d $2x + y = 10$ **e** $2x + 5y = 10$ **f** $5x + 2y = 10$

3 For each of the equations in **Q2 c**, **d** and **e**, rearrange to
   make y the subject.

> **Q2e hint**
>
> Divide by 5 to give $y = $ _____

4 Find the gradient of the line joining $(1, 4)$ and
   a $(3, 8)$ **b** $(3, -8)$ **c** $(-3, -8)$

5 The diagram shows four straight-line graphs and their equations.

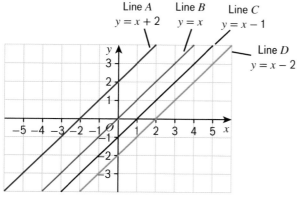

Line A Line B Line C
$y = x + 2$ $y = x$ $y = x - 1$

Line D
$y = x - 2$

What is the gradient of each line?

Key point

A **linear equation** produces a straight-line graph. The equation of a straight line is $y = mx + c$, where m is the gradient and c is the y-intercept.

6 Look at the lines in **Q5**.
   The y-intercept is the point where the line crosses the y-axis.
   a Write down the y-intercept for each of the lines A, B, C and D.
   b What do you notice about the equation of the line and the y-intercept?
   c Where would you expect $y = x + 4$ to cross the y-axis?

7 **a** For line A

    i work out the gradient, $m = \square$

    ii write the y-intercept, $c = \square$

    iii substitute the values of m and c into $y = mx + c$ to give the equation of line A.

  b Use the method in part **a** to work out the equation of line B.

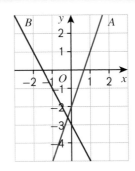

Q7b hint

The gradient is negative so
$y = -\square x + \square$

8 Work out the equations of lines A, B, C, D and E.

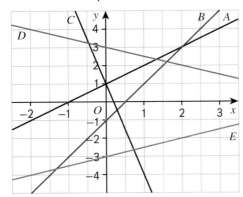

9 Here are the equations of three different lines

$$y = 6x + 6 \qquad y = 5x - 6 \qquad y = -6x + 6$$

  a Write down the equation of the straight line that is parallel to $y = 5x + 6$.

  b Write down the equation of a straight line that has the same y-intercept as $y = 5x + 6$.

  c Write down the equation of the line that passes through the point $(1, -1)$.

Q9c hint

Work out the value of y when $x = 1$. Which equation gives $y = -1$?

10 Which of these lines pass through

  a $(0, 0)$

  b $(0, -3)$?

$y = 3x$	$y = 3x - 3$	$y = 3x + 3$
$y = -3x$	$y = -3x + 3$	$y = x - 3$

Exam-style question

11 Does the point $(-2, -5)$ lie on the straight line with $y = 3x + 11$?

You must show how you get your answer. **(2 marks)**

Exam tip

It is not enough simply to show working. You must also write a sentence stating whether the point lies on the straight line.

12 A straight-line graph A has a gradient of 3 and passes through the point $(2, 1)$.

  a Substitute these values into $y = mx + c$ $y = \underset{\uparrow}{\square} x + \underset{\uparrow}{\square}$

                                                        m c

  b Solve the equation you wrote for part **a** to work out the value of c.

  c Copy and complete the equation of graph A.

Q12a hint

gradient $m = \square$
Point $(2, 1)$, $x = \square$, $y = \square$

13 In each of these you are given one point that lies on a line and the gradient of the line.
Use the method in **Q12** to work out the equation of the line.

  a $(4, 6)$ and $m = 2$

  b $(-1, 2)$ and $m = 4$

  c $(-2, 10)$ and $m = \frac{1}{2}$

  d $(5, -12)$ and $m = 3$

14 In each of these you are given two points that lie on a line.
First work out the gradient of the line (m).
Then use the method from **Q12** to work out the equation of the line.

  a $(2, 3)$ and $(4, 7)$

  b $(-6, 0)$ and $(2, 4)$

  c $(3, 0)$ and $(4, 2)$

  d $(-2, 4)$ and $(3, -1)$

15 **a** Draw a coordinate grid from -8 to 8 on both axes.

  b For the graph $y = 2x - 7$, write down

    i the y-intercept

    ii the gradient

  c Plot a point at the y-intercept.

  d From the y-intercept, draw a line with the given gradient.
Extend your line across the grid and label it with its equation.

16 Use the method in **Q15** to draw these lines from their equations on a coordinate grid from -8 to 8 on both axes.

  a $y = -3x + 5$ **b** $y = \frac{1}{2}x - 4$ **c** $y = 6 - x$

> **Q16c hint**
>
> The y-intercept may not always come at the end of an equation.

17 For each line, copy and complete

    i when $x = 0$, $y = \square$, $(0, \square)$

    ii when $y = 0$, $x = \square$, $(\square, 0)$

  a $x + y = 8$ **b** $x + y = 6$ **c** $x + y = 4$

18 Draw a coordinate grid from 0 to 8 on both axes.

  a Use the coordinates you found in **Q17** to plot each of the lines.

  b Write a sentence explaining what you notice about these lines.

19 Draw a coordinate grid from -8 to $+8$ on both axes.
Use the method in **Q17** to plot these lines.

  a $2x + y = 8$ **b** $3x + 2y = 6$ **c** $4x + 2y = 4$ **d** $2x + 4y = 6$

20 In **Q18** and **Q19**, you drew seven lines. For each line

  a rearrange its equation to make y the subject

  b use the equation to write down the gradient and y-intercept

  c look back at your lines in **Q18** and **Q19** to check the gradients and y-intercepts are correct.

9.5 Real-life graphs

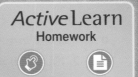
- Draw and interpret graphs from real data.

Warm up

1 **Fluency** Work out
   a 5 kg at 70p per kg
   b 4 hours at £30 per hour
   c the cost per hour when £90 is paid for 2 hours' work.

   > **Q1a hint**
   > 70p per kg means 70p for each kg.

2 Write the value each arrow is pointing to.

   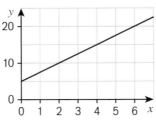

3 For this graph, what is
   a the y-intercept
   b the gradient
   c its equation?

4 Draw the graph of $y = \frac{1}{5}x - 4$.

---

5 A car uses 1 litre of fuel for every 10 kilometres it travels.
   a Copy and complete the table.

Fuel used (litres)	0	1	2	3	4	5
Distance travelled (km)	0	10				50

   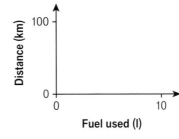

   b Copy these axes on graph paper.
     Draw a graph of fuel used against distance.
   c Use your graph to work out how much fuel is used on a 60 kilometre journey.
   d **Problem-solving** At the start of a 100 kilometre journey there are 27 litres of fuel in the tank. How many litres will be left at the end of the journey?

6 A water company charges a flat rate of £30 per month plus £2 per cubic metre of water used.
   a Copy and complete the table.

Water used (m³)	0	1	2	3	4	5	6	7	8	9	10
Cost (£)	30										50

   b Draw a graph of water used against cost. Put water used (m³) on the x-axis and cost (£) on the y-axis.
   c **Problem-solving** Lianne's water bill is £41 this month. Use your graph to find the number of cubic metres of water she used.
   d **Problem-solving** How many litres of water has Lianne used?

   > **Q6d hint**
   > 1 m³ = 1000 litres

7 **Reasoning** The graph shows the cost of buying different quantities of tomatoes.

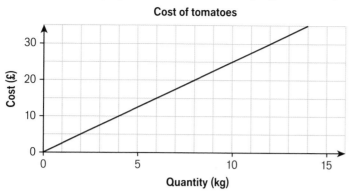

Cost of tomatoes

a What is the cost of

  i 10 kg of tomatoes

  ii 1 kg of tomatoes?

b Copy and complete: gradient $= \dfrac{\text{distance up or down}}{\text{distance across to the right}} = \dfrac{\boxed{}}{\text{quantity (kg)}}$

c Work out the gradient of the line.

d **Reflect** Copy and complete this sentence.
The gradient of the line represents cost in £ per _____.

Key point

You can use graphs to compare trends and relationships between two or more sets of data.

---

8 **Reasoning** The graphs show the costs of grapes at different supermarkets.

Supermarket A: Green grapes £4 per kg
Supermarket B: Organic grapes £5 per kg
Supermarket C: Black grapes £3 per kg

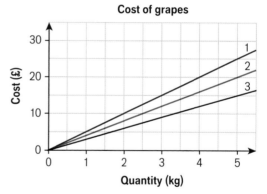

Cost of grapes

a Match each line on the graph with the correct supermarket.

b Copy and complete this statement.
The steeper the slope the .... the unit cost.

9 **Reasoning** An electrician charges a £60 callout fee and £30 per hour.

a Copy and complete the table of values.

Hours worked	0	1	2	3	4	5
Total cost (£)	60					210

b Draw a graph of hours worked against total cost.
Put hours worked on the x-axis and total cost on the y-axis.

c Work out the equation of the line.

d Write a sentence stating

  i what the gradient represents

  ii what the y-intercept represents

10 The graph shows the costs of hiring two
    different plumbers.
    Plumber A charges a callout fee and
    £35 per hour.
    Plumber B has no callout fee but charges an
    hourly rate.

    a Match each line on the graph with the
      correct plumber.
    b How much is Plumber A's callout fee?
    c What is Plumber B's hourly rate?
    d Work out the difference in cost between
      Plumber A and Plumber B for 7 hours' work.
    e Alice thinks her plumbing work will take about
      22 hours.
      Explain which plumber will be cheaper.

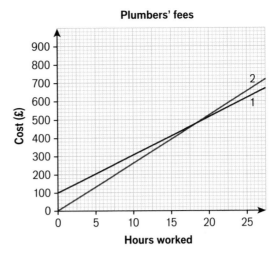

Plumbers' fees

11 **Problem-solving / Reasoning**
    The graph shows two different ways to pay for
    golf at a golf club.
    a Explain in words what each option means.
    b John expects to play about 45 games of golf
      in a year. Which option should he choose?

Golf club fees

12 Bob delivers soil.
    You can use this graph to find the delivery cost for different distances.

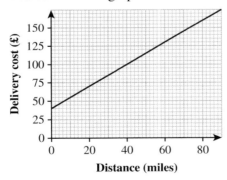

Distance (miles)

Exam tip

Read the question carefully.
Part **b** is not asking for the cost
of a delivery 30 miles away.
You need to look at the costs
of two deliveries 30 miles apart.

For each delivery, there is a fixed charge for the distance.

a How much is the fixed charge?   **(1 mark)**

Bob makes 2 deliveries of soil.
The distance for 1 delivery is 30 miles more than the other delivery.

b Work out the difference in cost between the two deliveries.   **(2 marks)**

13 **Problem-solving / Reasoning** A company works out the monthly pay for sales staff using the equation $y = \dfrac{x}{10} + 600$, where y is the monthly wage (£) and x is the value of the sales made in the month.

   a Draw the graph for the equation.

   b What is the monthly pay when no sales are made?

   c Pete's sales average £12 000 per month.
     He is offered a different job where the monthly pay is £2000.
     In which job would his average monthly pay be higher? Explain.

Q13a hint

Plot the y-intercept at 600.
Draw the line with gradient $\frac{1}{10}$.

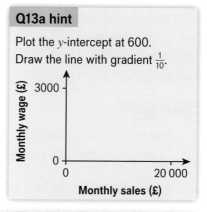

Exam-style question

14 The graph shows the cost of getting work done by three different companies.

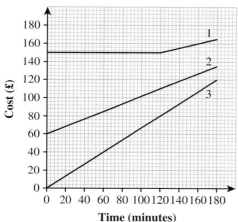

Exam tip

For part **c**, you cannot see 8 hours on the graph. But you have the information to work out the cost of 8 hours' work by each company.

Here are the charges of the three companies.
Company A: Callout charge of £60 and every hour costs £25.
Company B: No callout charge but every hour costs £40.
Company C: Callout charge of £150 which includes the first 2 hours and then every hour costs £15.

   a Match each line on the graph with the letter of the company it represents.

Ned uses company A for 2 hours of work.

   b Find the total cost.

Fiona needs 8 hours of work done.

   c Explain which company would be the cheapest for her to use.
     You must give reasons for your answer.   **(4 marks)**

9.6 Distance–time graphs

- Use distance–time graphs to solve problems.
- Draw distance–time graphs.
- Interpret rate of change on graphs.

Active Learn
Homework

Warm up

1   Fluency   Which of these are measures of speed?

   4 km   45 min   40 seconds   400 m/s   450 m   45 km/h   40 m/min

2   A car is travelling at 40 km per hour.
   a   How far will it travel in   **i** 1 hour   **ii** 3 hours   **iii** 30 minutes?
   b   How long will it take to travel   **i** 200 km   **ii** 10 km   **iii** 5 km?

Key point

A **distance–time** graph represents a journey. The vertical axis represents the distance from the starting point. The horizontal axis represents the time taken.

Example

Jen walks 500 metres in 5 minutes, then arrives at the bus stop. She waits 5 minutes for the bus. She travels 3000 metres on the bus and gets off 16 minutes after she left home.

a   Draw a distance–time graph for her journey.

b   Work out the average speed in km/h of Jen's walk.

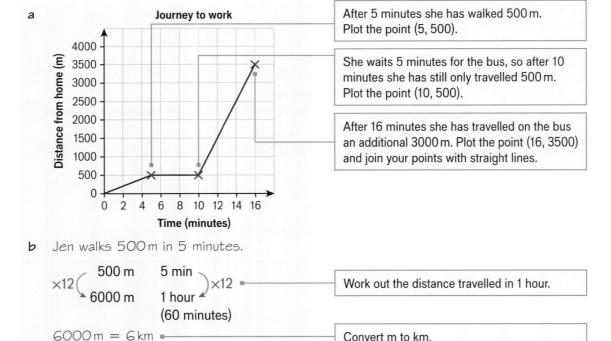

a

After 5 minutes she has walked 500 m. Plot the point (5, 500).

She waits 5 minutes for the bus, so after 10 minutes she has still only travelled 500 m. Plot the point (10, 500).

After 16 minutes she has travelled on the bus an additional 3000 m. Plot the point (16, 3500) and join your points with straight lines.

b   Jen walks 500 m in 5 minutes.

$\times 12 \left(\begin{array}{l} 500\,m \\ 6000\,m \end{array} \right. \left. \begin{array}{l} 5\,min \\ 1\,hour \\ (60\,minutes) \end{array} \right) \times 12$

Work out the distance travelled in 1 hour.

6000 m = 6 km

Convert m to km.

Average walking speed = 6 km/h

---

3 Problem-solving Satbir leaves the house at 9 am. She cycles 5 km to town. It takes her 15 minutes. She goes to one shop for 30 minutes. Then she cycles 2 km further to her friend's house. The journey to her friend's house takes her 10 minutes.

a Draw a horizontal axis from 9 am to 10 am and a vertical axis from 0 to 8 km.

b Use your axes to draw a distance–time graph for Satbir's journey.

c Work out the average speed in km per hour for Satbir's

  i cycle to town   **ii** cycle to her friend's house

d Reasoning Compare the average speeds to decide in which part of the journey she was cycling uphill.

> **Q3b hint**
> How many 15 minutes are there in 60 minutes?

Key point

Average speed = $\dfrac{\text{distance travelled}}{\text{time taken}}$

---

4 Problem-solving This distance–time graph shows Isaac's journey on his bicycle.

a How far did Isaac ride his bike on the first part of the journey?

b How long did the first part of his journey take?

c What was his average speed on the first part of the journey?

d How many minutes did Isaac rest for?

e What was his average speed for the last part of the journey?

f Reflect Write a sentence explaining how you can tell, by looking at the graph, when Isaac was travelling fastest.

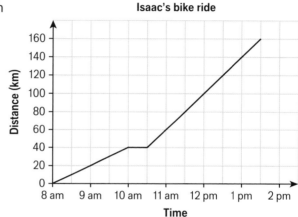

Isaac's bike ride

Key point

On a distance–time graph the gradient represents the speed of the journey.

---

5 Problem-solving This graph shows a fire engine's journey to and from a fire.

a Work out the average speed of the fire engine in km/h on the way to the fire.

b For how long was the fire engine at the fire?

c Work out the average speed of the fire engine in km/h on the way back to the fire station.

d The fire engine travels to the fire and back again. How far did the fire engine travel altogether?

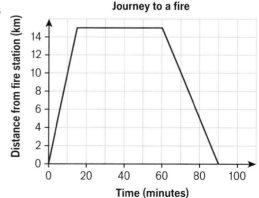

Journey to a fire

6 Reasoning On a distance–time graph, what does

a a horizontal line represent

b a negative gradient represent?

7 Problem-solving

Mike walks to the doctor's surgery. He stops to talk to a friend on the way home. The graph shows his journey.

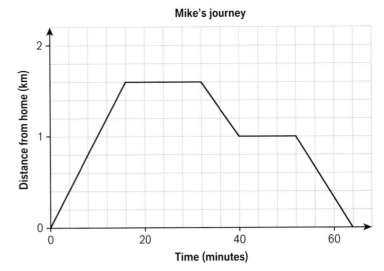

Mike's journey

a How far is the doctor's surgery from Mike's home?

b How long does Mike spend in the surgery?

c How long does he spend talking to his friend?

d Reasoning On which part of the journey is Mike walking fastest?

8 Reasoning Bob leaves home at 10 am and cycles 5 km to the sports centre.
It takes him $\frac{1}{4}$ of an hour. He spends $2\frac{1}{2}$ hours at the sports centre.
He starts to cycle home on the same route and after 4 km stops at a friend's house.
He arrives at 1 pm and stays $1\frac{1}{2}$ hours. He gets home at 2.35 pm.

a Draw a distance–time graph to show Bob's journey.

b On which part of the journey was he cycling the fastest?

9 Problem-solving The graph shows two bus journeys.
One bus leaves Southampton at 9 am and travels to Winchester. The other bus leaves Winchester at 9 am and travels to Southampton.

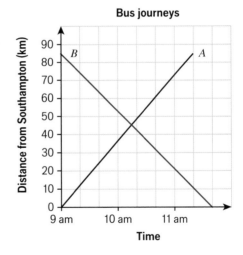

Bus journeys

a Which line shows the Winchester to Southampton bus?

> **Q9a hint**
>
> The vertical axis shows the distance from Southampton.

b At what time does each bus get to its destination?

c Reasoning Which bus journey is faster?

10 Problem-solving Sasha and Kelly go hill walking along the same route.

The graph shows their journeys.

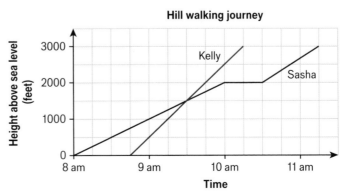

Hill walking journey

a How high do they climb?

b How long does it take Sasha to reach the top?

c How long does it take Kelly to reach the top?

d Reasoning At what time does Kelly overtake Sasha?

11 Jon drove from Junction 10 to Junction 11 on a motorway.

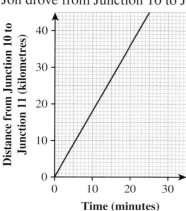

Q11 hint

You could plot Carol's journey on the graph to compare the speeds. Or you could work out Jon's speed from the graph.

The travel graph shows Jon's journey. Carol also drove from Junction 10 to Junction 11 on the same motorway. She drove at an average speed of 105 km/h.
Who had the faster average speed, Jon or Carol?
You must explain your answer. **(4 marks)**

Key point

A **rate of change graph** shows how a quantity changes over time.
Velocity means speed in a particular direction.
A **velocity–time graph** shows how velocity changes over time.
It has time on the x-axis and velocity on the y-axis.

12 Reasoning Match each velocity–time graph to one of the sentences.

A 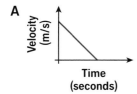 **B** 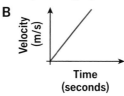 **C**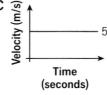

 a Which graph starts from **i** rest (0 m/s) **ii** more than 5 m/s?
 b Which graph shows velocity that is **i** steady **ii** steadily decreasing?

Key point

On a **velocity–time graph** the gradient represents the acceleration.

13 Problem-solving The graph shows the movement of a particle over time.

 a What was the particle's
  i starting speed
  ii acceleration in m/s^2?
 b Use your answers to parts **a** and **b** to write the equation of the line.

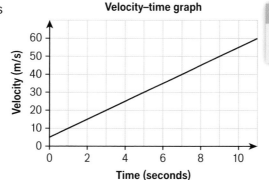

Q13a hint

Starting speed is the speed when time = 0

9.7 More real-life graphs

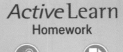 *Active* Learn
Homework

- Draw and interpret a range of graphs.
- Understand when predictions are reliable.

Warm up

1 Fluency When was the person travelling faster?

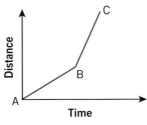

2 a What type of correlation does this scatter graph show? Choose from the words in the box.

negative correlation
positive correlation
no correlation

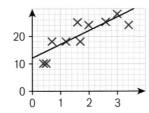

b Find the equation of the line of best fit.

---

3 Reasoning Containers A, B and C are filled at a constant rate. This means the same amount of water is poured into each container every second.

  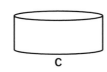

A   B   C

a Which container will fill fastest?

b The graph shows the depth of water for the three containers. What do the gradients of the lines represent?

c Match each container to a line on the graph.

4 Reasoning This graph shows the depth of water for two containers, A and B, with sloping sides filling at the same constant rate.
The depth in a container with sloping sides makes a curved graph.

a Which container will take longer to fill?

 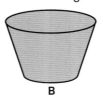

A   B

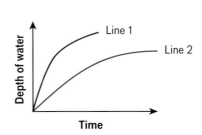

b Match each container to a line on the graph.

5 **Problem-solving** Sketch a graph of depth of water against time for each of these containers filled with water at a constant rate.

a

b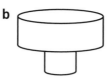

Q5 hint

Use the graphs in **Q3** and **Q4** to help you.

6 The two containers have been filled with water at the same rate. Match each container to a line on the graph.

A   B

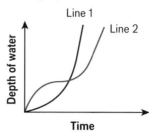

Line 1
Line 2
Depth of water
Time

7 **Reflect** Write a sentence explaining what you notice about container **a** in **Q5** and container A in **Q6**, and the graphs showing depth of water against time.

8 **Future skills** The graph shows pressure against temperature for a gas at a constant volume.

a Estimate the temperature when the pressure is 84 cm of mercury.

b Estimate the pressure when the temperature is −80 °C.

c Copy and complete these sentences

  i Pressure is increasing at a rate of _____ cm of mercury for every 100°C.

  ii Pressure is increasing at a rate of _____ cm of mercury/°C.

d Work out the gradient of the line.

e Is the increase in pressure the same for every °C increase in temperature?

f Would the graph still be a straight line if the answer to part **e** was no? Explain.

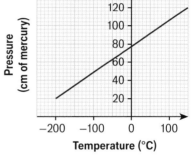

Pressure–temperature graph

Pressure (cm of mercury)

−200 −100   0   100
Temperature (°C)

Exam-style question

9 Oil flows out of a tank at a constant rate.
The graph shows how the depth of oil in the tank varies with time.

  a Work out the gradient of the straight line.
    (2 marks)

  b Write down a practical interpretation of the value you worked out in part **a**. **(1 mark)**

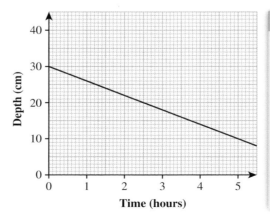

Depth (cm)

Time (hours)

Q9 hint

Part **b** is asking you to use your calculated value from part **a** to explain what is happening to the oil in the tank. Begin your answer with 'The depth of oil is _____'.

Exam-style question

10 Alex draws this time series graph to show the percentage of patients early or on time for appointments at the hospital each month from January to April.

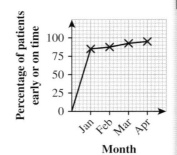

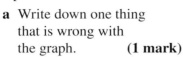

a Write down one thing that is wrong with the graph. **(1 mark)**

Exam tip

When asked to find something wrong with a graph, look at:

• the scales on the axes
• the labels on the axes
• the position of the points
• whether all the points should be joined.

One or more of these is likely to be wrong.

b What percentage of patients were late for appointments in January? **(1 mark)**

c Describe the trend in the percentage of patients being early or on time for appointments. **(1 mark)**

11 **Reasoning** The scatter graph shows the hours spent on study and test scores for some students.

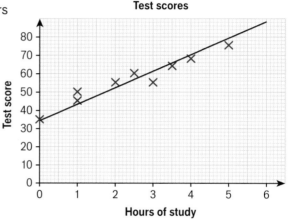

a Describe the correlation between hours of study and test scores.

b Use the line of best fit to predict the test score for someone who studies for 4.5 hours.

c Work out the equation of the line of best fit.

d Use your equation to predict the test score for someone who studies for 4.5 hours.

e **Reflect** Which gives a more accurate prediction: the graph (used to answer part **b**) or your equation (used to answer part **d**)? Explain.

f Explain why you cannot estimate the hours of study for someone with a test score of 20.

12 The table shows the average household size in the UK from 1961 to 2011.

Year	1961	1971	1981	1991	2001	2011
Average household size	3.1	2.9	2.7	2.5	2.4	2.3

a Use this information to predict the average household size in the year 2021.

b How reliable are your results? Explain any assumptions you have made.

Q12a hint

Plot a graph of the data. Put year on the x-axis and average household size on the y-axis. Draw a line of best fit through your plotted points and extend the line.

9 Check up

ActiveLearn
Homework

Algebraic straight-line graphs

1 a Copy and complete this table of values for the equation $y = 2x - 1$.

x	−2	−1	0	1	2
$y = 2x - 1$					

  b Draw a coordinate grid from −5 to +5 on both axes. Draw the graph of $y = 2x - 1$.

  c Use the graph to find the value of y when $x = -0.5$.

2 Work out the gradient of this line.

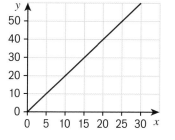

3 Find the midpoint of the line segment PQ where P is (2, 5) and Q is (6, 3).

4 Write the equation of lines A, B, C and D.

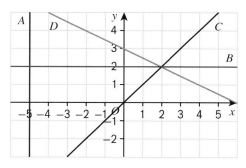

5 $y = 3x + 4$ $\qquad$ $y = x + 3$ $\qquad\qquad$ $y = 4x + 3$ $\qquad\qquad$ $y = 3x - 7$

  a Which of these lines are parallel? **b** Which of these lines have the same y-intercept?

  c Which of these lines is the steepest? **d** Which of these lines pass through the point (1, 7)?

Distance–time graphs and scatter graphs

6 The graph shows Rhoda's walk.

  a For how long did Rhoda stop to talk to a friend?

  b Between what times was Rhoda walking fastest? Explain how you know.

  c What was her average speed in km/h on the first part of the journey?

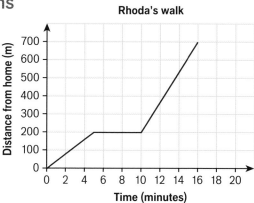

Rhoda's walk

7 Which graph shows the depth of water against time for filling this bath at a constant rate?

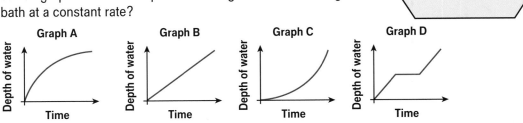

Graph A

Graph B

Graph C

Graph D

8 The scatter graph shows the marks for 10 students in their Spanish and French exams.

a Find the equation of the line of best fit.

b Use your equation to predict the French mark of a student who gets 48 in Spanish.

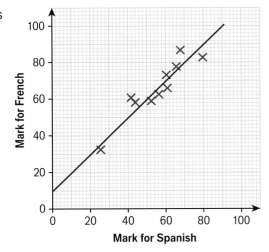

Real-life graphs

9 **Reasoning** The graph shows the cost of hiring two handymen, Fred and Joe.
Both men have a fixed charge for callout and then a rate per hour.

a How much does Joe charge for callout?

b What is Fred's hourly rate?

c Estimate the difference in their total charges for work that takes 7 hours.

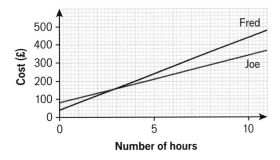

10 **Reflect** How sure are you of your answers? Were you mostly

Just guessing ☹   Feeling doubtful ☺   Confident ☺

What next? Use your results to decide whether to strengthen or extend your learning.

Challenge

11 Write down the coordinates of points to make geometrical shapes.
How many of these can you find?
  triangle, square, rectangle, parallelogram, rhombus, trapezium, kite

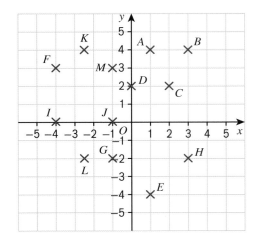

9 Strengthen

*Active*Learn
Homework

Algebraic straight-line graphs

1 a Copy and complete this table of values for the function $y = 3x$.

x	0	1	2	3	4
$y = 3x$	0	3			

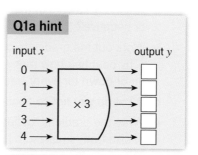

Q1a hint

input x output y

0 →
1 →
2 → ×3
3 →
4 →

b Write down the five pairs of coordinates from your table.

c Draw a grid with x-axis from 0 to 5 and y-axis from 0 to 15. Plot the coordinates on the grid.

d Join the points with a straight line. Extend it to the edge of the grid.

e Label the line $y = 3x$

f Find the point on the graph where $x = \frac{1}{2}$ then read across to the y-axis to find the value of y when $x = \frac{1}{2}$.

g Find the point on the graph where $y = 4$ then read down to the x-axis to estimate the value of x when $y = 4$.

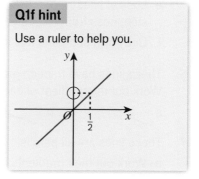

Q1f hint

Use a ruler to help you.

2 a Copy and complete this table of values for the function $y = 4x - 3$.

x	0	1	2	3	4
$y = 4x - 3$	-3	1			

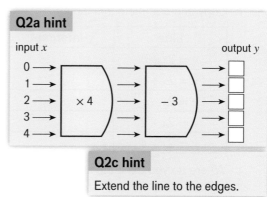

Q2a hint

input x output y

0 →
1 →
2 → ×4   −3
3 →
4 →

b Draw a grid with x-axis from 0 to 5 and y-axis from -5 to $+20$. Plot the coordinates on the grid.

c Join the points with a straight line.

d Label the line $y = 4x - 3$

e What are the coordinates of the point where the graph crosses the y-axis?

Q2c hint

Extend the line to the edges.

3 a Write down three points with an x-coordinate of 1 and a y-coordinate of between -5 and $+5$.

b Draw a grid with x- and y-axes from -5 to $+5$. Use your answer to part **a** to draw the graph of $x = 1$.

c Use the method in parts **a** and **b** to draw these lines on the same grid

   i $x = 4$ **ii** $x = -2$ **iii** $y = 3$ **iv** $y = -1$

4 The steepness of a graph is called the gradient.

   a A line that goes upwards from left to right has a
      _____ gradient.

   b A line that goes downwards from left to right has a
      _____ gradient.

Q4 hint

Positive gradients

Negative gradients

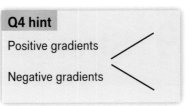

5 **a** Are the gradients of these lines positive
     or negative?

   b Work out the gradients of lines A, B, C, D and
     E by counting how many squares the line goes
     up or down for every 1 across to the right.

   c What do you notice about lines B and D?

   d Write the gradients of lines A, B and E in order
     of size from smallest to largest.

   e Look at lines A, B and E. Write the lines in order of
     steepness from least steep to steepest.

   f Copy and complete this sentence: The _____ the gradient the steeper the line.

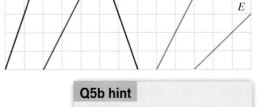

Q5b hint

Gradient of $C = -\square$

6 These lines have fraction gradients. Decide if each gradient is positive or negative.
   Work out the gradient of each line.

   a   **b**   **c**

7 These lines are all parallel.

   a Work out their gradients.
     What do you notice?

   b Copy and complete this table.

Line	y-intercept
$y = 2x + 3$	
$y = 2x + 1$	
$y = 2x$	
$y = 2x - 2$	
$y = 2x - 4$	

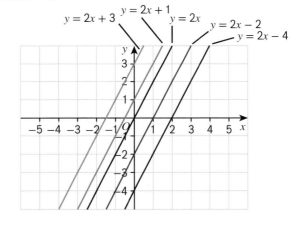

   c Where do you think the line $y = 2x - 5$ will cross the y-axis?

8 $y = 4x + 2$, $y = 2x + 4$

   a Which line is parallel to the lines in **Q7**?
     Write a sentence explaining how you know.

   b Which line has the same y-intercept as $y = 4x + 4$?
     Write a sentence explaining how you know.

Q8b hint

The y-intercept is the value
where the line crosses the y-axis.

9 **a** Work out the gradient of
     these lines.

   b Find the y-intercept.

   c Write the equation of
     each line.

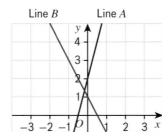

Q9c hint

$y = \square x + \square$

  gradient y-intercept

10 Use the method in **Q9** to write the equation of this line.

Q10 hint

For the gradient, look carefully at the scales on the axes.

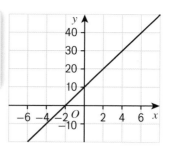

11 a Draw and label the line segment AB with start point $A(1, 3)$ and end point $B(5, 1)$.

b Label the midpoint of the line M.

c Copy and complete the table to work out the coordinates of M.

d Check your answer matches the midpoint M on your line.

	x-coordinate	**y-coordinate**
A	1	3
B	5	1
M	$\dfrac{1+5}{2} = \square$	$\dfrac{\square+\square}{2} = \square$

12 Follow the method in **Q11** to work out the midpoints of these line segments.

a From $C(2, 9)$ to $D(8, 3)$

b From $E(0, 4)$ to $F(4, 6)$

c From $G(-2, 7)$ to $H(4, -3)$

d From $J(3, 8)$ to $K(2, 5)$

Q12 hint

Coordinates can include fractions.

Distance–time graphs and scatter graphs

1 The graph shows Ali's car journey to his friend's house and home again.

a Follow the graph with your finger. Does your finger always

   i move across showing the time increasing?

   ii move up showing the distance from home is increasing?

b How can you tell from the graph when Ali is at his friend's house?

c How long does it take Ali to get to his friend's house?

d How long does Ali spend at his friend's house?

e How far from Ali's home does his friend live?

f How long does it take Ali to get home from his friend's house?

g Which part of the journey was fastest?

h Ali left home at 7.30 pm. At what time did he get home again?

i Copy and complete to work out Ali's average speed in miles per hour for the first part of his journey.

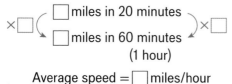

Average speed $= \square$ miles/hour

2 The graph shows sales of ice cream and temperature on each of 10 days.

Ice cream sales

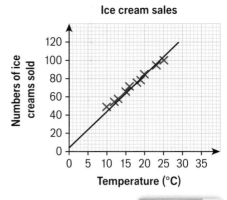

a For the line of best fit

   i write down the *y*-intercept

   ii work out the gradient

   iii copy and complete the
      equation: $y = \boxed{}x + \boxed{}$
               ↑   ↑
       gradient *y*-intercept

b Use your equation to estimate the number of ice creams sold when the temperature was 27 °C.

> **Q2b hint**
>
> $x = 27°$

Real-life graphs

1 **Reasoning** The graph shows costs to have documents translated. The translators charge a fixed fee, and then a rate per word.

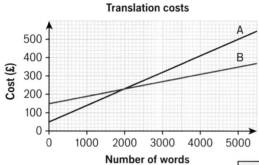

a Copy and complete this table.

b Use the graph to work out who is cheaper for

   i fewer than 2000 words

   ii more than 2000 words

	Translator A	Translator B
Fixed fee		
Cost for 1000 words		
Rate per word		
Cost for 1500 words		
Cost for 2500 words		

2 Water is poured into each container at a steady rate. Match each container to a statement, then match each container to a graph.

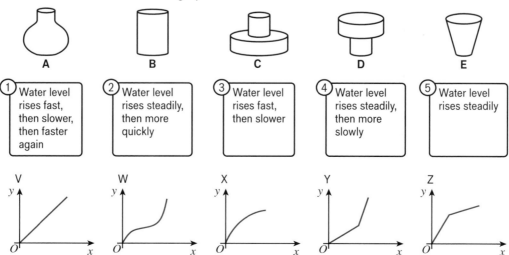

9 Extend

1 **Problem-solving** Here is some information about the gradients of three hills.

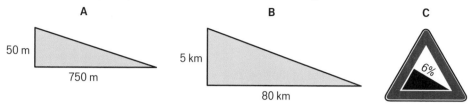

A 50 m / 750 m

B 5 km / 80 km

C 6%

Which hill is the steepest? You must show working to explain your answer.

Exam-style question

2 The diagram shows the distance travelled by a particle (A) over a period of time.

Another particle (B) travelled at a speed of 15 m/s.

Which particle moved faster?

You must show your working.

(3 marks)

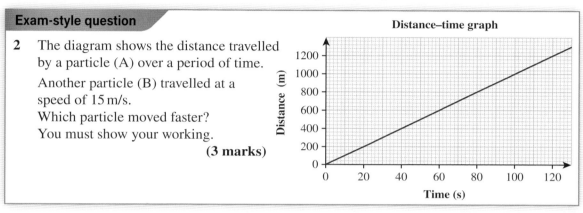

Distance–time graph

Distance (m) vs Time (s)

3 The graph shows some charges for a car park.

   a How much is parking for 3 hours?

   b The open circles shows that the lower limit of each bar is not included in the cost. How much is parking for 4 hours?

   c What is the cost per hour if you park for 5 hours?

Cost of parking

Cost of parking (£) vs Time parked (hours)

4 **Reasoning** Sam has a landline telephone.

He pays £16 per month plus 12p per minute for calls.

Sam uses the phone for about 80 minutes each month.

He sees this offer: £20.50 per month plus 6p per minute for calls.

   a Should he change to the rate in the offer?

   b **Reflect** How could you see at a glance when each payment option was better value?

5 **Problem-solving** The table shows the total area of ice (in km²) on Mount Kilimanjaro between 1912 and 2011.

Year since 1900	12	62	75	84	93	100	111
Area of ice (km²)	11.4	7.32	6.05	4.82	3.8	2.9	1.76

If this trend continues, by what year will all the ice have melted?

Q5 hint

Draw a scatter graph with a line of best fit to show the trend. The **trend** is the general direction in which something is developing.

6 The lines $y = -x$, $x = -4$ and $y = 0$ drawn on a centimetre-squared grid enclose a triangle. What is the area of the triangle? **(4 marks)**

7 **a** Plot the points (2, 1), (2, 3), (0, 3) and (−2, −1) on a coordinate grid.

b Join them to make a quadrilateral. What is the name of the quadrilateral?

c What is the equation of the line of symmetry of the quadrilateral?

8 **a** Make y the subject of the formula $2x + y = 7$.

b What is the gradient and y-intercept of the line $2x + y = 7$?

c The equation of another line is $y + 2x - 3 = 0$.
Show that this line is parallel to the line $2x + y = 7$.

> **Q8c hint**
>
> Rearrange the formula into the form $y = mx + c$

9 The graph shows the movement of a particle over time.

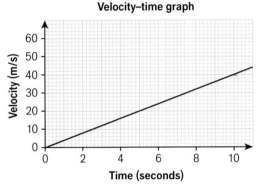

Velocity–time graph

a What is the equation of the line?

b What is the speed of the particle after 10 seconds?

c How far did the particle travel in the first 10 seconds?

> **Q9c hint**
>
> The enclosed area gives the distance travelled. Find the area of the triangle shown.
>
>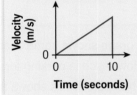

10 **Reasoning** The graph shows the population of a town from 1951 to 2011.

a In what year was the population lowest?

b Gary says, 'From 2001 to 2011 the population increased by more than 14%.'
Is Gary correct? You must show your working.

> **Q10b hint**
>
> Increase the 2001 population by 14% to check Gary's statement.

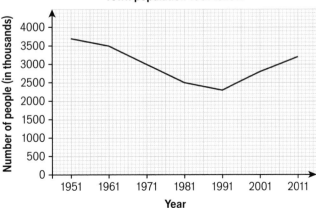

Town population 1951 to 2011

11 Draw an octagon on a coordinate grid.
Write down the equations of each of its sides.

9 Test ready

Summary of key points

To revise for the test:

- Read each key point, find a question on it in the mastery lesson, and check you can work out the answer.

- If you cannot, try some other questions from the mastery lesson or ask for help.

Key points

1 On the line with equation $y = 1$ the y-coordinate is always 1. The line is parallel to the x-axis.
On the line with equation $x = 3$ the x-coordinate is always 3.
The line is parallel to the y-axis. → **9.1**

2 A line segment has a start and end point.
The **midpoint** of a line segment is the point exactly in the middle of the line. → **9.1**

3 A table of values can be used to show x- and y-values for the equation of a line. → **9.2**

4 The steepness of a graph is called the **gradient**. To find the gradient work out how many units the graph goes up or down for each unit it goes across to the right.
Gradients can be positive (/) or (\) negative. → **9.3**

5 A **linear equation** produces a straight-line graph. The equation of a straight line is $y = mx + c$, where m is the gradient and c is the y-intercept. → **9.4**

6 You can use graphs to compare trends and relationships between two or more sets of data. → **9.5**

7 A **distance–time** graph represents a journey. The vertical axis represents the distance from the starting point. The horizontal axis represents the time taken. → **9.6**

8 Average speed $= \dfrac{\text{distance travelled}}{\text{time taken}}$ → **9.6**

9 On a distance–time graph the gradient represents the speed of the journey. → **9.6**

10 A **rate of change graph** shows how a quantity changes over time.
Velocity means speed in a particular direction.
A **velocity–time graph** shows how velocity changes over time.
It has time on the x-axis and velocity on the y-axis. → **9.6**

11 On a **velocity–time** graph the gradient represents the acceleration. → **9.6**

Sample student answers

On the grid below, draw the graph of $y = -2x + 1$ for values of x from -2 to 3.　　　**(3 marks)**

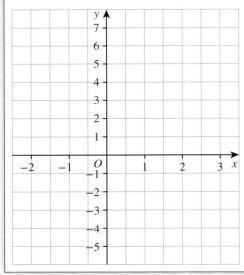

Student A

x	-2	0	3
y	5	1	-5

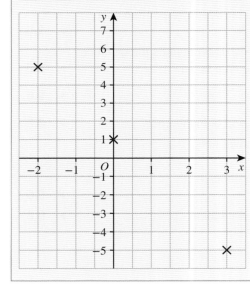

Student B

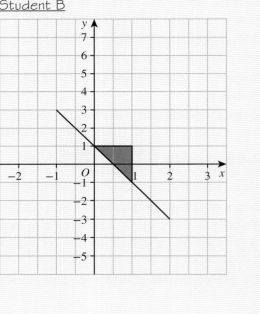

Neither student has given a perfect answer. Explain what each student has done wrong.

9 Unit test

*Active*Learn
Homework

1 **a** What is the equation of line *A*? **(1 mark)**

b Draw a coordinate grid from −5 to +5 on both axes. On this grid, draw and label the line with equation $y = 1$. **(1 mark)**

c On your grid from part **b**, draw and label the line with equation $y = x$. **(1 mark)**

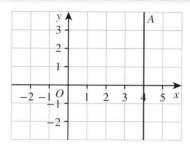

2 *A* has coordinates (0, 1). *B* has coordinates (4, 5). Work out the midpoint of the line *AB*. **(2 marks)**

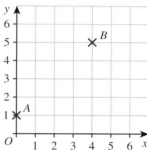

3 **a** Copy and complete the table of values for $y = 4x − 1$. **(1 mark)**

x	−1	0	1	2	3
y		−1			11

b Draw a grid from −1 to 3 on the *x*-axis and from −6 to 18 on the *y*-axis. Plot the graph of $y = 4x − 1$. **(2 marks)**

4 Use the graph of $y = 4x − 1$ you drew in **Q3**. Estimate the value of

a *y* when $x = 2.4$ **(1 mark)**

b *x* when $y = 1.6$ **(1 mark)**

5 From the list of equations, write down

a the equation of the steepest line **(1 mark)**

b a pair of parallel lines **(1 mark)**

c a line that slopes downwards from left to right **(1 mark)**

d a line with a negative gradient **(1 mark)**

e a line that crosses the *y*-axis at 2 **(1 mark)**

f a line that passes through (3, −1) **(1 mark)**

$$y = 2x + 3$$
$$y = 5x + 3$$
$$y = -\tfrac{1}{2}x + 6$$
$$y = 3x + 1$$
$$y = 2x - 5$$
$$y = 0.5x + 2$$

6 Find the equation of each line. **(4 marks)**

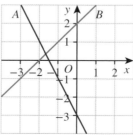

7 Find the equation of the straight line passing through the points $A(−1, −3)$ and $B(3, 5)$. **(3 marks)**

8 Draw a coordinate grid with x- and y-axes from -6 to $+6$.
Plot the graph of $2y + x = 6$. **(3 marks)**

9 The graph shows Paul's journey from his home to town.

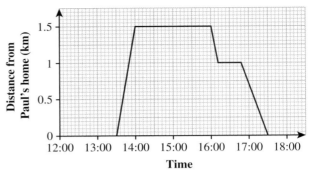

a At what time did Paul leave his home? **(1 mark)**

b The distance from Paul's home to town is 1.5 km. What was Paul's speed on the journey to town? **(2 marks)**

c For how long did Paul stay in town? **(1 mark)**

d Describe fully Paul's return journey home. **(2 marks)**

e For which part of the return journey was Paul travelling the fastest? You must give a reason for your answer. **(2 marks)**

10 The graph shows the relationship between the speed of sound and the height above sea level.

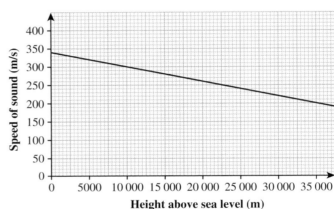

a What is the speed of sound at 5000 m above sea level? **(1 mark)**

b Copy and complete the statement.
As height above sea level increases, the speed of sound _____. **(1 mark)**

c Use the graph to estimate the rate of change of speed with height above sea level. **(2 marks)**

11 Sketch a graph of depth of water against time for filling this container at a constant rate. **(2 marks)**

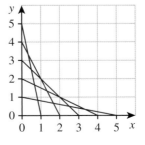

(TOTAL: 40 marks)

12 Challenge Five straight lines are drawn like this to make a pattern that forms a curve.

a Copy the grid and the lines.

b Write down the equation of each line.

c Draw the line $y = -x + 5$ then reflect the pattern in this line.

d Write down the equation of each line in the reflected pattern.

e What do you notice about your answers to **b** and **d**?

13 Reflect In this unit, which was easier, 'plotting and drawing graphs' or 'reading and interpreting graphs'? Copy and complete this sentence to explain why:
I find _____ graphs easier, because _____

10 Transformations

Prior knowledge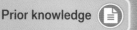

10.1 Translation

- Translate a shape on a coordinate grid.
- Use a column vector to describe a translation.

Active Learn
Homework

Warm up

1 **Fluency** Choose the correct translation to move from point *A* to point *B*.

| 5 left, 2 up | 5 right, 2 up | 3 right, 2 down |

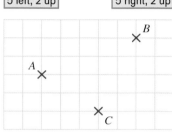

2 Find the missing letters.

  a *A* is translated 3 right, 2 down to ☐.

  b *B* is translated 2 left, 4 down to ☐.

  c *C* is translated 3 left, 2 up to ☐.

3 Draw a grid and describe how to move between each pair of points.

  a (1, 2) and (3, 6)

  b (0, 5) and (2, 3)

  c (−2, 3) and (5, 8)

---

4 Copy and complete this sentence to describe the translation from shape *A* to shape *B*.

Shape *A* is translated _____ to shape *B*.

> **Q4 hint**
>
> Consider how each vertex (corner) of the shape moves:
> ☐ left or ☐ right,
> ☐ up or ☐ down.

a

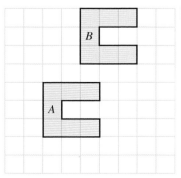

b

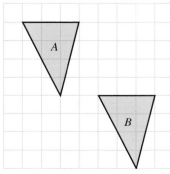

c

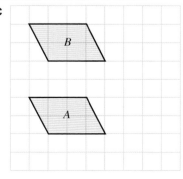

Example

Translate shape P by the column vector $\begin{pmatrix} 7 \\ -1 \end{pmatrix}$.   ← $\begin{pmatrix} 7 \\ -1 \end{pmatrix}$ means 7 right, 1 down.

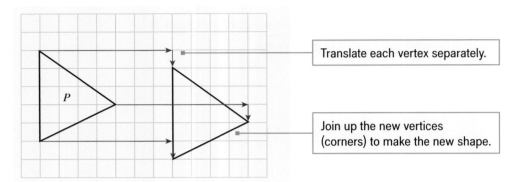

Translate each vertex separately.

Join up the new vertices (corners) to make the new shape.

5 Write the column vector that translates each shape A to shape B in **Q4**.

6 Copy the diagram.

   a Translate shape A by $\begin{pmatrix} 2 \\ 3 \end{pmatrix}$.

   b Translate shape B by $\begin{pmatrix} -2 \\ -1 \end{pmatrix}$.

   c Translate shape C by $\begin{pmatrix} 1 \\ -2 \end{pmatrix}$.

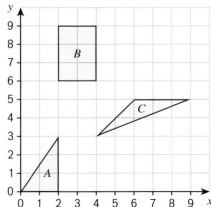

7 Here is Austin's answer to **Q6a**.
Write a sentence explaining what he did wrong.

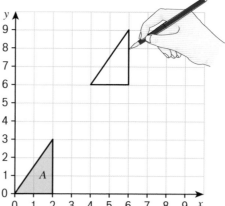

Key point

A **transformation** (such as a translation) transforms a shape to a different position.
The original shape maps to an **image**. The **image** is the shape after a transformation.

---

8 Copy the grid.
Translate shape *A* by each of
these vectors.

a $\begin{pmatrix} 2 \\ 1 \end{pmatrix}$ Label the image *B*.

b $\begin{pmatrix} -3 \\ -2 \end{pmatrix}$ Label the image *C*.

c $\begin{pmatrix} 0 \\ -5 \end{pmatrix}$ Label the image *D*.

d $\begin{pmatrix} 7 \\ -3 \end{pmatrix}$ Label the image *E*.

e $\begin{pmatrix} 5 \\ 0 \end{pmatrix}$ Label the image *F*.

f $\begin{pmatrix} 3 \\ -4 \end{pmatrix}$ Label the image *G*.

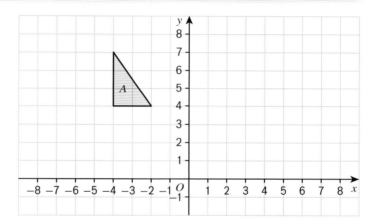

9 **Reasoning** Write the column
vector that maps shape

a *A* to *B*   **b** *A* to *C*

c *A* to *D*   **d** *A* to *E*

e *B* to *D*   **f** *C* to *D*

g *D* to *C*   **h** *E* to *B*

i **Reflect** Write a sentence
explaining what you notice
about the column vectors that
translate shape *C* to shape *D*
and shape *D* to shape *C*.

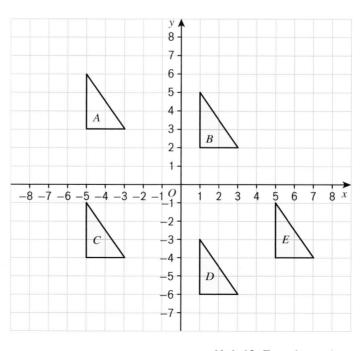

Exam-style question

10 Describe fully the single transformation that maps triangle P to triangle Q.

(2 marks)

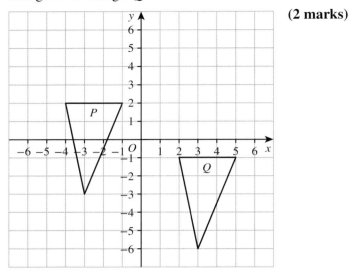

Exam tip

'Describe fully' means write the type of transformation (translation) and the vector.

11 **Problem-solving / Reasoning** Copy the grid and triangle S.

Triangle R is translated by vector $\begin{pmatrix} -2 \\ 2 \end{pmatrix}$ to give triangle S.
Draw triangle R.

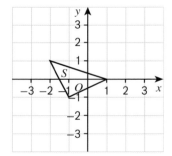

12 **Reflect** Suzie draws this diagram to help her remember how to use a column vector.

$\begin{pmatrix} 3 \\ 4 \end{pmatrix} \begin{matrix} \rightarrow \\ \uparrow \end{matrix}$

$\begin{pmatrix} -2 \\ -5 \end{pmatrix} \begin{matrix} \leftarrow \\ \downarrow \end{matrix}$

a Do you think it is a good diagram? Explain.
b What other diagram would you draw?

10.2 Reflection

- Draw a reflection of a shape in a mirror line.
- Draw reflections on a coordinate grid.
- Describe reflections on a coordinate grid.

Active Learn
Homework

Warm up

1 Fluency Which pairs of lines on this grid are perpendicular?

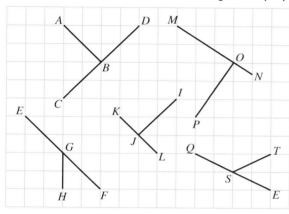

2 Here is a pair of axes.
Four lines are drawn on the axes.
Which line is

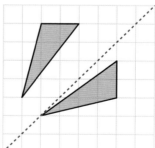

  a the x-axis   **b** $x = 1$ **c** $y = -x$
  d the y-axis   **e** $y = 1$ **f** $y = x$?

3 Which reflection is correct? Explain why.

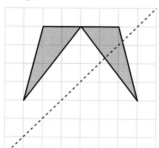

4 Copy each diagram. Draw the reflection of the shape in the mirror line.

  a 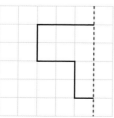   **b** 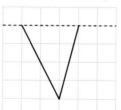   **c**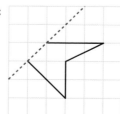

Unit 10 Transformations   **305**

5 Copy the diagram.

   a Reflect shape A in the x-axis.
     Label the image B.

   b Reflect shape A in the y-axis.
     Label the image C.

   c Reflect shape B in the y-axis.
     Label the image D.

   d **Reasoning** David says, 'If I reflect a point in the x-axis, its x-coordinate stays the same. If I reflect a point in the y-axis, its y-coordinate stays the same.' Is he correct? Give examples.

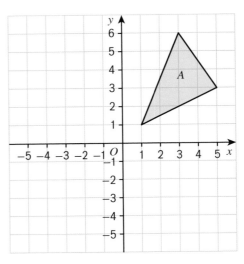

6 Draw a coordinate grid with x-axis from -5 to 5 and y-axis from -8 to 6.
Copy shape A from **Q5** onto your grid.

   a Draw the line $x = 1$.

   b Reflect shape A in the line $x = 1$. Label the image X.

   c Draw the line $y = -1$.

   d Reflect shape A in the line $y = -1$. Label the image Y.

7 Copy the diagram.

   a Reflect shape P in the line $y = 2$.
     Label the image Q.

> **Q7a hint**
>
> Draw the line $y = 2$ first.

   b Reflect shape Q in the line $x = -1$.
     Label the image R.

   c **Reasoning** Shona starts with shape P and reflects it first in $x = -1$ and then in $y = 2$. Does she get the same final image?

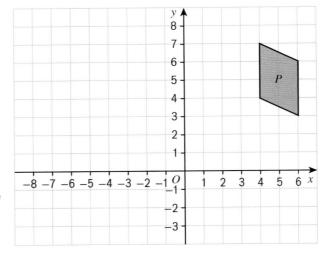

8 **Reasoning** Toby is asked to reflect shape S in the line $y = 1$, and label the image T.
Shape S and its image T overlap. Explain why.

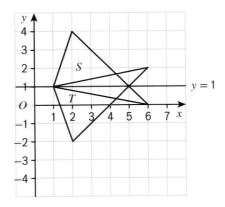

9 Copy the diagram.

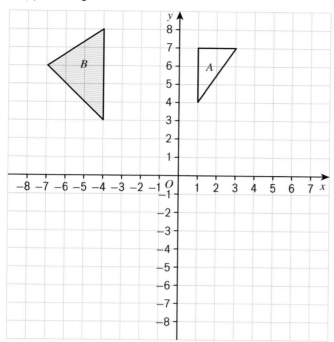

Q9 hint

Count the perpendicular distance of each vertex from the mirror line, then count the same again the other side of the line.

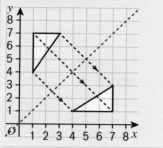

a Reflect shape A in the line $y = x$. Label the image A'.

b Reflect shape B in the line $y = -x$. Label the image B'.

10 Reflect Write down three steps for reflecting a shape.

Key point

To describe fully a single transformation involving a **reflection** on a coordinate grid, you must state it is a reflection and give the equation of the **mirror line**.

11 a Write down the equation of the mirror line halfway between the vertices of the shape A and its image B.

b Copy and complete to describe fully the single transformation that maps shape A to shape B: Reflection in the line

_____ .

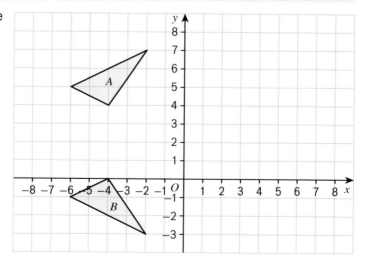

12 Reasoning Describe fully the single transformation that maps shape

a A to B **b** C to F **c** D to F **d** E to F

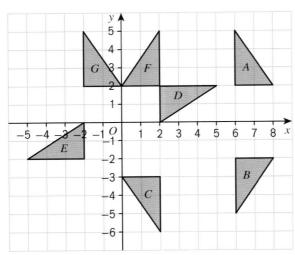

13 Reasoning Isabel looks at the shapes in **Q12**.
She writes the single transformation that maps shape F to G:

Reflection in the line y.

a What mistake has Isabel made?

b Write the correct description of the full transformation.

14 Describe fully the single transformation that maps shape A to shape B.

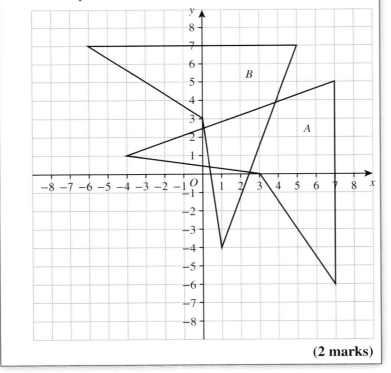

(2 marks)

10.3 Rotation

- Rotate a shape on a coordinate grid.
- Describe a rotation.

Active Learn
Homework

Warm up

1 Fluency How many degrees are there in   **a** a full turn   **b** a half turn   **c** a quarter-turn?

2 For each turn, write the number of degrees and whether the direction is clockwise or anticlockwise.

a   **b**   **c**

Key point

You **rotate** a shape by turning it around a point called the **centre of rotation**.

Example

Rotate the shape 90° anticlockwise about the point (1, 2).

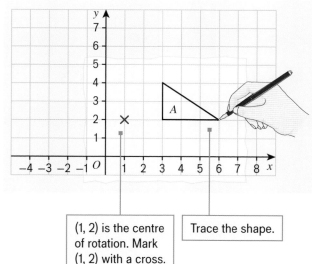

(1, 2) is the centre of rotation. Mark (1, 2) with a cross.

Trace the shape.

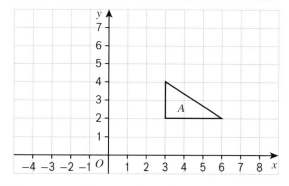

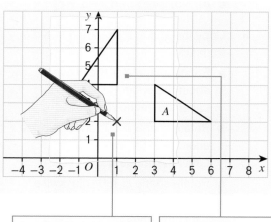

Rotate the tracing paper 90° anticlockwise about (1, 2).

Lift up the tracing paper and draw the image on the grid.

Unit 10 Transformations   309

3 Copy each diagram. Use the centre of rotation marked to draw the image of each shape after the rotation given.

a

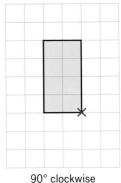

90° clockwise

b

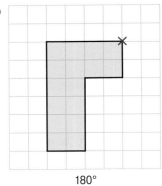

180°

c

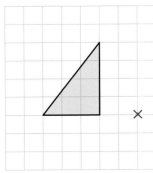

90° anticlockwise

d

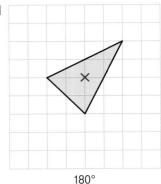

180°

4 **Reflect** Write a sentence comparing

a your rotations for **Q3a** and **Q3b** with your rotation for **Q3c**

b your rotations for **Q3a** and **Q3b** with your rotation for **Q3d**

Q4 hint

Look at the centres of rotation.

5 Copy the grid and shape *A* only. Rotate shape *A* 90° clockwise about each centre of rotation.

a $(0, 0)$ **b** $(2, 0)$

c $(0, -5)$ **d** $(-3, -3)$

6 Copy the grid and shapes *B* and *C* from **Q5**.

a Rotate shape *B* 90° clockwise about the origin.
Label the image *B′*.

b Rotate shape *B* 90° anticlockwise about $(-2, 0)$.
Label the image *B″*.

c Rotate shape *C* 180° clockwise about $(0, 1)$.
Label the image *C′*.

d **Reflect** Nicky says for part **c**, you don't need to know the direction.
Is Nicky correct? Explain.

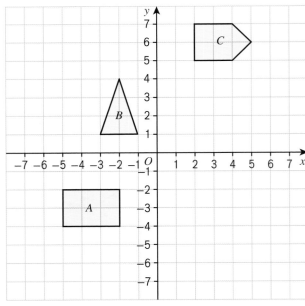

7 **Reasoning** Shape P is rotated to give shape Q.

a Trace shape P.
Rotate the tracing paper until the shape is the same way round as image Q.
What is the direction and what is the angle of rotation?

b Where is the centre of rotation?

c Copy and complete to describe fully the transformation that maps shape P to shape Q.
Rotation _____° clockwise about (☐, ☐).

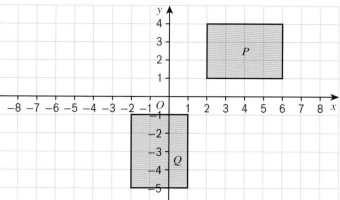

Q7b hint

Try holding the tracing paper at different centres of rotation.

Key point

To describe fully a single transformation involving a rotation on a coordinate grid, you must state it is a rotation and give the angle, direction and centre of rotation.
No direction is needed for a rotation of 180°.

8 **Reasoning** Describe fully the transformation that maps shape

a A to B

b A to C

c B to C

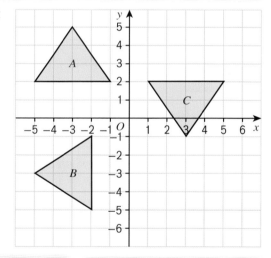

Exam-style question

9 Describe fully the single transformation that maps triangle A to triangle B.

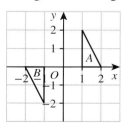

(3 marks)

Exam tip

You are allowed to use tracing paper in an exam to help you.

10 **Reflect** Look back at the reflections in Lesson 10.2 **Q12**.
How can you tell the shapes are reflected and not rotated?

10.4 Enlargement

- Enlarge a shape by a scale factor.
- Enlarge a shape using a centre of enlargement.

Warm up

1 **Fluency** What are the missing numbers?
Height of B = □ × height of A
Base of B = □ × base of A

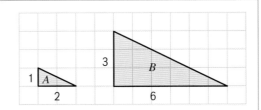

2 Work out

a $4 \times 1\frac{1}{2}$ **b** $6 \times \frac{1}{2}$ **c** 4×0.5 **d** 10×2.5 **e** $2 \times \frac{1}{4}$ **f** 1.2×20

Key point

To **enlarge** a shape you multiply all the side lengths by the same number.
The number that the side lengths are multiplied by is called the **scale factor**.

3 Copy the diagrams. Enlarge each shape by the scale factor.

a

b

c

Scale factor 3   Scale factor $1\frac{1}{2}$ Scale factor 0.5

Q3c hint

Triangle height is 4.
Enlargement height is
$4 \times 0.5 = □$

4 **a** **Reflect** Does enlargement always make a shape bigger?

b Describe what happens when the scale factor is

    i greater than 1   **ii** less than 1   **iii** equal to 1

Key point

When you enlarge a shape using a **centre of enlargement**, you multiply the distance from the centre to each vertex by the scale factor.

Example

Enlarge shape A by scale factor 2, using centre of enlargement (1, 1).
Label the image B.

Mark the centre of enlargement.

Count the squares from the centre of enlargement to each vertex.
Multiply all the distances from the centre by the scale factor.

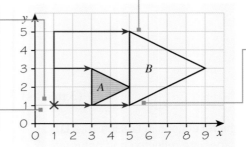

The distance to the top vertex changes from 2 up, 2 right to 4 up, 4 right.

The distance to the bottom vertex changes from 2 right to 4 right.

Check that the lengths of the image are twice as long as the original.

5 Copy each diagram. Enlarge each shape by the scale factor from the centre of enlargement.

a

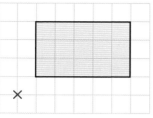

Scale factor 2

b

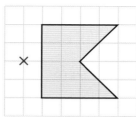

Scale factor 3

c

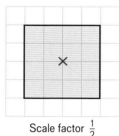

Scale factor $\frac{1}{2}$

d

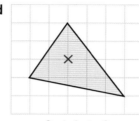

Scale factor 2

6 For each of your answers to **Q5**

   a draw lines from the centre of enlargement through the vertices on the original shape and across the grid. Check these lines go through the vertices of the image.

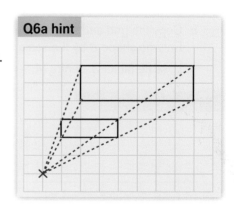

Q6a hint

   b check that

$$\frac{\text{each length on the}}{\text{original shape}} \times \frac{\text{scale}}{\text{factor}} = \frac{\text{corresponding length}}{\text{on the enlarged shape}}$$

   c **Reflect** Parts **a** and **b** are two different ways to check your answers when enlarging a shape by a scale factor.
   Which method do you prefer? Why?

7 Copy the diagram.
Enlarge the shape by scale factor 2 from the given centre of enlargement.

   a (1, 2)

   b (0, 3)

   c **Reflect** Describe how changing the centre of enlargement affects the image.

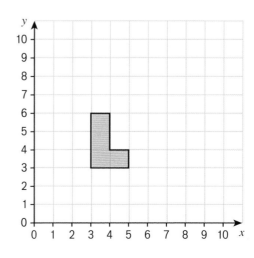

8

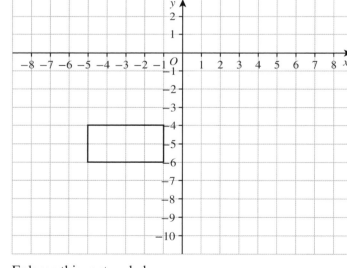

a Copy the grid and shape A only.
  Enlarge shape A by the scale factor from the centre of enlargement
  i scale factor 2, centre $(-4, 2)$ ii scale factor 2, centre $(-3, -3)$

b Copy the grid and shape B only.
  Enlarge shape B by the scale factor from the centre of enlargement
  i scale factor 3, centre $(4, -4)$ ii scale factor $\frac{1}{2}$, centre $(7, -10)$

Exam-style question

9

Exam tip

Check your answer using one of the methods in **Q6**.

Enlarge this rectangle by
a scale factor $1\frac{1}{2}$, using centre of enlargement $(-5, -4)$ **(2 marks)**
b scale factor $\frac{1}{4}$, using centre of enlargement $(-1, -6)$ **(2 marks)**

10 Problem-solving Raj enlarges this shape on a photocopier.
  a He enlarges it to 120%.
    Work out the length and width of this enlargement.
  b He enlarges the original shape to 80%.
    Work out the length and width of this enlargement.

15 cm

20 cm

10.5 Describing enlargements

- Identify the scale factor of an enlargement.
- Find the centre of enlargement.
- Describe an enlargement.

*Active*Learn
Homework

Warm up

1 Fluency Which shape is not an enlargement of A?

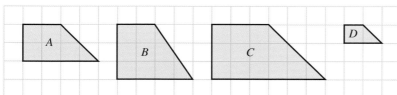

2 What is the missing number in these calculations?

  a $3 \times \square = 6$ **b** $4 \times \square = 12$ **c** $8 \times \square = 4$ **d** $2 \times \square = 5$

> **Q2 hint**
>
> The answer to part **c** is a fraction.

3 Simplify

  a $\frac{10}{5}$ **b** $\frac{3}{6}$ **c** $\frac{12}{8}$ **d** $\frac{18}{6}$ **e** $\frac{2}{8}$

4 **a** How many squares long is
    i the object
    ii the image?

  b Use your answers to part **a** to write a calculation:

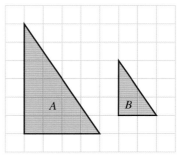

    length of object $\times$ scale factor = length of image

    $\square \times$ scale factor = $\square$

  c Use your calculation in part **b** to work out the scale factor.

5 Shape B is an enlargement of shape A.
Work out the scale factor of enlargement for each pair of shapes.

  a

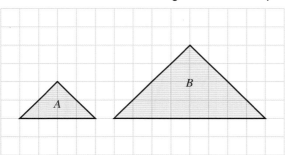

  b

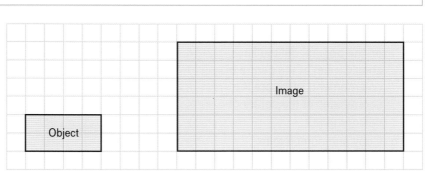

6 Reasoning

Triangle ABC is enlarged to triangle DEF.

a What is the scale factor of the enlargement?

b $AC = 5$ cm. Work out the length of DF.

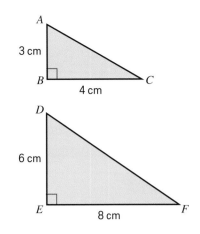

7 Reflect Ali says you can find the scale factor like this:

$$\text{scale factor} = \frac{\text{length of image}}{\text{length of object}}, \text{ then simplify the answer.}$$

Is Ali correct? Give examples using questions you have answered to explain.

Key point

To describe fully a single transformation involving an enlargement on a coordinate grid, you must state the scale factor and the coordinates of the centre of enlargement.

---

8 Shape B is an enlargement of shape A.

a What is the scale factor of the enlargement?

b Draw lines through each vertex on the image and the corresponding vertex on the original. All the lines should meet at the centre of enlargement. What are the coordinates of the centre of enlargement?

c Copy and complete to describe fully the enlargement that maps shape A to shape B.
Shape A is enlarged by scale factor _____ using centre of enlargement ($\square$, $\square$)

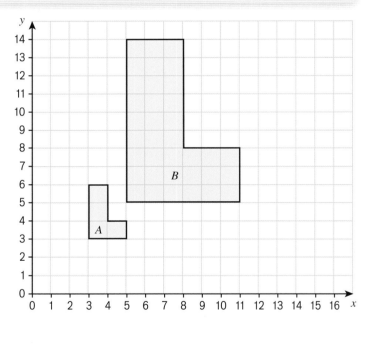

9 Reasoning Describe fully the single transformation that maps shape A on shape B.

a

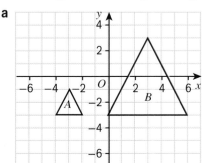

b

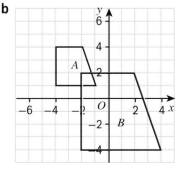

c

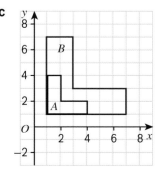

Exam-style question

10 Describe fully the single transformation that maps shape
A to shape *B*.

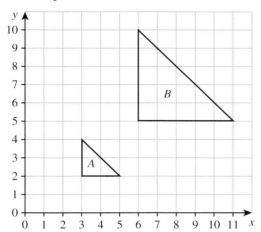

(3 marks)

11 Reasoning For each diagram
   i describe fully the transformation that maps shape *A* to shape *B*
   ii describe fully the transformation that maps shape *B* to shape *A*

a

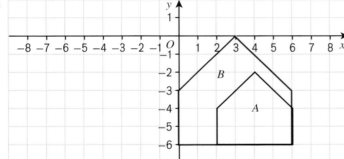

b

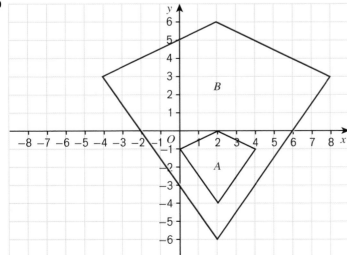

c A shape has been enlarged by scale factor 4.
   What scale factor returns the shape to its original size?

10.6 Combining transformations

- Transform shapes using more than one transformation.
- Describe combined transformations of shapes on a grid.

Warm up

1 Fluency Match each type of transformation to the information needed to describe it.

Translation   Reflection   Rotation   Enlargement

Equation of a line   Column vector

Scale factor and centre   Angle, direction and centre

2 Which type of transformation has been used to map triangle A to
   a triangle 1
   b triangle 2
   c triangle 3
   d triangle 4?

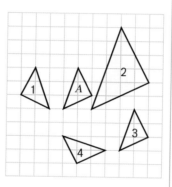

3 Reasoning What information is missing from each description?
   a Rotate shape A 90°, centre of rotation (1, 2).
   b Enlarge shape B scale factor 3.
   c Translate shape C 4 across.
   d Reflect shape D in the mirror line.

Key point

You can transform shapes using more than one transformation.

4 Copy the grid and shape R.
   a Translate shape R by $\begin{pmatrix} 4 \\ 3 \end{pmatrix}$.
      Label the image S.
   b Translate shape S by $\begin{pmatrix} -2 \\ 1 \end{pmatrix}$.
      Label the image T.
   c Reasoning Describe fully the single transformation that maps shape R to shape T.
   d Reflect Write a sentence describing what you notice about the two translations in parts **a** and **b** and the translation in your answer to part **c**.

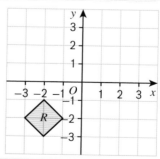

Q4d hint
$4 - 2 = \square$
$3 + 1 = \square$

Example

Triangle A is reflected in the line $y = 1$ to give triangle B.
Triangle B is reflected in the line $x = -1$ to give triangle C.
Describe fully the single transformation that maps triangle A to triangle C.

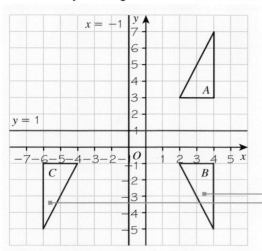

Draw triangles B and C on the diagram.

Triangle A is rotated $180°$ about $(-1, 1)$.

State the type of transformation that maps A to C.
Add *all* the information needed to 'describe fully' that transformation.

5   Copy the grid and shape A.
   a Reflect shape A in the y-axis.
      Label the image B.
   b Reflect shape B in the line $y = x$.
      Label the image C.
   c **Reasoning** Describe fully the single
      transformation that maps shape A to shape C.

6   **Reasoning** Copy the grid and shape D from **Q5**.
   Triangle D is reflected in the line $x = 1$
   to give image E.
   Triangle E is reflected in the line $x = -4$ to give
   triangle F.
   Describe fully the single transformation that maps
   triangle F to triangle D.

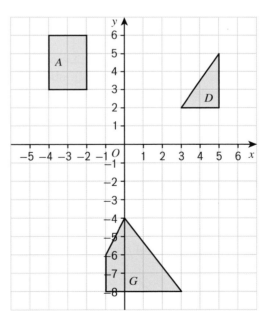

7   **Reasoning** Copy the grid and shape G from **Q5**.
   Shape G is reflected in the line $y = -2$ to give shape H.
   Shape H is reflected in the line $x = 2$ to give shape I.
   Describe fully the single transformation that maps shape I to shape G.

8   **Reflect** Which single transformation is the same as
   a reflecting in two parallel lines
   b reflecting in two perpendicular lines?

Q8 hint

Use your answers to **Q6** and **Q7** to help you.

9 **Reasoning** Copy the diagram.
Shape A is rotated $180°$ about
the origin to give shape B.
Shape B is reflected in the x-axis
to give shape C.
Describe fully the single
transformation that maps
shape A to shape C.

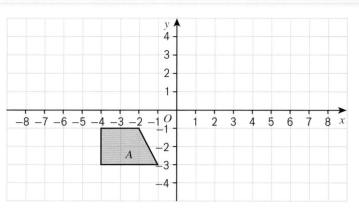

10 Copy the diagram.

  a Enlarge shape A by scale factor 2, centre $(7, 8)$.
  Label the image B.

  b Enlarge shape B by scale factor 0.5, centre $(5, 0)$.
  Label the image C.

  c **Reasoning** Describe fully the single
  transformation that maps shape A to shape C.

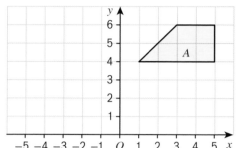

11 **a** **Reasoning** Shape A is transformed to shape B by a
  translation $\begin{pmatrix} -3 \\ -2 \end{pmatrix}$ followed by a reflection.

  Describe fully the reflection.

  b Chen says it is also possible to transform shape A to
  shape B by a rotation $180°$ about $(3, 0)$ followed by a
  reflection. Describe fully the reflection.

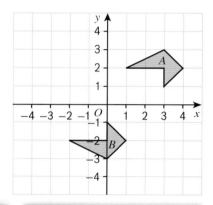

12 Shape S can be transformed to shape T by a rotation $90°$
clockwise about $(-3, -1)$ followed by a translation $\begin{pmatrix} a \\ b \end{pmatrix}$.
Find the value of a and the value of b.

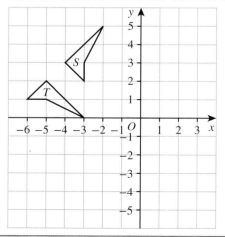

Q12 hint

Transform shape S by the
rotation. Then work out the
translation. The letters a and
b are used to represent the
numbers in the column vector
for the translation.

(3 marks)

10 Check up

Active Learn
Homework

Translations, reflections and rotations

1 Copy the grid and shape A only.

  a Reflect shape A in the x-axis.
   Label the image B.

  b Reflect shape A in the line $x = 2$.
   Label the image C.

  c Reflect shape A in the line $y = x$.
   Label the image D.

  d Translate shape A by the column vector $\begin{pmatrix} -5 \\ -1 \end{pmatrix}$.
   Label the image E.

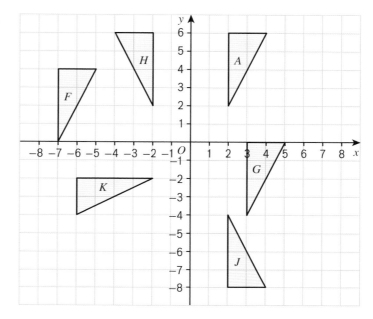

2 Look at the grid and shapes in **Q1**. Write down the vector that translates shape

  a A to F **b** A to G **c** F to G

3 Look at the grid and shapes in **Q1**. Describe fully the reflection that maps shape

  a A to H **b** A to J **c** A to K

4 Copy the grid and shape A.

  a Rotate shape A 90° clockwise about the origin.

  b Rotate shape A 90° anticlockwise about $(-2, -2)$.

5 Look at the grid and shapes in **Q4**.
Describe fully the transformation that maps shape

  a A to B

  b A to C

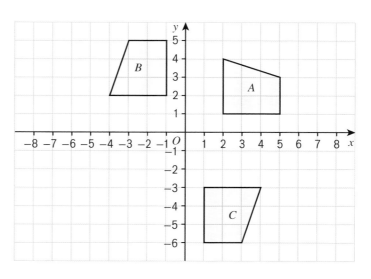

Enlargements

6 Copy the diagram. Enlarge the shape by scale factor 3 from the centre of enlargement.

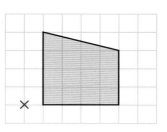

7 Copy the diagram.

    a Enlarge shape A by scale factor 2, centre of enlargement (1, 1).

    b Enlarge shape A by scale factor $\frac{1}{2}$, centre of enlargement (3, 3).

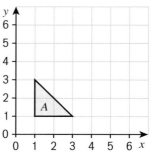

8 Describe fully the transformation that maps shape

    a A to B **b** B to C

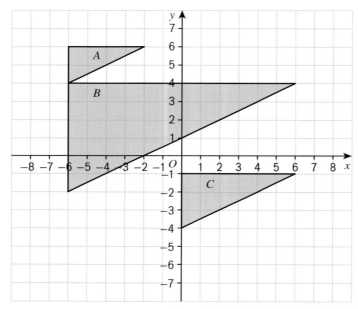

9 How sure are you of your answers? Were you mostly

Just guessing 😞   Feeling doubtful 😐   Confident 🙂

What next? Use your results to decide whether to strengthen or extend your learning.

Challenge

10 Copy the diagram.

    a Draw a reflection of the shape in the red mirror lines to make a shape with four lines of symmetry.

    b Create your own design using a different shape.

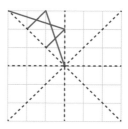

10 Strengthen

Active Learn
Homework

Translations, reflections and rotations

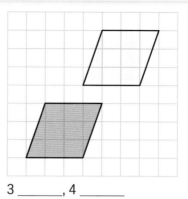

1 The yellow shape **translates** to give the orange shape.

  a Does the yellow shape move right or left to the orange shape?

  b Does the yellow shape move up or down to the orange shape?

  c Copy and complete the description below the grid of how the yellow shape translates to the orange shapes.

> **Q1c hint**
>
> **Translate** means 'slide'.

3 _____, 4 _____

2 Column vectors describe **translations**.
Write each of these as a column vector. The first two are done for you.

  a 2 right, 4 up $= \begin{pmatrix} 2 \\ 4 \end{pmatrix}$ **b** 2 left, 4 down $= \begin{pmatrix} -2 \\ -4 \end{pmatrix}$

  c 3 right, 2 up   **d** 3 left, 2 up

  e 2 left, 1 down   **f** 2 right, 1 down

  g 4 right, 3 down   **h** 3 left, 4 up

> **Q2 hint**
>
> $\rightarrow$ right is positive
> $\leftarrow$ left is negative
> $\uparrow$ up is positive
> $\downarrow$ down is negative

3 Choose the correct description for the column vector $\begin{pmatrix} 6 \\ -4 \end{pmatrix}$.

| 6 right, 4 up | 6 left, 4 up | 4 left, 6 up | 6 right, 4 down | 6 left, 4 down | 4 right, 6 down |

4 Copy the diagram.

  a Copy and complete the description of the **translation** that uses column vector $\begin{pmatrix} 6 \\ -4 \end{pmatrix}$

     6 _____, 4 _____

  b Translate each of the vertices (corners) of shape A by the column vector $\begin{pmatrix} 6 \\ -4 \end{pmatrix}$.
Label the new shape B.

  c Translate shape A by the column vector $\begin{pmatrix} -2 \\ 5 \end{pmatrix}$.
Label the image (new shape) C.

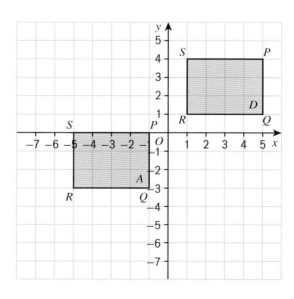

5 **a** Copy and complete this table.

	Coordinates of vertex			
	P	*Q*	*R*	*S*
Shape *A*	(−1, 0)	(−1, −3)		
Shape *D*	(5, 4)			

b Write the column vector that translates shape *A* to shape *D*.

Q5b hint

Check this column vector works for the other coordinates.
Shape *A*:   *P*(−1, 0)
                6 _____ ↓, ↓ up
Shape *D*:   *P* (5, 4)
6 _____, ☐ up is the

column vector $\begin{pmatrix} \square \\ \square \end{pmatrix}$

6 Copy the diagram.

a Shapes *P* and *Q* are both **reflected**. For shape *P*:

  i Label its vertices *A*, *B* and *C*.
  (It does not matter which vertex is which letter.)

  ii Reflect the points *A*, *B* and *C* in the mirror line *y* = 2.

  iii Join the reflected points. You have drawn a reflection of shape *P* in the line *y* = 2.

  iv Label the image *P′*.

b Follow the method in part **a** to reflect shape *Q* in the line *x* = −3. Label the image *Q′*.

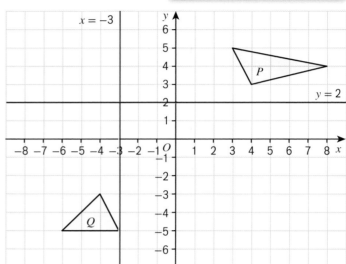

7 Shape *A* is **reflected** to give shape *B*. Copy the diagram.

a Draw in the mirror line for the reflection that maps *A* to *B*.
Write the equation of the line.

b A transformation moves a shape so that it is in a different position.
Copy and complete the description of the transformation that maps *A* to *B*:
Reflection in the _____

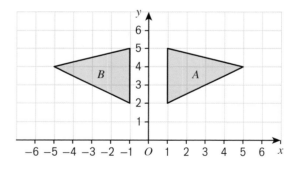

8 Copy the diagram. **Reflect** points *A*, *B* and *C* in the line *y* = *x*. Label the new points *A′*, *B′* and *C′*.

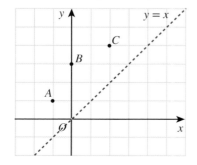

Q8 hint

For a diagonal line of symmetry, turn the page so that the mirror line is vertical.

9 Match each diagram to its description of a **rotation**.

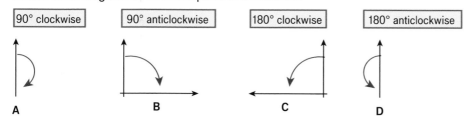

90° clockwise   90° anticlockwise   180° clockwise   180° anticlockwise

A   B   C   D

10 Copy each diagram.

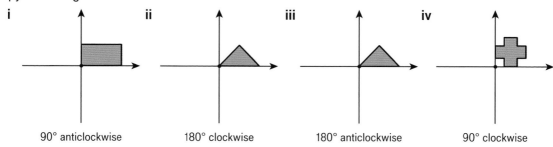

i   ii   iii   iv

90° anticlockwise   180° clockwise   180° anticlockwise   90° clockwise

For each diagram

a Trace the shape and the arrow.

b Hold the tracing paper on the dot with your pencil. This is called the **centre of rotation**.

c **Rotate** the shape in the given direction.
Stop when the arrow has turned through the correct angle.

d **Reflect** Look at your rotations for **ii** and **iii**. Is it necessary to give a direction, clockwise or anti-clockwise? Explain why.

11 Copy the diagram.

a Trace shape A.

b Hold the tracing paper at the origin, $(0, 0)$ with your pencil.
Rotate shape A 90° clockwise.

c Lift up the tracing paper and copy the rotated shape onto the coordinate grid.
Label the image B.

d Repeat steps **b** and **c** holding the tracing paper at $(0, 6)$.
Label the image C.

e **Reflect** Explain what changes when you hold the tracing paper in a different place.

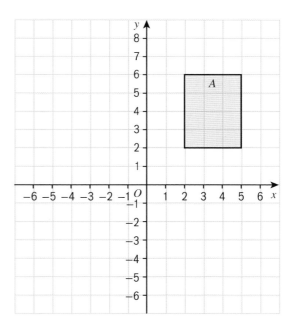

12 Use your diagram from **Q11**.

a **Rotate** shape A 180° about $(0, 0)$.
Label the image D.

b Rotate shape A 90° anticlockwise about $(-1, 2)$.
Label the image E.

Q12 hint

Follow similar steps to those in **Q11**.

13 a Trace shape A.

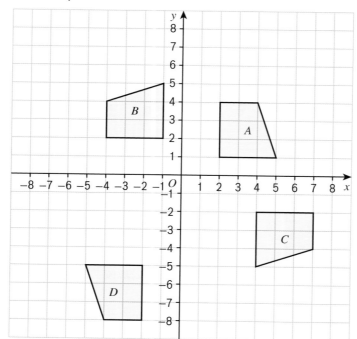

b Use your pencil to hold the tracing paper at different points.
Turn the tracing paper until shape A **rotates** to map onto shape B.
What is the centre of rotation that maps A to B?

c A **transformation** moves a shape so it is in a different position.
Copy and complete the transformation that maps A to B:
Rotation $\square°$ _____ about (__ , __).

d Follow similar steps to those in parts **b** and **c** to describe
fully the transformation that maps shape

i A to C **ii** A to D

> **Q13c hint**
>
> $\square°$ clockwise or
> $\square°$ anticlockwise?

14

y
3
A   B
2
1

−3 −2 −1 O   1 2 3 x
−1
−2
−3

a Explain how you know

i rectangle B is **not** a reflection of rectangle A

ii rectangle B is **not** a translation of rectangle A

b Explain how you can show that rectangle B is a rotation of rectangle A.

Enlargements

Shape A is drawn on a centimetre square grid.

Shape A   Shape B   Shape C   Shape D   Shape E

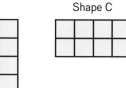

1 Which shapes are enlargements of shape A?

Q1 hint

Which of shapes B, C, D and E are the same shape (square) as A? It could be smaller: the word **enlargement** is used even when the new shape is smaller than the original shape.

2 a Shape F is an enlargement of shape A in **Q1**.
What have the side lengths of shape A been multiplied by to make shape F?
This is called the scale factor of enlargement.

b Work out the scale factor of any enlargements of shape A that you identified in **Q1**.

c Shape A is enlarged by scale factor 5.
Draw the enlargement.

Shape F

3 Copy each diagram.
Enlarge the shape by the scale factor.

a

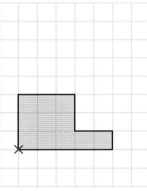

Scale factor 2

b

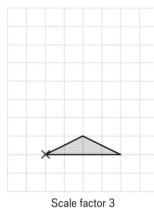

Scale factor 3

Q3b hint

Count the squares up and across between each vertex (corner).

$4 \times 3 = \boxed{}$ right

$1 \times 3 = \boxed{}$ up

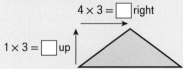

c

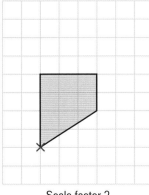

Scale factor 2

d

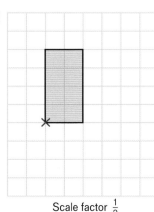

Scale factor $\frac{1}{2}$

4 Copy the diagram.

Use this method to enlarge rectangle $ABCD$ by scale factor 2, using centre of enlargement (0, 0).

a Draw lines up then across from (0, 0) A, B, C and D. (A is done for you.)

b Draw each line twice to extend by scale factor 2. Label the new points A', B', C' and D'. (A' is done for you.)

c Join the points A', B', C' and D' to enlarge rectangle $ABCD$ by scale factor 2 from (0, 0).

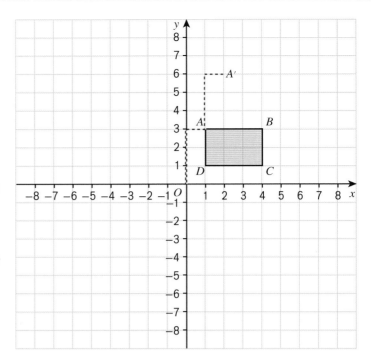

5 Use the same grid as you used for **Q4**.

a Mark the centre of enlargement (7, 0).

b Use the same method as in **Q4** to enlarge $ABCD$ rectangle by scale factor 2, using centre of enlargement (7, 0).

c **Reflect** Write a sentence stating how the centre of enlargement affects the image.

6 **Reasoning** Shape B is an enlargement of shape A.

a Work out the scale factor of the enlargement.

b Copy and complete this sentence.
Shape A is enlarged by scale factor _____.

Q6b hint

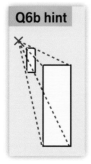

c Position your ruler so that it joins corresponding vertices (corners) on shapes A and B.
Repeat for all four vertices (corners).
What point does your ruler always go through?

d Use your answer to part **c** to copy and complete this sentence.
The centre of enlargement is ($\square$, $\square$).

e Use your answers to parts **b** and **d** to describe fully the transformation that maps shape A to shape B.

10 Extend

1 **Problem-solving** Describe fully at least three different transformations that map shape A to shape B.

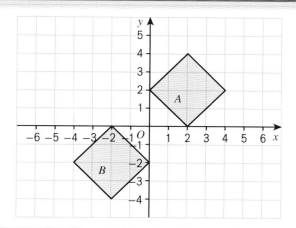

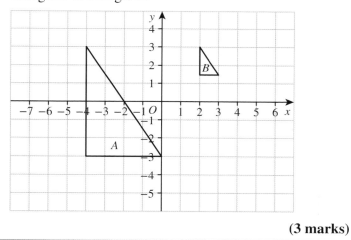
3 **Problem-solving** A regular hexagon $ABCDEF$ has centre O.
Describe fully the transformation that maps triangle ABO to triangle BCO.

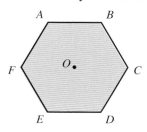

4 Reasoning

a What is the mathematical name of this shape?

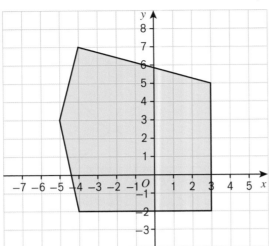

Q4a hint

Is it regular or irregular?

b Find its area.

c Copy the shape. Reflect it in the line $y = -x$.

d Does the reflection have the same area as the original shape? Explain.

Q4b hint

Trace the shape. Draw a vertical line to split the shape into a triangle and a trapezium.

5 Problem-solving

a A translation is described by the column vector $\begin{pmatrix} x \\ y \end{pmatrix}$.
Write the vector that would map the image back to the original shape.

b A ship sails from A to B, to C and then back to A.

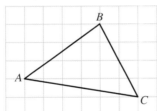

The vector from A to B is $\begin{pmatrix} 4 \\ 3 \end{pmatrix}$.

The vector from B to C is $\begin{pmatrix} 2 \\ -4 \end{pmatrix}$.

What is the vector from C to A?

Q5a hint

The vector $\begin{pmatrix} 4 \\ -2 \end{pmatrix}$ describes the translation from A to B. What vector describes the translation from B to A?

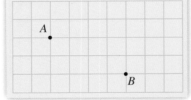

Exam-style question

6 A shape is translated twice using the column vector $\begin{pmatrix} a \\ b \end{pmatrix}$ followed by $\begin{pmatrix} c \\ d \end{pmatrix}$.

Write a vector using the letters a, b, c and d to describe the complete translation. **(1 mark)**

Exam tip

Try with small numbers for a, b, c and d.

7 Reasoning

a Rectangle A has length 4 cm and width 3 cm.
  Rectangle B is an enlargement of rectangle A and one of its sides has length 24 cm.
  What are the possible scale factors of the enlargement?

b Draw a rectangle with length 3 cm and width 2 cm. Label the rectangle A.
  i What is the perimeter of rectangle A?
  ii What is the area of rectangle A?
  iii Enlarge rectangle A by scale factor 2. Label the image B.
  iv What is the perimeter of rectangle B?
  v What is the area of rectangle B?

8 Reasoning When a shape is enlarged by scale factor x, what happens to its area?

9 Problem-solving This shape has an area of 4 cm^2.

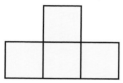

The shape is enlarged by scale factor 3.

a Predict the area of the enlargement.

b Draw the enlargement to check your answer to part **a**.

Exam-style question

10 Triangle ABC is drawn on a centimetre grid. A is the point $(2, 3)$.
B is the point $(6, 3)$. C is the point $(5, 6)$.
Triangle XYZ is an enlargement of triangle ABC with scale factor $\frac{1}{2}$ and centre $(0, 1)$.
Work out the area of triangle XYZ.

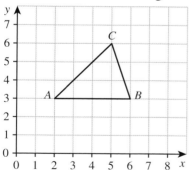

(3 marks)

Q10 hint

Work out the area of triangle ABC first.

10 Test ready

Summary of key points

To revise for the test:

- Read each key point, find a question on it in the mastery lesson, and check you can work out the answer.

- If you cannot, try some other questions from the mastery lesson or ask for help.

Key points

1 In a **translation**, all the points on the shape move the same distance in the same direction. You can use a **column vector** to describe a transformation. The top number describes the movement to the left or right, and the bottom number describes the movement up or down. For example:

$\begin{pmatrix} 3 \\ 2 \end{pmatrix}$ means 3 right, 2 up

$\begin{pmatrix} -4 \\ -5 \end{pmatrix}$ means 4 left, 5 down → **10.1**

2 Sometimes shapes are translated on coordinate grids. → **10.1**

3 A **transformation** (such as a translation) transforms a shape to a different position. The original shape maps to an **image**. The **image** is the shape after a transformation. → **10.1**

4 To describe fully a single transformation involving a **reflection** on a coordinate grid, you must state that it is a reflection and give the equation of the **mirror line**. → **10.2**

5 You **rotate** a shape by turning it around a point called the **centre of rotation**. → **10.3**

6 To describe fully a single transformation involving a rotation on a coordinate grid, you must state that it is a rotation and give the angle, direction and centre of rotation. → **10.3**

7 No direction is needed for a rotation of 180°. → **10.3**

8 To **enlarge** a shape you multiply all the side lengths by the same number. The number that the side lengths are multiplied by is called the **scale factor**. → **10.4**

9 When you enlarge a shape using a **centre of enlargement**, you multiply the distance from the centre to each vertex by the scale factor. → **10.4**

10 To describe fully a single transformation using an enlargement, you must state the scale factor and the centre of enlargement. → **10.5**

11 You can transform shapes using more than one transformation. → **10.6**

Sample student answers

1 Describe fully the single transformation that maps triangle P to triangle Q.

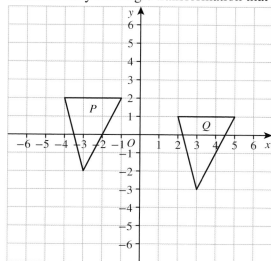

(2 marks)

Student A

Translation $\begin{pmatrix} 3 \\ -1 \end{pmatrix}$

Student B

Translation $\begin{pmatrix} 6 \\ -1 \end{pmatrix}$

a Which student has the correct answer? Explain why.

b What could the student draw on the diagram to help ensure that they get the correct answer?

2 Describe fully the single transformation that maps shape A to shape B.

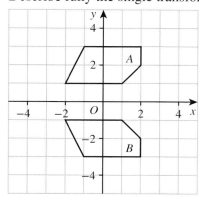

(2 marks)

Shape B is a reflection of shape A about the origin (0, 0).

Why do you think this student is likely not to gain full marks?
Write an answer that would gain full marks.

10 Unit test

ActiveLearn
Homework

1 Shape B is an enlargement of shape A.
   Write the scale factor of the enlargement.   **(1 mark)**

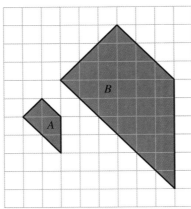

2 Copy shape S. Enlarge shape S by scale factor $\frac{1}{2}$ using the marked centre of enlargement.
                                                                (2 marks)

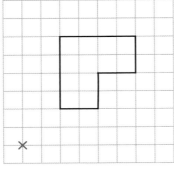

3 Copy the diagram.
   a Translate shape A by vector $\begin{pmatrix} -5 \\ 6 \end{pmatrix}$.
       Label the image B.
                            (1 mark)
   b Reflect shape A in the line $y = 1$.
       Label the image C.
                            (2 marks)
   c Rotate shape A 90° anticlockwise about $(1, -1)$.
       Label the image D.
                            (3 marks)

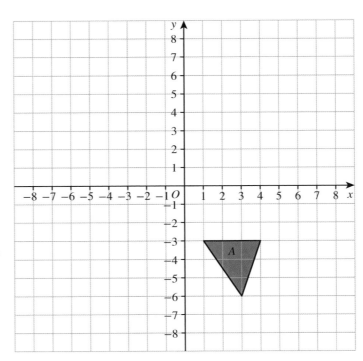

4 Look at the diagram.

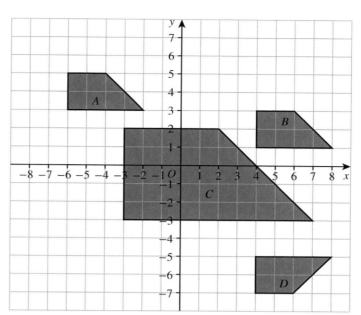

a Describe fully the transformation that maps shape *A* to shape *B*. **(2 marks)**

b Describe fully the transformation that maps shape *A* to shape *C*. **(2 marks)**

c Describe fully the transformation that maps shape *B* to shape *D*. **(2 marks)**

5 Copy the diagram and shape *A* only.
Enlarge shape *A* by scale factor 2,
centre of enlargement (6, 5). **(3 marks)**

6 Bella is asked to describe fully the
transformation that maps shape *A* to
shape *B*.
She writes, 'Rotation 90° clockwise.'
What is missing from Bella's
description? **(1 mark)**

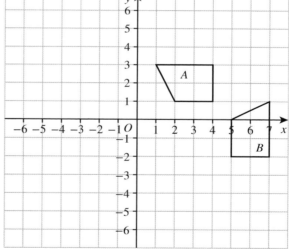

7 Describe fully the transformation that maps triangle *A* to triangle *B*. **(2 marks)**

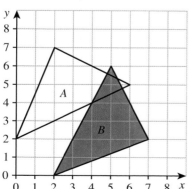

8 Copy the diagram.

  a Reflect triangle P in the line $y = x$.
    Label the image Q.
    (2 marks)

  b Reflect triangle Q in the line $y = -x$.
    Label the image R.
    (2 marks)

  c Describe fully the single transformation that maps triangle P to triangle R.
    (2 marks)

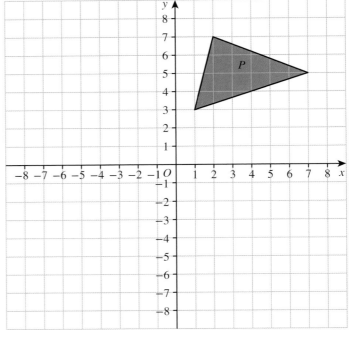

9 Triangle A is enlarged by scale factor 2 from centre $(0, 0)$.
  Work out the area of the enlarged triangle.
    (3 marks)

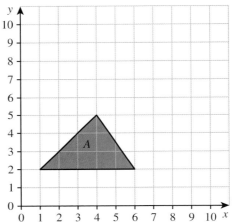

(TOTAL: 30 marks)

10 **Challenge** Design your own combined transformations GCSE exam question.

  a Draw a coordinate grid with axes labelled from -8 to 8.

  b Draw a shape and label it A.

  c Write the first transformation and carry it out on the grid.
    Label the image B.

  d Write the second transformation and carry it out on the grid.
    Label the image C.

  e If there is a single transformation that maps shape A to shape C, write it down.

  f Repeat the steps using different types of transformations.

11 **Reflect**

  a What are the four types of transformation?

  b List the pieces of information that are needed to describe fully a single transformation of each type.

Mixed exercise 3

1   A shape is translated by the vector $\begin{pmatrix} 0 \\ -3 \end{pmatrix}$.

Does the shape move

A up   **B** down   **C** left   **D** right?

2   **Reasoning** Felix is asked to complete a table of values for
$y = -2x + 1$ and to draw its graph. Here are Felix's table and graph:

x	-2	-1	0	1	2
y	-5	-3	1	-3	-5

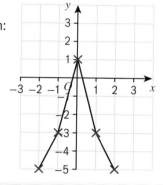

a How can you tell from looking at Felix's graph
that it is incorrect?

b Draw the correct graph of $y = -2x + 1$.

Exam-style question

3   Here is a diagram of part of a wall in Evie's house.

4 m

2 m

Evie is going to cover this part of the wall with square tiles.
The square tiles have sides of 20 cm.
Work out the number of tiles Evie needs.   **(3 marks)**

---

4   **Reasoning** Here are six graphs.

a Match each of these equations
to the correct graph.

  i $y = x + 2$

  ii $y = -x - 4$

  iii $y = \frac{1}{2}x + 2$

  iv $y = -x$

  v $y = \frac{1}{2}x - 2$

b Write a possible equation for the
graph not used in part **a**.

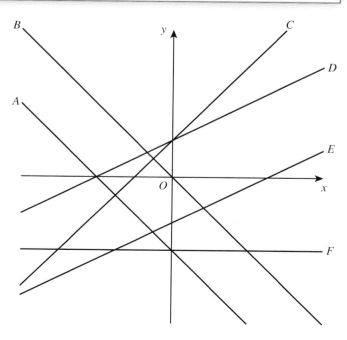

5 The diagram shows shape A.
All the measurements are in centimetres.

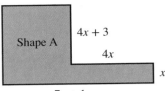

a Find an expression, in terms of x, for the perimeter of shape A. **(3 marks)**

A square has the same perimeter as shape A.

b Find an expression, in terms of x, for the length of one side of this square. **(1 mark)**

6 **Problem-solving** The midpoint of a line segment from $(3, 2)$ to (p, q) is $(-2, 5)$.
Work out the values of p and q.

7 Jack makes a wooden building block.
The block is in the shape of a cuboid.

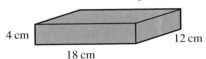

Jack is going to make a new block.
The new block will also be in the shape of a cuboid.
The cross section of the cuboid will be a 6 cm by 6 cm square.

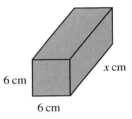

Jack wants the new cuboid to have the same volume as the 4 cm by 18 cm by 12 cm cuboid.
Work out the value of x. **(3 marks)**

8 **Problem-solving** The diagram shows shape P.

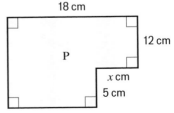

The area of shape P is 271 cm^2.
Work out the value of x.

9 **Reasoning** Toby is asked to enlarge a triangle with vertices $A(0, 5)$, $B(6, 5)$ and $C(4, 1)$ with scale factor $1\frac{1}{2}$ from centre of enlargement $(8, 7)$.
This is Toby's diagram.

a What mistake has Toby made?

b Copy triangle ABC and draw the correct enlargement with scale factor $1\frac{1}{2}$ from centre of enlargement $(8, 7)$.

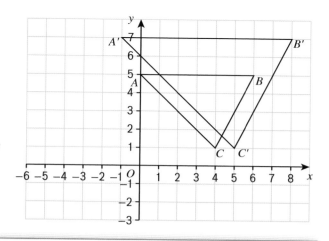

10 Jay drives 60 miles from his home to a meeting.
Here is the travel graph for Jay's journey to the meeting.

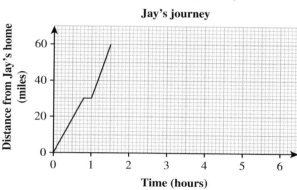

Jay's meeting lasts for $1\frac{1}{2}$ hours.
He then drives home at a steady speed of 40 miles per hour with no stops.
Copy and complete the travel graph to show this information. **(2 marks)**

11 A company designs these containers.
Each of the 5 faces of the container is made of glass.

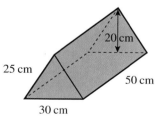

The container is in the shape of a prism.
The cross-section of the prism is an isosceles triangle with height 20 cm.
Work out the total area of glass needed to make the container. **(3 marks)**

12 Problem-solving Katy draws an equilateral triangle with sides of length 12 cm.
She then draws a square with the same perimeter as the triangle.
Work out the area of Katy's square.

13 Problem-solving The diagram shows the plan of Sean's patio, which is to be covered in decking.

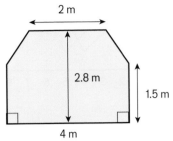

2 m

2.8 m

1.5 m

4 m

The decking is sold in packs, each covering 1.7 m².
A pack of decking normally costs £36.
Sean gets a discount of 25% off the cost of the decking.
Sean has £175 to spend on decking.
Does Sean have enough money to buy all the decking he needs? Explain why.

14 Reasoning The diagram shows triangle A drawn on a grid.

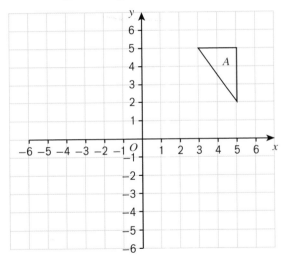

a Copy the diagram. Reflect triangle A in the y-axis to get triangle B and then translate triangle B by $\begin{pmatrix} 1 \\ -6 \end{pmatrix}$ to get triangle C.

b Ewan says, 'This also works the other way around. Translate triangle A by $\begin{pmatrix} 1 \\ -6 \end{pmatrix}$ to give triangle T and then reflect triangle T in the y-axis to get the same triangle C as in part **a**.'
Is Ewan correct? Show how you know on your diagram.

Answers

UNIT 1 Number

1.1 Calculations

1 a B   b A and D

2 a 4, −8   b −6, −7

3 a 6   b 6   c 9

4 a RS/GT; RS/WT; RS/YT; BS/GT; BS/WT; BS/YT
    b 6

5 a A1/S1; A1/S2; A1/S3; A2/S1; A2/S2; A2/S3; A3/S1;
    A3/S2; A3/S3; A4/S1; A4/S2; A4/S3
    b I listed the combinations of the first art activity with each
    of the sports activities, then repeated this for each of the
    other art activities.
    c 12

6 $2 \times 3 - 7$; $2 \times 7 - 3$; $2 - 3 \times 7$; $2 - 7 \times 3$;
    $3 \times 2 - 7$; $3 \times 7 - 2$; $3 - 2 \times 7$; $3 - 7 \times 2$;
    $7 \times 2 - 3$; $7 \times 3 - 2$; $7 - 2 \times 3$; $7 - 3 \times 2$

7 −1, 11, −19, −19;
    −1, 19, −11, −11;
    11, 19, 1, 1

8 A, E, F(6)

9 B, C, D, A

10 a 66   b 0   c 1

11 He should have divided 30 by 5 before adding 25.

12 a ≠   b =   c ≠   d ≠
    e 2 of: $8 - 4 \times 2 < (8 - 4) \times 2$; $(3 + 4)^2 > 3^2 + 4^2$
    f No, they are all either equal to or not equal to.

13 a 3, 27   b 4, 44   c 7
    d 8   e 12   f 14

14 a −5   b ÷2   c ×3   d +4

15 a $89 + 598 = 687$ b $46 \times 11 = 506$
    c $3168 \div 12 = 264$

16 C

17 a 72   b 5   c 20

18 33

19 a 25   b 343   c 2025   d 11

1.2 Decimal numbers

1 376.8   b 37.68   c 0.376

2 a 1 cm = 10 mm   b 1 m = 100 cm
    c 1 km = 1000 m   d £1 = 100p

3 a 0.19   b 19   c 190   d 1967

4 a 7.3   b 3.5   c 0.5   d 12.1
    e 9.0

5 a i 4.03   ii 16.17   iii 0.13
    iv 11.86   v 703.43
    b i 4.026   ii 16.172   ii 0.135
    iv 11.862   v 703.429   v 703.429

6 a 3.5 cm   b 6.85 m   c 25.325 km

7 583 mm

8 a £8.98 making a total of £35.92
    b 2 decimal places, to give the answer to the nearest penny.
    c $4 \times £8.98 = £35.92$

9 a 10, 8   b 10, 18   c 150

10 a 6   b 0.6   c 0.6   d 0.06
    e The answer is always 6×10^n.

11 a 7.2   b 7.2, 0.72   c 0.072   d −0.072
    e 0.072

12 There are many answers, such as 0.06×0.1, 0.03×0.2,
    0.15×0.04, etc.

13 a 25.5   b 39.96   c 21.3   d 47.52
    e 29.322
    f The number of digits after the decimal point in each
    answer is equal to the sum of the number of digits in
    each calculation. Yes, this works for the calculation in the
    worked example.

14 The answer to 14.3×0.96 should be less than 14.3, but
    14.688 is more than 14.3.

15 Total cost $= 10 \times £2.40 + 5 \times £3.50 = £24 + £17.50 = £41.50$.
    She will only get £8.50 in change, so she is wrong.

16 a 2   b 0.2   c 20   d 2
    e The answers are 2 or $2 \div 10$ or 2×10.

17 a 4.9   b 9.8   c 4.9
    d To estimate, round the first number to the nearest easy
    whole number, e.g. nearest ten or nearest multiple of 5. For
    example, $25 \div 5 = 5$, $30 \div 3 = 10$ and $35 \div 7 = 5$.

18 a 15.6   b 150   c 4000   d 130

19 16

20 5 bricks and 2 bags of mortar

1.3 Place value

1 a $\div 10 \times 2$ b $\div 100 \times 2$

2 a 1300   b 2 600 000   c 34.5   d 1 500 000

3 $\frac{40}{20}$, 2, 30

4 $7.6 \times 2.1 = 15.96$, so $15.96 \div 7.6 = 2.1$ and
    $15.96 \div 2.1 = 7.6$

5 a 7 000 000   b 4 600 000   c 10 100 000
    d 2 450 000   e 3 125 000   f 5 500 000

6 a 62 million   b 34.1 million   c 4.25 million
    d 58.42 million
    e 16.325 million
    f 74.3 million

7 a 11 050 000   b 10 550 000

8 a 6.437 km   b 431 m   c 650 000 km
    d 24 μm   e 6000

9 a 561.8   b 0.003 47   c 49 000

10 a 70   b 76 400   c 340 000   d 530 000

11 a 28, 280 000   b 2 400 000
    c 4   d 2000   e 510, 1020

12 a £225 000   b £234 999 (or £234 999.99)

13 a $275 \times 421 \approx 300 \times 400 = 120 000$
    b $\frac{876}{29} \approx \frac{900}{30} = 30$
    c $\frac{41 \times 482}{1182} \approx \frac{40 \times 500}{1000} = 20$
    d $\frac{284 \times 10.34}{4.52} \approx \frac{300 \times 10}{5} = 600$
    e $\frac{5.21 \times 3.84}{6.72} \approx \frac{5 \times 4}{7} \approx \frac{21}{7} = 3$
    f $\frac{9.83 \times 3.24}{7.65} \approx \frac{10 \times 3}{8} \approx \frac{32}{8} = 4$

14 a 7 **b** 1

15 Q13a 115 775 **Q13b** 30.206...
Q13c 16.719... **Q13d** 649.681...
Q13e 2.977... **Q13f** 4.163...
Q14a 7.194... **Q14b** 1.166...

16 a 4 **b** 2.5

17 a 5.715 **b** 4.5 **c** 45

18 a 5920 **b** 5733 **c** 5624

19 a 1.26 **b** 1.256 258 235

1.4 Factors and multiples

1 2

2 a i 3, 6, 9, 12, 15, 18, 21, 24, 27, 30
ii 4, 8, 12, 16, 20, 24, 28, 32, 36, 40
b 18 **c** 56

3 a 4 or 6 **b** 5 **c** 36

4 a $1 \times 36 = 36$, $2 \times 18 = 36$, $3 \times 12 = 36$, $4 \times 9 = 36$,
$6 \times 6 = 36$
b 1, 2, 3, 4, 6, 9, 12, 18, 36

5 a 1, 2, 4, 7, 14, 28 **b** 1, 2, 4, 8, 16, 32
c 1, 2, 4, 5, 8, 10, 20, 40

6 5, 15

7 a 31, 37
b Alice is incorrect. There are 2 prime numbers between 30 and 40 (31, 37). There are also 2 prime numbers between 20 and 30 (23, 29).

8 a For example, $3 \times 5 = 15$.
b Students' own answers, e.g. I disproved the statement using an example.

9 13, 23; or 19, 29; or 31, 41; or 37, 47; or 43, 53; or 61, 71; or 73, 83; or 79, 89

10 a 8 **b** 7 **c** 12 **d** 8

11 a 12 **b** 35 **c** 12 **d** 24
e 60 **f** 40
g Students' own answers, e.g. I wrote the first 5 multiples of each number unless I could see that one number was already a multiple of the other.

12 a HCF: 6, LCM: 36 **b** HCF: 20, LCM: 120
c HCF: 15, LCM: 150

13 E.g. 16 and 32

14 E.g. 5 and 8

15 9

16 100 seconds or 1 minute 40 seconds

17 2.4 seconds

1.5 Squares, cubes and roots

1 a 1, 4, 9 **b** 1, 8, 27

2 a 81 **b** 125

3 a 4.63 **b** 0.848 **c** 52.4 **d** 110

4 a 29.16 **b** 8.9 **c** 226.981 **d** 4.82

5 a 0.5 **b** 0.2

6 a 36 **b** 81 **c** 121
d The square of a number is the same as the square of the negative of the number.

7 a ±6 **b** ±9

8 a 64 **b** −64 **c** 125 **d** −125
e The cube of a negative number is a negative number.

9 a −4 **b** −5

10 a 17.64 **b** 50.7 **c** 21.9 **d** 8.26
e 4.960

11 a 1.6 **b** 4.7

12 12.96 m^2

13 a 7 m
b A rug cannot have a side less than 0 m in length.

14 a 12 **b** −17 **c** 19 **d** −3
e 22 **f** 11 **g** 8

15 a Adam added 2 and 3 and divided 50 by 5 to get 10; he should have divided 50 by 2 before adding 3.
b The correct answer is 28 ($50 \div 2 + 3 = 25 + 3$).

16 a 10.665 **b** 198.207 **c** 7.24

17 1.22

18 a 7 and 8 **b** 6 and 7
c 9 and 10 **d** 3 and 4

19 a 8 **b** 8
c They both equal 8. This suggests that $\sqrt{x} \times \sqrt{y} = \sqrt{x \times y}$.

20 15

21 a 3 **b** 5
c This suggests that $\sqrt{x} \times \sqrt{x} = x$.

22 a 5.83 **b** 15
c No, as $\sqrt{25} + \sqrt{9} = 8$
Yes, as $\sqrt{25} \times \sqrt{9} = 15$

23 0.999 301 242 4

24 1.0

25 a $2\sqrt{3}$ **b** 3.46 to 2 d.p.

26 a $\sqrt{11}$ **b** $\sqrt{8} = 2\sqrt{2}$ **c** $\sqrt{45} = 3\sqrt{5}$ **d** $\sqrt{8} = 2\sqrt{2}$
e $\sqrt{75} = 5\sqrt{3}$

1.6 Index notation

1 a 9 **b** 8 **c** 72

2 a −1 **b** −3 **c** 4
d 4 **e** 900

3 a 6.5 cm **b** 2.75 m

4 $10^1 = 10$
$10^2 = 10 \times 10 = 100$
$10^3 = 10 \times 10 \times 10 = 1000$
$10^4 = 10 \times 10 \times 10 \times 10 = 10\,000$
$10^5 = 10 \times 10 \times 10 \times 10 \times 10 = 100\,000$
$10^6 = 10 \times 10 \times 10 \times 10 \times 10 \times 10 = 1\,000\,000$

5 2, 4, 8, 16, 32

6 a 3^6 **b** 5^3 **c** 4^5

7 a $2 \times 2 \times 2 \times 2 \times 2 \times 2$
b $5 \times 5 \times 5 \times 5$
c $7 \times 7 \times 7 \times 7 \times 7$

8 a 4^5 **b** $2^3 \times 3^2$ **c** $2^2 \times 5^3$

9 a = **b** = **c** ≠ **d** ≠

10 a i 6561 **ii** 6561 **iii** 6561 **iv** 6561
b The indices add up to 8.
c E.g. $3^4 \times 3^4$

11 a 3^7 **b** 5^5 **c** 7^5 **d** 8^9

12 Ross has multiplied the indices instead of adding them together.

13 a 27 **b** 3^3
c i $\dfrac{3 \times 3 \times 3 \times 3 \times 3 \times 3}{3 \times 3 \times 3} = \dfrac{3^6}{3^3} = 3^3$
ii $3^5 \div 3^3 = \dfrac{3^5}{3^3} = \dfrac{3 \times 3 \times 3 \times 3 \times 3}{3 \times 3 \times 3} = 3^2$
d To divide powers of the same number, subtract the indices.

14 a 4^4 **b** 5^3 **c** 7^2 **d** 2^6

15 a 2^3 **b** 4^2 **c** 5^2 **d** 3^3

16 Multiply the indices.

17 a 3^{10} **b** 5^8 **c** 6^{15} **d** 7^{18}

18 a $100 = 10^2$ **b** $100^4 = (10^2)^4 = 10^8$
   c $100^6 = (10^2)^6 = 10^{12}$
   d The powers of 10 are always double the powers of 100.
   e 10^6

19 $10^3 = 1000$
   $10^2 = 100$
   $10^1 = 10$
   $10^0 = 1$
   $10^{-1} = \frac{1}{10} = 0.1$
   $10^{-2} = \frac{1}{100} = \frac{1}{10^2} = 0.01$
   $10^{-3} = \frac{1}{1000} = \frac{1}{10^3} = 0.001$

20 a 10^2 **b** 10^{-1} **c** 10^{-3} **d** 10^4

21 a 1 million $= 1\,000\,000 = 10^6$
   b 1 billion $=$ 1 thousand million $= 1\,000\,000\,000 = 10^9$
   c 1 trillion $=$ 1 thousand billion $= 1\,000\,000\,000\,000 = 10^{12}$

22

Prefix	Letter	Power	Number
tera	T	10^{12}	1 000 000 000 000
giga	G	10^9	1 000 000 000
mega	M	10^6	1 000 000
kilo	k	10^3	1000
deci	d	10^{-1}	0.1
centi	c	10^{-2}	0.01
milli	m	10^{-3}	0.001
micro	μ	10^{-6}	0.000 001
nano	n	10^{-9}	0.000 000 001
pico	p	10^{-12}	0.000 000 000 001

23 a 1000   **b** 1 000 000   **c** 0.000 001
   d 0.000 000 000 001

24 10^3 gigabytes

1.7 Prime factors

1 3, 5

2 $2^3 \times 3^4$

3 60

4 a 4   **b** 36

5 a
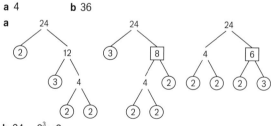

   b $24 = 2^3 \times 3$
   c No, it doesn't matter which factors you start with; you get the same answer in different numbers of steps.

6 a $2 \times 3 \times 3$ **b** $2 \times 3 \times 5$
   c $2 \times 2 \times 2 \times 7$ **d** $2 \times 2 \times 2 \times 3 \times 3$

7 D

8 D

9 a HCF $= 12$; LCM $= 480$
   b HCF $= 12$; LCM $= 216$
   c HCF $= 30$, LCM $= 600$

10 a 20   **b** 600

11 Students' own answers.

12 $3 \times 221 = 663$

13 $90 = 2 \times 3 \times 3 \times 5$

14 15 chocolates per box.

1 Check up

1 S = small, M = medium, G = large,
   A = Americano, L = Latte, E = Espresso
   SA, SL, SE, MA, ML, ME, GA, GL, GE

2 a 31   **b** 29   **c** -5 **d** -7
   e 15   **f** 0.75

3 a $35 \times 24 = 840$ **b** $(32 + 7) \div 3 = 13$

4 a 15.96   **b** 5

5 a $500 \div 50 = 10$
   b $(800 \times 100) \div 40 = 2000$
   c $(90 - 30) \div (9 + 3) = 5$

6 a 77.608   **b** 178

7 a -0.035 **b** 32

8 a 22.72747766   **b** 22.7

9 15.822

10 a $5^3 = 125$ **b** $6 \times 6 \times 6 \times 6 \times 6 = 6^5$
   c $10^3 = 1000$ **d** $\sqrt[3]{64} = 4$

11 a 72   **b** 5   **c** 4

12 a 7^7 **b** 7^2 **c** 7^8 **d** 7^4

13 7, 23; or 11, 19; or 13, 17

14 a 6   **b** 40

15 a $72 = 2^3 \times 3^2$, $96 = 2^5 \times 3$ **b** HCF $= 24$ **c** LCM $= 288$

17 Students' own answers.

1 Strengthen

Calculations

1

2 a 6.3   **b** 15.1   **c** 0.4   **d** 12.0

3 a 11.26   **b** 9.07   **c** 0.64   **d** 28.98

4 a 8.046   **b** 14.173   **c** 0.057   **d** 21.814

5 a 47.823; 40   **b** 0.00572; $\frac{5}{1000}$
   c 432 650; 400 000   **d** 0.6718; $\frac{6}{10}$

6 a 50   **b** 500   **c** 6000   **d** 9000

7 a 14.1   **b** 7.2   **c** 0.0432   **d** 0.0061

8 a 20   **b** 200   **c** 40   **d** 28
   e 23   **f** 50   **g** 10   **h** 6
   i -2 **j** -6

9 a 21   **b** 9   **c** 15   **d** 2

10 a A   **b** C

11 a B   **b** C

12 a 20   **b** 1.85   **c** 15.75   **d** 120
   e 90   **f** 650

13 a $\approx 500 \times 400 = 200\,000$ **b** $\approx \frac{600}{30} = 20$
   c $\approx \frac{300 \times 40}{60} = 200$ **d** $\approx \frac{20 - 5}{3 + 2} = 3$

14 a 226.8   **b** 2268   **c** 2.268   **d** 42

15 a 4.486 089 611   **b** 4.49

Powers and roots

1 a $2^3 = 2 \times 2 \times 2 = 8$ **b** $4^3 = 4 \times 4 \times 4 = 64$
   c $2 \times 2 \times 2 \times 2 = 2^4$
   d $2^2 \times 3^3 = 2 \times 2 \times 3 \times 3 \times 3$

2 a i $\sqrt[3]{64} = 4$ because $4^3 = 4 \times 4 \times 4 = 64$.
  ii $\sqrt[3]{1000} = 10$ because $10^3 = 10 \times 10 \times 10 = 1000$.
  iii $\sqrt[3]{8} = 2$ because $2^3 = 8$.
  iv The cube root of 27 is 3.
  b i 5 **ii** 9 **iii** 1 **iv** 12
3 a i 16 **ii** 16 **iii** 20
  b The two square roots of 16 are $+4$ and -4.
  c i ±3 **ii** ±5 **iii** ±12
4 a $2^{3+4} = 2^7$ **b** 5^5 **c** $4^{2+3} = 4^5$
  d 3^9 **e** $6^{2+3+4} = 6^9$ **f** 10^{10}
5 a $7^{6-3} = 7^3$ **b** $4^{8-2} = 4^6$ **c** 3^3 **d** 6^3
6 a $6^2 \times 6^2 \times 6^2 = 6^6$ **b** $2^5 \times 2^5 = 2^{10}$
  c 3^{12}
7 a 2 **b** 11 **c** 7
8 a 4 **b** 6 **c** 57

Factors, multiples and primes

1 a 4, 5, 6, ... 24
  b 5, 7, 11, 13, 17, 19, 23 **c** Prime numbers
2 a 5, 19; or 7, 17; or 11, 13
  b Yes. Possible pairs include 5 and 19, 7 and 17; or 11 and 13.
3 a 1, 2, 4, 5, 10, 20 **b** 1, 2, 3, 5, 6, 10, 15, 30
  c 1, 2, 5, 10 **d** 10 **e** 14
4 a 6: 6, 12, 18, 24, 30, 36, 42, 48, 54, 60
  b 7: 7, 14, 21, 28, 35, 42, 49, 56, 63, 70
  c 42 **d** 40
5 a

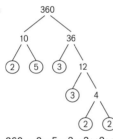

  b $360 = 2 \times 5 \times 3 \times 3 \times 2 \times 2 = 2^3 \times 3^2 \times 5$
  c i $144 = 2^4 \times 3^2$
    ii $396 = 2^2 \times 3^2 \times 11$
    iii $450 = 2 \times 3^2 \times 5^2$
    iv $72 = 2^3 \times 3^2$
    v $84 = 2^2 \times 3 \times 7$
6 a $72 = 2^3 \times 3^2$; $84 = 2^2 \times 3 \times 7$

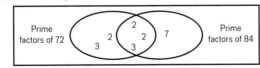

  b 12 **c** 504

1 Extend

1 $2^3 = 2 \times 2 \times 2 = 8$
  $2^2 = 2 \times 2 = 4$
  $2^1 = 2$
  $2^0 = 1$
  $2^{-1} = \frac{1}{2^1} = \frac{1}{2}$ or 0.5
  $2^{-2} = \frac{1}{2^2} = \frac{1}{4}$ or 0.25
  $2^{-3} = \frac{1}{2^3} = \frac{1}{8}$ or 0.125
2 a 0.0048 **b** 12 **c** 20.6 **d** 20.6
3 56

4 a 2 and 25 $(= 27)$ **b** 7 and 1 $(= 8)$
5 a $540 = 2^2 \times 135$ **b** $540 = 3^3 \times 20$
  c $540 = 3^2 \times 60$
6 a 48 **b** 6
7 a $70 + 40 + 20 + 20 + 20 + 20 + 50 + 50 + 80 + 10 =$
  $80 + 70 + (2 \times 50) + 40 + (4 \times 20) + 10 = 380$
  $380 \div 10 = 38$
  This estimated mean is higher than Sally's mean.
  b 38 minutes 30 seconds
8 $(2 + 4) \times 2 = 12$; $2 + 4 \times 2 = 10$; 12 is nearer 13.23 than 10 is, so $(1.6 + 3.8) \times 2.45 = 13.23$
9 6
10 She has rounded down both the cost of a ticket and the number of days she will travel.
11 5, 6, 7, 8 or 9
12 1 p.m.
13 6
14 0.17 lies between 0 and 1 so answer must be greater than 14.314
15 225
16 5
17 a 1 trillion is 10^{12}, so this is correct.
  b E.g. $10^{11} \times 10^1$ or $(10^6)^2$

1 Test ready

Sample student answers

  a They have correctly calculated the number of postcards needed.
  b But the cost of the postcards is incorrect. It should be £43.20. They have not compared this cost with £50.

1 Unit test

1 £12.60
2 Two pay £4.12, one pays £4.11
3 a 13 **b** 38 **c** 12
4 £52.88
5 16
6 2100
7 250
8 a 625 **b** 0.125 **c** 0.4 **d** 4
9 a 4.48 **b** 32 **c** 0.014
10 a 9^8 **b** 9^4 **c** 9^2 **d** 9^{15}
11 4 postcards, 2 greetings cards
12 2 times
13 a $75 = 3 \times 5 \times 5$, $90 = 2 \times 3 \times 3 \times 5$
  b i 15 **ii** 450
14 a 1.69842422 **b** 1.7 (2 s.f.)
15 a 9 **b** 12
16 Students' own answers.

UNIT 2 Algebra

2.1 Algebraic expressions

1 a $2a$ **b** $3h$ **c** x **d** $-3s$
2 a $7a$ **b** $11r$ **c** $-4x$ **d** $-s$
3 14 cm
4 a $9y + 12$ **b** $3t + 7$ **c** $7p + 4q$
  d $4k + 3t$ **e** $2x^2 + 9$ **f** $m^2 + 2$
  g $9r^2 + r$ **h** $x - y + 3$ **i** $2c + 6d - 10$

5 **a** $10x^2+4x+14$ **b** $-6x-5xy+3xz$

6 Ben is correct. Sam incorrectly added the 10 to the $4a$ and $-a$ but these are not all like terms.

7 **a** $3m$ **b** $40c$ **c** $8t$ **d** cd

    e $5ab$ **f** $3st$ **g** $\frac{h}{3}$ **h** $\frac{a}{b}$

8 **a** $4t$ **b** $2x^2$ **c** $-3a+4b+6$

9 **a** $x+6$ **b** $x-7$ **c** $n+5$ **d** $y-3$

    e $12y$ **f** $3m$ **g** $4k$ **h** $\frac{y}{2}$

    i $\frac{d}{2}$ **j** $\frac{t}{4}$

10 **a** B   **b** C   **c** A   **d** D

11 **a** $y-2$ **b** $5y$ **c** $y+3$ **d** $8y+1$

12 **a** $2n$ **b** $4n$ **c** $10n$ **d** xn

13 $4x$

14 $2b+2h$

15 $20b+32t$

2.2 Simplifying expressions

1 **a** 20   **b** -24 **c** -2 **d** -4

2 **a** $3y$ **b** lm **c** $\frac{h}{4}$ **d** $8n$

3 **a** 7^2 **b** 4^3 **c** 5^6 **d** 2^6

4 **a** 3^2 **b** x^2 **c** 4^3 **d** y^3

5 **a** 2^6 **b** x^6 **c** y^{11} **d** z^6

6 **a** $y\times y$ **b** $x^2\times x$ **c** $y^2\times y$

    d e.g. $x^4\times x^3$ **e** $x^5\times x^2$

7 **a** 5^3 **b** y^3 **c** 9^5 **d** x^5

8 **a** a^4 **b** z^4 **c** x^3 **d** g^5

9 **a** $\frac{1}{n^3}$ **b** $\frac{1}{x}$ **c** $\frac{1}{z^2}$

    d $\frac{1}{t}$ **e** $\frac{1}{k^3}$ **f** $\frac{1}{m^2}$

    g $\frac{1}{a^3}$ **h** $\frac{1}{s}$

10 **a** ef **b** $\frac{a}{b}$ **c** $\frac{d^2}{c}$

    d g^2h^3 **e** g^2h **f** gh

    g $\frac{y^3}{x}$ **h** $\frac{y^2}{x}$

11 **a** $4tw$ **b** $3km$ **c** $6st$ **d** $10nq$

    e $12ab$ **f** $14rs$ **g** $36abc$ **h** $18xyz$

    i $-42mn$ **j** $4a^2$ **k** $4t^5$ **l** $48rs^2$

    m $-16a^2b$ **n** $-24a^2b$ **o** $16a^2b$

12 **a** $3n^3$ **b** $6n^3$ **c** $6n^4$ **d** $6n^5$

    e $8a^5$ **f** $8b^9$ **g** $12c^3d$ **h** $12c^3d^4$

13 **a** $3b$ **b** $-4a$ **c** $2z$ **d** $4m$

    e $2n$ **f** $2p$ **g** $\frac{m}{2}$ **h** $\frac{p}{3}$

    i $-2e$ **j** $\frac{3t}{4}$ **k** $\frac{2f}{3}$ **l** $\frac{d}{2}$

14 **a** $7x$ **b** $10ry$

15 **a** $28gh$ **b** p^2 **c** $3x$

16 Jessie is correct.

17 **a** $3a$ **b** $4c^2$ **c** $\frac{1}{4}$ **d** $-5z$

    e $-\frac{p}{6}$ **f** $\frac{3m^2}{4}$ **g** $\frac{4}{n}$ **h** -4

    i Because the powers of x are the same in the numerator and denominator in each expression, so they cancel.

18 **a** x **b** $2x$ **c** x^2 **d** $2x$

    e x^2 **f** $2xy$

19 **a** x^6 **b** y^9 **c** z^{10}

    d $8x^3$ **e** $9x^2$ **f** $25y^2$

    g $25x^4$ **h** $25x^6$ **i** $16a^2$

    j $16a^4$ **k** $16a^4b^2$ **l** $64a^6b^3$

20 **a** $2st^2$ **b** $3uv^2$ **c** $4a^3$ **d** $\frac{2c}{d^2}$

    e $\frac{2}{cd^2}$ **f** $5ef$ **g** $\frac{5}{e}$ **h** $\frac{e}{2f}$

21 **a** n^6 **b** $8a^3b^6$ **c** $3y^3$

2.3 Substitution

1 **a** 2   **b** 42   **c** -7 **d** $-\frac{1}{4}$

2 **a** $\frac{x}{2}$ **b** $2x$ **c** $x-2$ **d** $x+2$

3 **a** 14   **b** 31   **c** 4   **d** 14

4 **a** 7   **b** 3   **c** 10   **d** 20

    e 16   **f** 50   **g** -8 **h** 5

5 36

6 **a** 12   **b** 18   **c** 36   **d** 48

    e 2   **f** -45 **g** -3 **h** 3

7 **a** -20 **b** -7 **c** -6 **d** 3

    e 15   **f** 50   **g** 22   **h** 13

8 **a** $n-20$ **b** $10n$ **c** $\frac{n}{5}$

9 **a** $4n+5$ **b** $2n-6$ **c** $\frac{3n}{4}$ **d** $\frac{n}{2}+4$

    e $\frac{n+4}{2}$ **f** Matty: $\frac{6}{2}+4=7$, Lisa: $\frac{6+4}{2}=5$

10 $6n^2$

11 **a** $c+p$ **b** 8

12 **a** $n-1$ **b** $\frac{n-1}{2}$ **c** 2

13 **a** $3x$ **b** $2y$ **c** $3x+2y$ **d** £2.85

14 **a** k represents each kg over the 20 kg limit.

    b 59.50, 68.50, 86.50, 100

2.4 Formulae

1 **a** 9   **b** 16   **c** 25

2 **a** 29   **b** 2   **c** 75   **d** 34

3 $r+4$

4 £290

5 **a** 17   **b** 2   **c** 17   **d** 32

6 **a** 20   **b** 45

7 **a** $27\,\text{cm}^2$ **b** $75\,\text{cm}^2$ **c** $300\,\text{cm}^2$

8 **a** 25 mph   **b** 35 mph   **c** 60 mph   **d** 350 mph

9 **a** 3 hours   **b** 4 hours   **c** 2 hours

10 Formula; term, expression

11 **a** Expression   **b** Formula   **c** Expression

    d Expression   **e** Formula   **f** Formula

12 **a** $n+10$ **b** $C=n+10$

13 **a** $12x$ **b** $n=12x$

14 **a** $C=x-15$ **b** $C=3x$ **c** $C=\frac{x}{2}$

15 $C=hx$ where C represents the charge.

16 **a** $C=35+20h$ **b** £95

17 $N=\frac{y}{m}$

18 **a** $D=nr$ **b** £56   **c** $S=\frac{t}{2}$ **d** £15

    e £71

    f No; you can use any two different letters to represent the number of hours and hourly rate.

19 **a** $C=3+4k$ **b** £23

20 $L=4x+3$

2.5 Expanding brackets

1 **a** -12 **b** 42 **c** $3b$ **d** $12a$
e x^2 **f** $2a^2$ **g** $-2y^2$ **h** $-6y$

2 **a** $5x$ **b** $4y$

3 **a** $-3y$ **b** $a-3b$ **c** $3m^2+2m$

4 **a** $3\times5+3\times4=3(5+4)$
b $2\times x+2\times7=2(x+7)$
c $6(y+3)=6y+18$
d $3(z+4)=3z+12$

5 **a** $3x+6$ **b** $5a+35$ **c** $4t+36$ **d** $6y+18$

6 **a** $10t-60$ **b** $4x-20$ **c** $7n-14$ **d** $-5m+15$

7 **a** $-2x-6$ **b** $-3a-15$ **c** $-3k+6$ **d** $-2y+2$

8 **a** $6w+3$ **b** $10x+15$ **c** $12t+20$
d $6m+2$ **e** $6m-2$ **f** $10c-5$
g $10d-15$ **h** $-6p-2$ **i** $-12g-15$
j $-12r+15$ **k** $6x-2$ **l** $-2-6x$

9 $12x-18$

10 Benton is correct. Hannah has made the common mistake of forgetting that the negative sign is part of the number term.

11 **a** x^2+x **b** r^2+4r **c** g^2-2g
d $-h^2+3h$ **e** $2m^2-m$ **f** $3t^2-2t$
g $-5a^2-4a$ **h** $2b^2-7b$ **i** $-7c-2c^2$

12 $5x^2+15x$

13 **a** $4(n+3)$ **b** $10(n+5)$ **c** $10(n-5)$
d $2(n+1)$ **e** $2(n-1)$

14 **a** $2(n+3)$
b She could have doubled the number and then added 6.

15 **a** $5(3n+1)=15n+5$ litres
b 305 litres

16 **a** $x+2$ **b** $\frac{x+2}{2}$ **c** 8

17 **a** $12n+25$ **b** $C=12n+25$
c $E=36n+75$ **d** £1155

18 **a** 70 represents the first 70 litres that are used, 5 represents the additional five minutes of water flow, and $x+2$ represents the rate of water flow during these 5 minutes.
b 110 litres

19 **a** 6 **b** 8 **c** 45
d 121 **e** -8 **f** 100
g 36 **h** 6 **i** -16

20 **a** 36 **b** 400 **c** 1 **d** 16

21 **a** $3t+14$ **b** $5m-4$ **c** a^2-6a **d** $-c^2+c$
e $-2m+1$ **f** $10x+14$ **g** $16e-2$

22 **a** $5x+8$ **b** $3n+29$ **c** $k-17$ **d** $5a+6$
e $7d-18$ **f** $7r+7$

23 $12x+9$

24 Students' own answers, e.g. In maths, brackets are used to group terms together, and operations inside brackets should be done first.

2.6 Factorising

1 **a** 2 **b** 3 **c** 10

2 **a** $3x+12$ **b** $5x-30$ **c** x^2+x **d** $2x^2-2x$

3 **a** 3 **b** 4 **c** $a+2$ **d** 12

4 **a** $1\times20, 2\times10, 4\times5$
b $1\times10t, 2\times5t, 5\times2t, 10\times t$
c 10

5 **a** 3 **b** 4 **c** 6

6 **a** $8(y+2)$ **b** $7(m-3)$ **c** $6(y+4)$ **d** $5(2t-5)$

7 **a** Emily
b Phil has not taken the HCF out of the expression. Factorising an expression requires you to 'completely factorise'.

8 **a** $9(x+2)$ **b** $6(w-2)$ **c** $5(3a+2)$ **d** $4(3-5t)$
e $4(2m+5)$ **f** $2(5-2p)$ **g** $3(2-3k)$ **h** $6(v+4)$

9 **i** $12x+20$ **ii** $2(x+4)$

10 **a** $8x+12=4(2x+3)$ **b** $6y+4=2(3y+2)$
c $6z-10=2(3z-5)$

11 **a** d **b** a **c** b **d** x
e n **f** y **g** $3y$ **h** $3y$

12 **a** $6y$ **b** $5a$ **c** $10a$ **d** $2x$
e $3c$ **f** $2x$

13 **a** $n(n+1)$ **b** $k(k-1)$ **c** $r(1-r)$
d $t(1+t)$ **e** $x(4x+1)$ **f** $x(4x+3)$
g $2x(2x+3)$ **h** $2t(s+2)$ **i** $5b(a-4)$
j $5b(a-4b)$ **k** $5a(4b-a)$ **l** $5a(3a-b)$

14 **a** A **b** B **c** B **d** A

15 **a** $x^2(x+1)$ **b** $7x(x^2-3)$ **c** $3x^2(3+4x)$
d $2y(3y^2-1)$ **e** $n(4-n)$ **f** $2n(2+n)$
g $v^2(3v+1)$ **h** $v^2(3v+2)$
i When the terms inside the brackets have no common factors.

16 $5c(2-c)$

17 **b** $2t+4\equiv2(t+2)$ **d** $5t+7\equiv7+5t$

18 **a** $4a+a\equiv5a$ **b** $0.5a\equiv\frac{a}{2}$
c $3(a+4)\equiv3a+12$ **d** $a+2\equiv2+a$

19 **a** Formula **b** Identity **c** Expression
d Identity **e** Expression **f** Formula

20 **a** $\neq$ **b** $\neq$ **c** $\equiv$ **d** $\equiv$

21 Term, identity

2.7 Using expressions and formulae

1 **a** $x-4$ **b** $\frac{x}{2}$ **c** $2x$
d $\frac{x}{3}$ **e** $10x$

2 **a** 8 **b** 8 **c** 5 **d** 7

3 **a** $3r$ **b** $r+20$ **c** $C=4r+20$

4 **a** $5n+2$ **b** $2(n-1)$ **c** $\frac{3n}{2}$
d n^2 **e** n^3 **f** $\sqrt{n}$

5 **a** $3x$ **b** $4y$ **c** $3x+4y$
d $N=px+by$

6 **a i** $m-20$ **ii** $2(m-20)$
b $4m-60$ **c** 180

7 10.8

8 **a** -30 **b** 160

9 2

10 1

11 **a** 15 **b** 80

12 **a** 25 **b** 5

13 10

14 **a** $59\,°F$ **b** $23\,°F$
c $25\,°C=77\,°F$ so Monday was hotter.

15 **a** 120 minutes **b** $M=40w+20$
c 180 minutes **d** 5.20 pm

2 Check up

1 **a** $5a-5b$ **b** m^2-8m **c** $20mn$ **d** $4a^2b$
e $10c^3$ **f** $-8x$ **g** $\frac{15a}{b}$ **h** $9n^8$

2 a 18   **b** 2   **c** 4   **d** −5
   e 32   **f** 30   **g** 64   **h** 16

3 a $x + 5$ **b** $\frac{4x}{5}$ **c** $5(x + 4)$

4 a $2a + 2$ **b** $15f + 10$ **c** $6y^2 - 2y$ **d** $-6a - 10$

5 a $6(x + 2)$ **b** $4x(x + 4)$ **c** $3y(3y^2 + 7)$
   d $5y(3x - 1)$

6 a $\neq$ **b** $\equiv$ **c** $\equiv$ **d** $\neq$

7 a Formula   **b** Identity   **c** Expression

8 a 8 km/h   **b** 23

9 a $T = b + p$ **b** 35

10 a −30   **b** 8

12 a $x^2 + 3x$, $2x^2 + 4x$, $x^2 - 2x$, $3x^2 - x$
   b $7x^2 + 4x$ **c** $x(7x + 4)$

2 Strengthen

Expressions and substitution

1 a $6d + 4e$ **b** $15x + 10y$ **c** $2r + 8s$ **d** $4p - 9q$
   e $-3v - 10w + 4$ **f** $2g + 4h + 10$
   g $10x^2$ **h** $8a^2 + 2a$

2 a $6a$ **b** $4n$ **c** $-4k$ **d** $-2y$

3 a ab **b** fg **c** $2fg$ **d** $3fh$

4 a $10a$ **b** $18y$ **c** $40st$ **d** $6st$
   e $-12pq$ **f** $-12pq$

5 a $2a$ **b** $2c$ **c** $-6b$ **d** $\frac{x}{2}$

6 a $n + 4$ **b** $n - 6$ **c** $2n$ **d** $\frac{n}{3}$
   e $2n + 4$

7 a a^2 **b** $2a^2$ **c** a^5 **d** q^5
   e $2q^5$ **f** $6n^4$ **g** $6n^5$ **h** $15n^6$

8 a 4   **b** a **c** st **d** $\frac{1}{3}$
   e $\frac{x^2}{y}$ **f** $\frac{a}{b}$

9 a x^3 **b** a^4 **c** w^3 **d** 1

10 a $3x^4$ **b** $3f^4$ **c** $3d^2$ **d** $\frac{3}{s}$

11 a $4n^6$ **b** $25t^4$ **c** $16x^6$ **d** $9p^{10}$

12 a 4   **b** 20   **c** 24   **d** 16   **e** 8
   f 12   **g** 75   **h** 100   **i** 2

13 a 7   **b** 5   **c** 49   **d** 18

14 a 2   **b** 6   **c** 8   **d** 4
   e −8   **f** −4   **g** −12   **h** 8

Expanding and factorising

1 a $3x + 6$ **b** $4m + 4$ **c** $5n + 15$ **d** $6h + 24$

2 a $5x - 15$ **b** $2x - 2$ **c** $4k - 20$ **d** $3n - 6$

3 a $6a + 3$ **b** $12b - 4$ **c** $10c + 15$ **d** $15r - 10$

4 a $b^2 + 4b$ **b** $t^2 + 2t$ **c** $2d^2 - 5d$ **d** $2f^2 - f$

5 a $2(b + 3)$ **b** $3(c - 3)$ **c** $2(4 - t)$ **d** $5(1 + 3x)$

6 a b **b** $5d$

7 a $3a(a - 3)$ **b** $4x(4x + 3)$ **c** $5a(a + 3b)$
   d $2q(q^2 + 4)$ **e** $12(7a - 1)$ **f** $a(5a + b)$
   g $y(y^2 + y)$

8 a $\equiv$ **b** $\neq$ **c** $\neq$ **d** $\equiv$

Writing and using formulae

1 a 12   **b** 60   **c** 500

2 a i £100   **ii** £350   **iii** $3a$ **iv** ad
   b $T = ad$ **c** £75

3 a £30   **b** $hx + t$ **c** $E = hx + t$ **d** £70

4 a 14   **b** 35   **c** 4   **d** 11   **e** −6

2 Extend

1 Students' own answers, e.g. $10x + 15x$, $5x \times 5$,
   $5(5x + 1) - 5$, $\frac{50x}{2}$

2 a $100x$ **b** $1000x$

3 a −24   **b** 16   **c** 128   **d** 74

4 a $x^3y + x^3$ **b** $x^2y - xy^2$ **c** $an^3 + an$
   d $2b^2c - 2bc^2$ **e** $3d^3e + 3de^2$ **f** $6gh^2 + 3g^2h$
   g $10p^2q + 15pq^2$ **h** $12st^2 - 20s^2t$

5 a $rs(r + 1)$ **b** $rs(r + s)$ **c** $rs(2r - s)$
   d $2rs(r + 2s)$ **e** $jk(3 - jk)$ **f** $3jk(1 + 2jk)$
   g $2gh(2g - 5)$ **h** $2gh(3g + 5h)$

6 a $7(1 - 3p)$ **b** $3xy(y + 3x)$

7 a $-7(t + 3)$ **b** $-a(a + 5)$
   c $-6f(f + 2)$ **d** $-2r(1 + 2r)$

8 a $3y + 2x$ **b** $T = ay + bx$

9 $3x - 25$

10 6936

11 a £30   **b** The second electrician

12 a a^9 **b** a^{14} **c** a^{m+n} **d** a^3
   e a^2 **f** a^{m-n}

13 5

14 5

15 a n^6 **b** n^6 **c** x^8 **d** $8y^3$
   e $8y^6$ **f** a^{3m} **g** a^{2m} **h** a^{mn}

16 4

17 a 3.0   **b** 0.6

18 a 134   **b** −73

19 a −120   **b** 9   **c** 5   **d** 5.4

20 Students' own answers, e.g. $\frac{qp}{2} = 12$, $\sqrt{prs} = 6$, $qp + s^2 = 28$,
   $\frac{r^2p}{s^2} = 27$

2 Test ready

Sample student answers

a It is a good idea to write out what you know first, then it is clear to see what the girls' individual expressions are.

b So that you don't forget to multiply both terms in the bracket by what is outside.

c You may still get marks for the method even if you make an error with the final answer.

2 Unit test

1 a $5x$ **b** $20m^2$ **c** $4b$ **d** $3xyz$

2 a $4p + 32$ **b** $x^2 - 3x$ **c** $20n - 10$

3 a $12(7a - 1)$ **b** $3a(a - 3)$ **c** $a^2(a + 1)$

4 a $18x - 6$ **b** $5(m + 4)$

5 a 6   **b** 3   **c** 24

6 a $\neq$ **b** $\equiv$

7 a $5x$ **b** $3x - 4$

8 a $n + 9$ **b** $2(n + 5)$

9 $1000t$

10 a $2a + 3c$

11 a Formula   **b** Expression   **c** Identity

12 $E = 1600$

13 a $I = \frac{P}{V}$ **b** 5 amps

14 a £274   **b** $gp + hv$ **c** $C = gp + hv$ **d** £332

15 $4y + 12$

16 a t^6 **b** $27x^6$ **c** $2a^2b$

17 a $x = 3$ **b** $m = 3$

18

$7a+6b$	$2a+b$	$3a+8b$
$7b$	$4a+5b$	$8a+3b$
$5a+2b$	$6a+9b$	$a+4b$

19 Students' own answers.

UNIT 3 Graphs, tables and charts

3.1 Frequency tables

1 a 8   **b** 11

2 a 122   **b** 140

3

Time (seconds)	Tally	Frequency
$12.8 < t \leqslant 13.0$	卌 |	6
$13.0 < t \leqslant 13.2$	||||	4
$13.2 < t \leqslant 13.4$	卌	5

4 Friday should be 17, not 12.

5 a 26   **b** 46

6 Missing value is 3.

7

Shoe size	Tally	Frequency
2–4	卌	5
5–7	卌 卌	10
8–10	||||	4
11–13	|	1

8 a 25, 25.5   **b** 25.5, 26   **c** 26, 26.5, 27

9 a

Age (years)	Tally	Frequency
$15 \leqslant y < 20$	||||	4
$20 \leqslant y < 25$	卌 ||	7
$25 \leqslant y < 30$	卌 |||	8
$30 \leqslant y < 35$	|||	3

b $30 \leqslant y < 35$ **c** 19

10 a Peru, Ghana, India

b Any suitable data collection sheet.

11 a Continuous

b Any suitable grouped frequency table. e.g. Length split as $12 \leqslant y < 14, 14 \leqslant y < 16, 16 \leqslant y < 18, 18 \leqslant y\ 20$

12 Any suitable data collection sheet.

13 Any suitable data collection sheet. e.g. Hours split as 11–14, 15–18, ...

3.2 Two-way tables

1 a 60, 24   **b** 16:25, 7.50 pm

2 a 130 minutes or 2 hours 10 minutes

b 125 minutes or 2 hours 5 minutes

c 11 hours 15 minutes

3 a 4 hours 15 minutes   **b** 13:15

4 a 23 minutes

b Yes. She leaves at 9.50 am and arrives at Hatfield station at 10.03 am. She takes the 10.04 am train, arriving at Kings Cross at 10.41 am. She arrives at work at 10.51 am.

5 a 42 miles   **b** 42 miles   **c** 45 miles

6 a 478 km   **b** $336 - 228 = 108$ km

c Exeter and Edinburgh

7 $5 + 26 + 24 = 55$

8 $24 - 9 = 15$

9 Any suitable two-way table.

10 Any suitable two-way table.

11 a 10   **b** 8

c

	Small	Medium	Large	Total
Pine	7	12	4	23
Oak	10	6	8	24
Yew	3	8	2	13
Total	20	26	14	60

12

	Piano	Guitar	Trumpet	Total
Year 7	20	11	6	37
Year 8	4	7	12	23
Total	24	18	18	60

13 25

14 35

3.3 Representing data

1 a i A   **ii** C

b 24°C

2 a Action   **b** 14   **c** 45

3 a

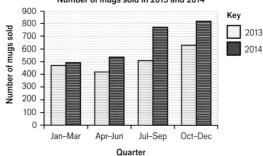

Number of mugs sold in 2013 and 2014

b Oct–Dec 2014

c Students' own answers, e.g. More mugs were sold in 2014 than in 2013.

4 a 100   **b** 380

c Students' own answers, e.g. Website sales had the biggest increase, because both shop and craft fair sales decreased.

5 a

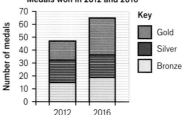

Medals won in 2012 and 2016

b Students' own answers, e.g. In 2016 the team won more medals than in 2012, or, The greatest improvement in 2016 was in the number of gold medals won.

6 a

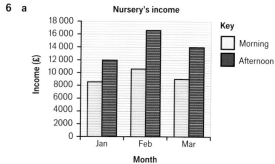

Nursery's income

Key
Morning
Afternoon

b

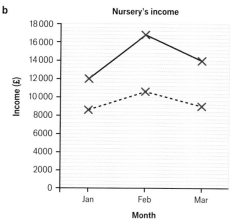

Nursery's income

Key ----- Morning ——— Afternoon

7 a, b, c

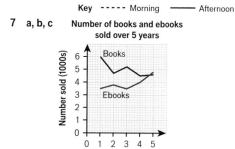

Number of books and ebooks sold over 5 years

d Students' own answers, e.g. Book sales are decreasing, ebook sales are increasing.

8 a 5   **b** 9   **c** 16

9

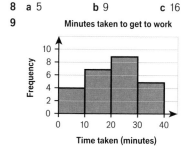

Minutes taken to get to work

10

Minutes taken to get to work

11

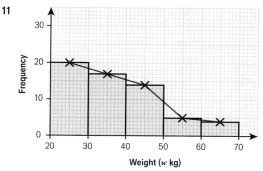

12 a Students' own answers, e.g. A dual bar chart.
  b Students' own answers, e.g. Histogram or frequency polygon.

3.4 Time series

1 a 1 600 000   **b** 0.16 million

2 a 45 metres   **b** 32 seconds
  c Students' own answers, e.g. No, as the line is not straight.

3 a

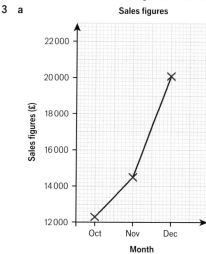

Sales figures

b Lucy is not correct. Her sales figures are increasing but NOT at a steady rate, i.e. the differences between the months are not the same.

c Students' own answers, e.g. Her sales figures are rising at an increasing rate, so are likely to increase in January.

4 a

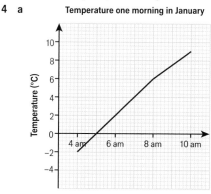

Temperature one morning in January

b 7.6°C

5

Year	2016	2017	2018	2019
Number of visitors (thousands)	4.75	4.4	5.2	3.9

6 a

Month	March	April	May
Money (£)	5800	2600	4500

b

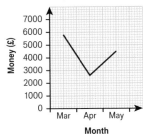

Money received each month

7 a

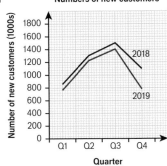

Numbers of new customers

b Students' own answers, e.g. The number of customers increases from Q1 to Q3 then decreases in Q4.

8 a

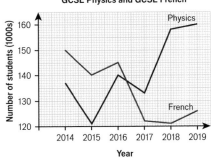

Numbers of students taking GCSE Physics and GCSE French

b Overall trend is that numbers taking French GCSE are decreasing, but numbers taking Physics GCSE are increasing.

c Sally is correct. The numbers taking GCSE Physics have been greater than the numbers taking GCSE French for the last three years shown (2017, 2018, 2019). There has also been a big difference in the numbers for each subject in 2018 and 2019, and there is likely to remain a big difference even if the French numbers continue to increase slightly, as they did in 2018 to 2019.

9 Students' own answers, e.g. The first label on the y-axis should be 5.5, not 5, or data is missing for Q4, or there should be straight lines between data points, not a curve.

3.5 Stem and leaf diagrams

1 a 123, 123, 125, 125, 132, 135, 143, 146
   b 1.4, 3.2, 3.9, 3.9, 4.1, 6.7, 7.5, 8.4

2 a 6   **b** 34 m
3 a 19   **b** 9   **c** 39 mm
4

```
19 | 2 8
20 | 5
21 |
22 | 0 0
23 | 0 5 8 8 8
```
Key: 19 | 2 means 192 seconds

5

```
0 | 8
1 | 0 2 3 5 7 8
2 | 0 1 2 2 2 3 3
3 | 1 3 4 5
4 | 4 5 6
```
Key: 4 | 4 means 44 minutes

6

```
0 | 9
1 | 4 6 6 7
2 | 0 1 2 4 6 7
3 | 0 3 8 8
```
Key: 1 | 4 means 14 marks

7

```
3 | 4 9 9 9
4 | 6 7 7 8 9
5 | 1
```
Key: 5 | 1 means 5.1 metres

8 a 75 bpm   **b** 107 bpm   **c** 23 bpm
   d Students' own answers, e.g. Malek is correct. More males than females had a heart rate of more than 90 bpm. Nine males did (92, 93, 95, 97, 98, 99, 101, 103 and 107) whereas only 4 females did (91, 96, 98 and 99).
   e Students' own answers, e.g. There are more figures in the last two lines of the diagram on the male side than on the female side.

9 a

```
Abbey Hotel    |   | Balmoral Hotel
            2  | 0 |
      9 7 5 1  | 1 |
        9 8 3 1 | 2 | 6
9 9 9 9 7 6 6 5 3 3 2 | 3 | 4 4 7
      9 8 7 7 5 0 | 4 | 0 0 5 5 6 9
              8  | 5 | 0 0 0 0 1 3 6 6 7
                 | 6 | 2 3 3 4 5 7
                 | 7 | 0 1 5
```
Key: For Abbey Hotel 9 | 1 means 19 years old
      For Balmoral Hotel 7 | 0 means 70 years old

b Students' own answers, e.g. She is correct. Balmoral Hotel has several residents over 50 years old, whereas Abbey Hotel only has one resident over the age of 50.

10

```
Boys      |    | Girls
        9 | 14 | 6 8
    6 5 3 | 15 | 1 1 2 5 6 7
  5 5 5 3 | 16 | 4 9
      2 0 | 17 |
```
Key: For boys 6 | 15 means 156 cm
      For girls 14 | 6 means 146 cm

11 a

```
Dark       |   | Light
9 9 8 7 6 5 | 1 |
        2 0 | 2 | 7 8
            | 3 | 5 8
            | 4 | 1 3 4 5
```
Key: For dark 9 | 1 means 1.9 cm
      For light 3 | 8 means 3.8 cm

b Students' own answers, e.g. They grew better in the light because those seedlings grew taller.

3.6 Pie charts

1 a $\frac{1}{4}$ **b** $\frac{1}{2}$

2 360

3 a, b, c Accurate drawing of a circle with radius 4 cm and an angle of 60° at the centre.

4 a $\frac{1}{4}$

b Students' own answers, e.g. The percentage of householders who thought waste collection was poor has reduced from 50% in 2018 to 25% in 2019.

c Students' own answers, e.g. Yes, because the proportion of householders who answered poor in 2014 has fallen and the proportion who answered good has increased.

5 a 25%

b 45° (allow 43–47°) which equates to 12.5% of the students (accept 12–13%).

c Students' own answers, e.g. Twice the yes vote is 50%. Don't know is more than 50%, so Bethan is correct.

6 a 5 games.

b No; Team A won 10 games and Team B won 14 games.

7 a 90   **b** 4°

c Hot chocolate 80°; milkshake 60°, coffee 100°, tea 120°

d Accurate pie chart drawn with hot chocolate 80°, milkshake 60°, coffee 100°, tea 120°.

8 Accurate pie chart drawn with France 144°, Spain 108°, Germany 45°, Italy 63°.

9 Accurate pie chart drawn with gym 144°, swimming 36°, squash 72°, aerobics 108°.

10 a 60   **b i** 12   **ii** 6

11 a 90°   **b** 32   **c** Non-fiction

12 a Accurate pie chart drawn with spotty 120°, plain 80°, striped 160°.

b

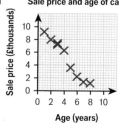
Kato's sales figures

c i The bar chart   **ii** The pie chart
iii Either

3.7 Scatter graphs

1 a, b, c

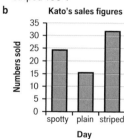
Number of bedrooms and bathrooms in four houses

2 a

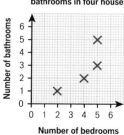
Sale price and age of cars

b Decreases

3 a

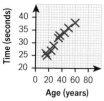
Time and age of 200 m runners

b Increases

4 Positive correlation (Q1), negative correlation (Q2), positive correlation (Q3).

5 a B   **b** C   **c** A

6 a Negative correlation   **b** 2

7 a

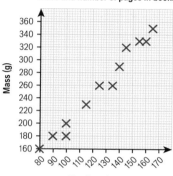
Mass and number of pages in books

b Increases; positive

8 a

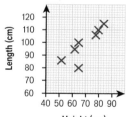
Height and length of sheep

b (65, 80)   **c** Positive correlation

9 a Negative

b Students' own answers, e.g. It is far away from the other points.

10 a

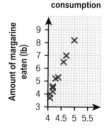
Divorces and margarine consumption

b Yes   **c** Students' own answers, e.g. No.

11 Arm length and leg length: positive correlation; possible causation

Exercise and weight: negative correlation; possible causation

Size of garden and running speed: no correlation

Hours of TV watched and shoe size: no correlation

3.8 Line of best fit

1 **a** Positive   **b** Negative

2 **a** 6   **b** 2.5

3 **a, b**

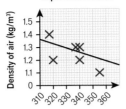

Experiment results

4 **a**

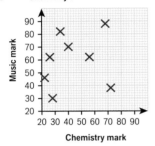

Chemistry and music exam results

b No correlation

c Students' own answers, e.g. No, a line of best fit needs a correlation, either positive or negative.

5 **a, b**

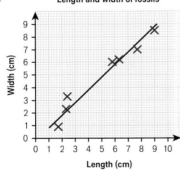

Length and width of fossils

c Allow 3.6–4.2 cm.

6 **a**

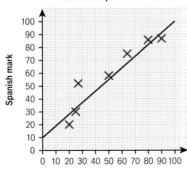

French and Spanish test results

b (27, 52) is an outlier. Students own explanations, e.g. These marks do not fit the pattern as the Spanish mark is a lot higher than the French mark.

c As the French mark increases the Spanish mark increases. Positive correlation.

d Allow 60–68 marks

7 **a**

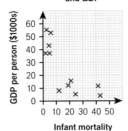

Infant mortality and GDP

b Students' own answers, e.g. The data supports her statement.

c Students' own answers, e.g. More data would allow for more confidence in any apparent correlation.

8 **a, c**

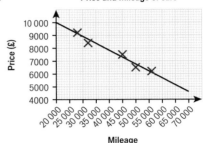

Price and mileage of cars

b Negative correlation

d i Allow £7600 to £8000   **ii** Allow £5000 to £5400

e Students' own answers, e.g. The estimate for 40 000 miles is more reliable because this is within the range of the data, whereas the estimate for 65 000 miles is an extrapolation for a point beyond the range of the data.

9 **a** (25, 70)   **b** Positive   **c** Allow 50 to 54

d Yes, there is a positive correlation between the marks for maths and science.

3 Check up

1

Height (m)	Tally	Frequency
$1.2 \leqslant h < 1.3$	|||	3
$1.3 \leqslant h < 1.4$	ⵏⵏ	5
$1.4 \leqslant h < 1.5$	||||	4

2 **a** Huntingdon   **b** 3 minutes   **c** 10:05

3 **a** 19

b

	Budget seats	Standard seats	Luxury seats	Total
Adult	15	17	19	51
Child	24	25	30	79
Total	39	42	49	130

4 Accurate pie chart drawn with perch 50°, bream 115° and carp 195°.

5 a

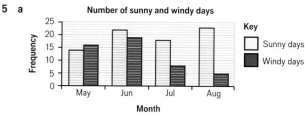

Number of sunny and windy days

Key: Sunny days, Windy days

b Students' own answers, e.g. There are more sunny days than windy days during the four-month period.

6 1 – Marc won 40 games and Caroline 39.

7 a 102 bpm   **b** $65 - 58 = 7$ bpm

8 a 5.5 cm   **b** Approx. 6.0 cm   **c** Approx. 2 hours 45 min

9 a (11, 50)   **b** Positive correlation   **c** Allow 64 to 66

11 Any suitable data collection sheet

3 Strengthen

Tables

1 a $9 \leqslant l < 11$

b

Length, l (cm)	Tally	Frequency
$7 \leqslant l < 9$	||	2
$9 \leqslant l < 11$	||	2
$11 \leqslant l < 13$	|||| |	6
$13 \leqslant l < 15$	||||	5

2 7 hours 5 minutes

3 a 34 minutes   **b i** 10:13   **ii** 10:29

4

	White	Black
Circle	3	4
Square	6	5

5 a Girls: 46, car: 33, bicycle: 30   **b** 22   **c** 25

d

	Walk	Car	Bicycle	Total
Boys	15	25	14	54
Girls	22	8	16	46
Total	37	33	30	100

6 a, b

	London	York	Total
Girls	19	24	43
Boys	23	14	37
Total	42	38	80

Graphs and charts

1 a

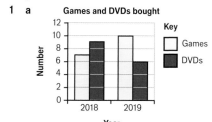

Games and DVDs bought

Key: Games, DVDs

b Students' own answers, e.g. In 2018 he bought more DVDs than games. In 2019 he bought more games than DVDs.

2 a

Daily hours of sunshine

Key: Majorca, Crete

b Students' own answers, e.g. In April and May, Majorca has more sunshine hours. In July and August, Crete has more.

3 a 30   **b** 12°   **c** 120°

d Green 60°, yellow 72°, orange 108°

e Accurate pie chart drawn with red 120°, green 60°, yellow 72°, orange 108°. All labelled correctly.

4 a 103, 105, 108, 110, 112, 113, 114, 117, 119, 121, 123, 125, 125, 125, 127

b

```
10 | 3 5 8
11 | 0 2 3 4 7 9
12 | 1 3 5 5 5 7
```
Key: 10 | 3 = 103 grams

Time series and scatter graphs

1 a

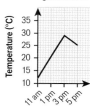

Temperature in a greenhouse

b 2 pm and 5 pm   **c** 27 °C

2

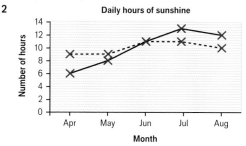

Daily hours of sunshine

Key: ---- Majorca —— Crete

3 a, b

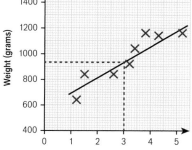

Lobsters – age and weight

c 930 kg

d i 1160–1180 kg   **ii** 800–820 kg

3 Extend

1 a i 20 **ii** 18

b

2 Students' own answers, e.g. The vertical axis is not labelled; the scale on the vertical axis does not have equal divisions; the shapes used for the bars are different; using 3D shapes for bars makes it difficult to read values.

3 He is not correct. $50 + 110 + 240 + 140 = £540$
$\frac{540}{12} = £45$, so he needs to pay £45 per month.

4 a

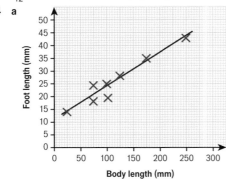

b Positive correlation

c Students' own answers, e.g. No, the point is outside of the trend of the data.

d Students' own answers, e.g. The point is outside the range of the scatter diagram.

5 a 58%   **b** 54%

c Students' own answers, e.g. No, the chart tells us the about the proportion of mice in the two areas. It does not tell us anything about the numbers of mice.

6 Accurate bar chart or line graph.

3 Test ready

Sample student answers

a Student A, because the data shows changes over a period of time so the graph should be a time series graph.

b Student A would get 1 mark for the suitable graph, but would not get the second mark because they did not label the vertical axis Temperature (°C). Student B got 0 marks because they did not label the vertical axis or draw a suitable graph.

3 Unit test

1 a 9 minutes   **b** 78 minutes or 1 hour 18 minutes

2 a 75

b False, the graph only specifies numbers not individuals.

3 a i 573 km   **ii** 662 km

b Caen and Nantes

4 a 29   **b** 5

c

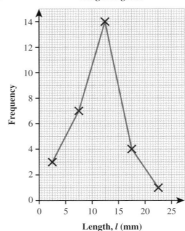

5 a (left to right:) 66%, 14%

b

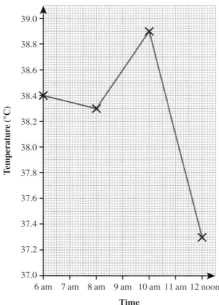

6 a

b 38.6 °C

c The patient's temperature fell slightly, then rose, then fell to below the starting temperature.

7

8 a

Jigsaw		**Percy**
9	4	
7 1 0	5	5 9 9
8 6 3	6	0 2
0	7	3 4
	8	1

Key: Jigsaw 9 | 4 means 49 points
Percy 6 | 0 means 60 points

b Jigsaw had a median of 60 points while Percy had a median of 61 points, so Percy had a higher average score. The range for Jigsaw was 21 while the range for Percy was 26, so Jigsaw's scores were more consistent.

9 51 students

10 The line of best fit does not fit the data; 1.7 is missing from horizontal axis.

11 Students' own answer.

12 Students' own answer.

UNIT 4 Fractions and percentages

4.1 Working with fractions

1 a 9   **b** 5

2 a >   **b** <

3 a $\frac{4}{6}$ **b** $\frac{12}{27}$ **c** $\frac{8}{20}$ **d** $\frac{4}{6} = \frac{8}{12}$

4 a $\frac{3}{4}$ **b** $\frac{2}{3}$ **c** $\frac{3}{5}$ **d** $\frac{2}{3}$

5 a 4   **b** 10   **c** 24   **d** 12

6 a $\frac{2}{4}, \frac{3}{4}$ **b** $\frac{7}{10}, \frac{6}{10}$ **c** $\frac{3}{24}, \frac{20}{24}$

7 Yes, because $\frac{1}{3} = \frac{3}{9}$ and $\frac{4}{9} > \frac{3}{9}$

8 $\frac{3}{5} = \frac{9}{15}, \frac{2}{3} = \frac{10}{15}$, so $\frac{2}{3}$ is larger.

9 Spinner A as on spinner A P(red) $= \frac{15}{24}$ while on spinner B P(red) $= \frac{14}{24}$

10 $\frac{8}{40} = \frac{1}{5}$. All the other fractions simplify to $\frac{4}{5}$

11 a $\frac{7}{12}, \frac{2}{3}, \frac{5}{6}$ **b** $\frac{1}{5}, \frac{1}{2}, \frac{3}{5}, \frac{2}{3}, \frac{3}{4}$

12 $\frac{4}{7}, \frac{7}{10}$

13 $\frac{3}{7}, \frac{1}{2}, \frac{7}{12}, \frac{5}{6}$

14 a Students' own answers, for example: When the numerators are the same, the smaller fraction has the greater denominator.

b Students' own answers, for example $\frac{8}{9}$ is larger because
$\frac{7}{8} = \frac{63}{72}, \frac{8}{9} = \frac{64}{72}$ and $\frac{64}{72} > \frac{63}{72}$

15 a $\frac{7}{8}$ **b** $\frac{11}{15}$ **c** $\frac{1}{4}$ **d** $\frac{5}{9}$

    e $\frac{1}{2}$ **f** $\frac{2}{3}$ **g** $\frac{1}{3}$ **h** $\frac{3}{5}$

16 a $\frac{7}{12}$ **b** $\frac{11}{12}$ **c** $\frac{3}{10}$

    d $\frac{1}{10}$ **e** $\frac{1}{12}$ **f** $\frac{7}{20}$

17 Change one denominator when one is a multiple of the other. Change both denominators when one is not a multiple of the other.

18 a $\frac{1}{2}$ **b** $\frac{3}{4}$ **c** $\frac{7}{12}$ **d** $\frac{2}{5}$

19 a $\frac{2}{3}$ **b** $\frac{2}{5}$ **c** $\frac{2}{7}$ **d** $\frac{3}{10}$

20 $\frac{7}{10}$

21 a $1\frac{3}{4}$ **b** $1\frac{1}{4}$ **c** $2\frac{3}{4}$

22 $\frac{8}{15}$

23 a $\frac{13}{24}$ **b** $\frac{11}{12}$ **c** $\frac{7}{12}$ **d** $\frac{11}{16}$

24 a $\frac{4}{7}$ **b** $\frac{2}{5}$ **c** $\frac{28}{63}$ **d** $\frac{3}{28}$

4.2 Operations with fractions

1 a 100   **b** 1000   **c** 60

2 a $\frac{1}{2}$ **b** $\frac{5}{8}$ **c** $\frac{1}{24}$

3 a $3\frac{1}{2}$ **b** $1\frac{2}{9}$ **c** $8\frac{1}{3}$ **d** $2\frac{1}{12}$

4 a $\frac{5}{2}$ **b** $\frac{19}{4}$ **c** $\frac{43}{6}$ **d** $\frac{21}{2}$

5 a 2   **b** 4   **c** 4

    d 28   **e** 15   **f** 64

6 a 150 kg   **b** 40 cm   **c** 1 m

    d 1 kg   **e** 36 minutes

7 45

8 £121

9 20

10 a Students' own answers.   **b** £30

    c Students' own answers, for example:
$(1 - \frac{2}{3}) \times £90 = \frac{1}{3} \times £90 = £90 \div 3 = £30$

11 490

12 Students' own answers, for example:
Total frequency $= 30 + 15 + 5 + 10 = 60$
Goats $= \frac{10}{60} = \frac{1}{6}$

13 a 150   **b** Answers between 220 and 260

14 a $\frac{8}{12} + \frac{9}{12} = \frac{17}{12} = 1\frac{5}{12}$ **b** $1\frac{7}{40}$

    c $1\frac{6}{35}$ **d** $1\frac{11}{24}$

15 Students' own answers

16 Students' own answers, for example:
$1 - \frac{7}{11} = \frac{4}{11}, \frac{11}{7} - 1 = \frac{4}{7}, \frac{4}{11} < \frac{4}{7}$ so $\frac{7}{11}$ is closer to 1.

17 a $5\frac{3}{4}$ **b** $13\frac{1}{4}$ **c** $11\frac{3}{4}$ **d** $5\frac{5}{6}$

    e $3\frac{7}{20}$ **f** $7\frac{1}{12}$ **g** $3\frac{5}{12}$ **h** $4\frac{13}{30}$

18 Yes, Crista's method works when adding mixed numbers.

19 a $4\frac{3}{4}$ **b** $1\frac{1}{2}$ **c** $2\frac{1}{2}$

    d $1\frac{5}{12}$ **e** $3\frac{1}{10}$ **f** $1\frac{7}{20}$

    g $1\frac{3}{7}$ **h** $\frac{3}{4}$ **i** $3\frac{5}{6}$

    j Students' own answers, for example: Yes, Crista's method works when subtracting mixed numbers but if the second fraction is greater than the first fraction, the result will be a negative fraction part.

20 Top to bottom: $7\frac{1}{4}, 1\frac{3}{4}$

4.3 Multiplying fractions

1 a and **c**; **b** and **d**; **e** and **f**

2 a $7\frac{1}{2}$ **b** $\frac{18}{5}$

3 **a** $\frac{7}{2} = 3\frac{1}{2}$ **b** $\frac{36}{3} = 12$ **c** 5 **d** 10

4 $5 \times \frac{2}{3} = 3\frac{1}{3}$, $\frac{1}{3}$ of an hour is 20 minutes, so Scott will finish at 12.20 am, after midday.

5 **a** £345

b Yes

Students' own answers, for example: Because a reduction of $\frac{1}{4}$ means it is $\frac{3}{4}$ of the original price.

6 **a** 4 **b** 4 **c** 3 **d** 1.5

7 **a** $\frac{1}{8}$ **b** $\frac{6}{35}$

8 **a** $\frac{6}{49}$ **b** $\frac{8}{63}$ **c** $\frac{1}{5}$ **d** $\frac{1}{2}$

9 Students' own answers, for example: Yes the rule works because every integer has a denominator of 1.

10 **a** $\frac{2}{5}$ **b** $\frac{7}{30}$ **c** $\frac{5}{14}$

d $\frac{2}{9}$ **e** $\frac{2}{9}$ **f** $\frac{13}{28}$

g Students' own answers, for example: It is easier to rearrange and simplify first to help keep numbers in the numerator and denominator smaller.

11 $\frac{1}{6}$

12 No. $\frac{3}{4} \times \frac{1}{3} = \frac{1}{4}$, $\frac{1}{4}$ kg = 250 g, Garcia only has $\frac{3}{4} - \frac{1}{4} = \frac{1}{2}$ kg = 500 g

13 **a** $5\frac{1}{4}$ **b** $12\frac{6}{7}$ **c** $17\frac{1}{2}$ **d** $35\frac{1}{5}$

14 Students' own answers

15 $7\frac{1}{2}$ hours

16 £180

4.4 Dividing fractions

1 **a** 4 **b** 2 **c** 3

2 **a** $\frac{15}{4}$ **b** $6\frac{1}{3}$

3 **a** 12 **b** 24 **c** $\frac{6}{11}$ **d** $\frac{10}{3} = 3\frac{1}{3}$

4 **a** $\frac{4}{3}$ **b** $\frac{2}{7}$ **c** 5 **d** $\frac{1}{4}$

e Students' own answers, for example: It gives the result 1.

5 **a** $4 \times 3 = 12$ **b** 25

c 96 **d** $5\frac{1}{3}$

6 2

7 **a** $5 \times \frac{2}{3} = \frac{10}{3} = 3\frac{3}{3}$ **b** 9

c $\frac{70}{11} = 6\frac{4}{11}$ **d** $\frac{80}{9} = 8\frac{8}{9}$

8 **a** $\frac{14}{5} = 2\frac{4}{5}$ **b** $\frac{50}{12} = 4\frac{1}{6}$

c $\frac{24}{16} = 1\frac{1}{2}$ **d** $\frac{48}{11} = 4\frac{4}{11}$

9 **a** 6 **b** $2\frac{1}{7}$

c $5\frac{1}{3}$ **d** 20 **e** 10

10 $80 \div \frac{4}{5} = 100$ cushions

In 4 weeks, he will make $4 \times 30 = 120$ cushions, so he does not have enough fabric.

11 30 is 30×1, not 10×1.

12 **a** $\frac{1}{6}$ **b** $\frac{1}{9}$ **c** $\frac{1}{5}$

13 **a** $\frac{5}{4} \div 2 = \frac{5}{8}$ **b** $\frac{7}{3} \div 5 = \frac{7}{15}$ **c** $\frac{15}{4} \div 8 = \frac{15}{32}$

14 Answers to Q13 checked using calculator.

15 1375 m

16 **a** $\frac{14}{35} = \frac{2}{5}$ **b** 24 minutes

17 **a** $\frac{3}{2} = 1\frac{1}{2}$ **b** $\frac{15}{4} = 3\frac{3}{4}$ **c** $\frac{14}{10} = 1\frac{2}{5}$

4.5 Fractions and decimals

1 7.00, 7, 7.000 000

2 **a** 0.3 **b** $\frac{2}{5}$ **c** $\frac{12}{25}$

3 **a** −0.43, 0.349, 0.39, 0.4 **b** 0.4

4 **a** 1.25 **b** $4.1\dot{6}$ **c** 0.6 **d** 1.5

5 **a** $\frac{5}{12}$ **b** $\frac{2}{3}$

6 0.125

7 0.28

8

Fraction	$\frac{1}{1000}$	$\frac{1}{100}$	$\frac{1}{10}$	$\frac{1}{8}$	$\frac{1}{5}$	$\frac{1}{4}$
Frequency	0.001	0.01	0.1	0.125	0.2	0.25

9 **a** $2 \times 0.2 = 0.4$ **b** 0.375 **c** 1.25

10 **a** 2.5 **b** 37.5 **c** 0.5

11 0.625, 0.7, 0.49, so $\frac{49}{100}$ is closest to $\frac{1}{2}$

12 **a** $\frac{61}{100}$ **b** $\frac{39}{50}$ **c** $\frac{229}{1000}$

d $\frac{9}{20}$ **e** $\frac{12}{125}$

13 **a** $0.6, \frac{5}{8}, 0.628, \frac{2}{3}$ **b** $-4.5, -\frac{17}{4}, \frac{5}{2}, 2.8$

14 **a** Ed pays £84.38; Sam pays £140.63 **b** Yes

c Round up

15 $\frac{8}{15}, \frac{7}{12}, \frac{2}{3}, \frac{17}{25}$

16 $\frac{150}{360} = \frac{15}{36} = \frac{5}{12}$

17 $\frac{18}{60} = \frac{3}{10} = 0.3$ hours

4.6 Fractions and percentages

1 **a** 100 **b** 1000

2 0.5

3 **a** $\frac{35}{100} = \frac{7}{20}$, 35% **b** 20%

c 30% **d** 25%

4 **a** 40 **b** 70

5 **a** $\frac{2}{25}$ **b** $\frac{6}{25}$ **c** $\frac{13}{20}$ **d** $\frac{24}{25}$

6 $\frac{7}{20}$

7 30% is $\frac{3}{10}$

8 **a** $\frac{14}{100} = 14\%$ **b** 12%

c 80% **d** 9%

9 $16\% = \frac{16}{100}$, $\frac{3}{20} = \frac{15}{100} = 15\%$

10 **a** 40% **b** 60%

c Students' own answer, for example: They add to 100%.

d $100\% - 40\% = 60\%$

11 **a** 30% **b** 40% **c** $6.\dot{6}\%$

d 35% **e** 5%

f Students' own answers, for example: I converted the fraction to denominator 100 when the denominator was a factor of 100.

12 Maths: he scored 70% for maths and 80% for spelling.

13 $\frac{64}{80} = 80\%$; Jill scored higher.

14 **a** $\frac{1}{3}$ **b** 33%

15 **a** $\frac{35}{75} = 46.7\%$

b Students' own answers, for example, Half of 35 is 17.5; Kia + Audi is 20.

16 93.5%

4.7 Calculating percentages 1

1 **a** £40 **b** £120 **c** £80 **d** £40
   e **a** and **d**, as $\frac{1}{4} = 25\%$

2 **a** 0.38 **b** 71 **c** 20 **d** 60
   e 120 **f** 0.061 **g** 0.084 **h** 0.005

3 **a** 40 **b** 45 **c** 90 **d** 224

4 **a** 1.5 years **b** 3.25 years

5 **a** 0.65 **b** 0.42 **c** 0.07

6 **a** Fraction–decimal–percentage triangle showing
   $\frac{1}{4}$, 25%, 0.25
   b Fraction–decimal–percentage triangle showing
   $\frac{4}{5}$, 80%, 0.8
   c Fraction–decimal–percentage triangle showing
   $\frac{3}{5}$, 60%, 0.6

7 **a**
   decimal $\longrightarrow$ $\boxed{\times 100}$ $\longrightarrow$ percentage

   b i 99% **ii** 9% **iii** 90% **iv** 0.9%

8 **a** 50%, 55%, 59%, 60%, 65%
   b 0.5%, 1%, 26%, 50%

9 0.033

10 **a** $\frac{1}{5}$ **b** 300

11 **a** 360 **b** 210 **c** 270 **d** 10

12 33 g

13 3 days

14 **a** 0.2 **b** 60

15 **a** 225 **b** 210 **c** 14

16 £8.75

17 **a** 1) 25% of 140 = 35, 140 − 35 = 105
       2) 100% − 25% = 75%, 75% of 140 = 105
    b Students' own answers.

18 75% of 1050.60 = 787.95
    787.95 ÷ 12 = £65.66
    The council are NOT charging correctly.

19 £285

20 **a** 1.3 **b** 1.2 **c** 3.5

21 50%

22 **a** £120 **b** 250 ml **c** £1750

23 £9200

24 **a** £82.50 **b** £24.38 **c** £132

25 £2140

26 **a** £20 **b** 1%

27 Same answers as Q24.

4.8 Calculating percentages 2

1 **a** 110% **b** 60% **c** 30%

2 **a** 0.2 **b** 3.2 **c** 0.03 **d** 0.235

3 **a** 21 **b** 150 **c** 175

4 **a** 39.6 **b** 532 **c** 505 **d** 2002

5 £41.20

6 130% of 240 is 312. Her wage was only £300.
   Wage has not increased with the cost of living.

7 **a** 36 **b** 330 **c** 423 **d** 4970

8 £180

9 Singh family's food bill is higher (£130 × 0.965 = £125.45
   compared with £120 × 1.02 = £122.40).

10 **a** £4.50 **b** £22.50 **c** £22.50
    d Students' own answers, for example, They are the same.

11 **a** 1.37 **b** 1.2 **c** 1.75 **d** 1.12
    e 1.5 **f** 1.05 **g** 1.875 **h** 1.075

12 **a** 2.12
    b Students' own answers, for example, An increase of 12%
       would have a multiplier of 1.12, so an increase of 112% has
       a multiplier of 2.12.

13 **a** 144 **b** 525 **c** 9375 **d** 3.6

14 **a** 0.6 **b** 0.65 **c** 0.94 **d** 0.988

15 **a** 120 **b** 325 **c** 122.2 **d** 3952
    e Students' own answers, for example, It cannot be
       less than 0.

16 £54.60

17 £8330

18 4% would be 1.04, not 1.4

19 £18.70

20 Paint R Us; 9 tins cost £13.14 at Paint R Us and £14.58 at
    Deco Mart.

21 £40

22 £1288

23 Quote 2 is £600 including VAT. Quote 1 is more expensive.

24 **a** £356.40
    b Sarah and Peter get the same answer.
       Sarah: £99 × 1.2 = £118.80 for one ticket;
       3 × £118.80 = £356.40 for 3 tickets.
       Peter: 3 × £99 = £297; £297 × 1.2 = £356.40.

25 Six winners

26 £1610.00

4 Check up

1 **a** $\frac{8}{15}$ **b** $\frac{5}{8}$ **c** $\frac{1}{4}$ **d** $\frac{19}{24}$

2 **a** $\frac{6}{25}$ **b** $\frac{4}{5}$ **c** $6\frac{2}{5}$ **d** $2\frac{4}{9}$

3 **a** $\frac{32}{3} = 10\frac{2}{3}$ **b** $\frac{3}{32}$ **c** $\frac{3}{2} = 1\frac{1}{2}$ **d** $\frac{1}{3}$

4 **a** $1\frac{1}{24}$ **b** $5\frac{7}{8}$ **c** $2\frac{1}{4}$ **d** $1\frac{5}{12}$

5 **a** $\frac{7}{1000}$ **b** $\frac{13}{40}$

6 $\frac{36}{60} = \frac{3}{5}$

7 0.44

8 **a** $\frac{1}{20}$ **b** $\frac{7}{25}$ **c** $\frac{3}{2}$

9 12%

10 58%, $\frac{3}{5}$, 0.62, $\frac{2}{3}$

11 $\frac{5}{12}$

12 **a** 13% **b** Greater (15%)

13 £9

14 **a** 18 cm **b** 18 km **c** £106.60

15 £3.30

16 £78

17 £325

18 £10 200

19 £43.75

21 **a** £2341.67 **b** £632.25

22 **a i** 24% **ii** 30% **iii** 16% **iv** 34%
    b *B*, *C* and *E*

4 Strengthen

Operations with fractions

1 a i $\frac{2}{4}$ ii $\frac{3}{4}$

 b i $\frac{2}{8}$ ii $\frac{5}{8}$

 c i $\frac{3}{9}$ ii $\frac{5}{9}$

 d i $\frac{2}{6}$ ii $\frac{1}{6}$

2 a $\frac{7}{24}$ b $\frac{13}{40}$ c $\frac{1}{6}$ d $\frac{7}{30}$

3 a $\frac{7}{12}$ b $\frac{19}{20}$ c $\frac{17}{24}$ d $\frac{3}{20}$

4 a $1\frac{1}{4}$ b $1\frac{2}{9}$ c $1\frac{1}{20}$ d $1\frac{5}{14}$

5 a $\frac{11}{6}$ b $\frac{5}{2}$ c $\frac{15}{6} - \frac{11}{6} = \frac{4}{6} = \frac{2}{3}$

6 a i $7\frac{7}{8}$ ii $2\frac{3}{8}$

 b i $8\frac{1}{4}$ ii $1\frac{1}{4}$

 c i $1\frac{3}{8}$ ii $8\frac{3}{8}$

 d i $1\frac{1}{8}$ ii $6\frac{3}{8}$

7 a $\frac{4}{3} = 1\frac{1}{3}$ b $2\frac{6}{7}$ c 9

8 a $\frac{1}{6}$ b $\frac{2}{5}$ c $\frac{28}{55}$ d $\frac{16}{25}$

9 a $\frac{3}{2}$ b $\frac{3}{5}$ c $\frac{4}{7}$ d $\frac{3}{5}$

10 a $\frac{4}{3}$ b $4 \times \frac{4}{3} = \frac{4 \times 4}{3} = \frac{16}{3} = 5\frac{1}{3}$

11 a $6\frac{3}{4}$ b $5\frac{5}{7}$ c $9\frac{4}{5}$

12 a $7 \times \frac{5}{1} = 35$ b $3 \times \frac{3}{2} = \frac{3 \times 3}{2} = \frac{9}{2} = 4\frac{1}{2}$

13 a $\frac{1}{10}$ b $\frac{1}{12}$ c $\frac{1}{15}$ d $\frac{1}{14}$

14 a $\frac{1}{4} \times \frac{15}{2} = \frac{15}{8} = 1\frac{7}{8}$ b $\frac{3}{2} \times \frac{1}{3} = \frac{3}{6} = \frac{1}{2}$

Percentages, decimals and fractions

1 a 0.125 b 0.625 c 0.32 d 0.04

2 a $\frac{3}{5}$ b $\frac{3}{100}$ c $\frac{1}{4}$

 d $\frac{1}{200}$ e $\frac{33}{1000}$ f $\frac{1}{8}$

3 a $\frac{1}{50}$ b $\frac{1}{20}$ c $\frac{12}{25}$

 d $\frac{39}{50}$ e $\frac{8}{5}$ f $\frac{9}{5}$

4 a 28% b 82% c 12%

5 a $0.8 = 80\%$ b $0.4 = 40\%$ c $0.375 = 37.5\%$

6 a $\frac{3}{8}$ b $\frac{17}{25}$ c $\frac{6}{60} = \frac{1}{10}$ d $\frac{3}{20}$

 e Students' own answers, for example: Because it is a fraction of 1 hour, and 1 hour = 60 minutes.

7 48%

8 a 16% b 48% c 25% d 20%

9 14.3%

10 a 0.4 b 0.7 c 0.12

 d 0.47 e 0.02 f 0.025

Calculating percentages

1 a £8 b 7 c 22

2 a 24 b 14 c 132

3 a 3 b 10 c 187.5

4 a 5.52 b 329 c 1.8 d 3.75

5 a £37.50 b £350

6 a 4 b 44

7 a 33 b 30 c 324 d 3150

8 £1440

9 a 7 b 63

10 a 54 b 72 c 16 d 950

4 Extend

1 a $-\frac{1}{14}$ b $-\frac{5}{14}$ c $-\frac{7}{8}$ d $-\frac{1}{28}$

2 30

3 0.8

4 750 square miles

5 28.6%

6 £4800

7 a £15

 b $\frac{1}{5} \times 30\% \times £50 = £30$, which is more than $5\% \times \frac{1}{2} \times £800 = £20$.

8 35%

9 Bethany saves $\frac{1}{8} = 12.5\%$. Bethany saves more.

10 Yes; 10% of 8.9 is 0.89; $8.9 + 0.89 = 9.79$ but the percentage has gone up to 10.3%

11 Option 1: $22\,000 + 37\,000 = 59\,000$

 $59\,000 \times 3\frac{1}{2} = 206\,500$

 Option 2: $222\,000 + 33\,000 = 255\,000$

 $\frac{1}{4}$ million $= 250\,000$

 Only option 2 is suitable.

12 2 hours 20 minutes

13 £2.53

14 Yes. $(£72\,000 \times 60\%) \div 54 = £800$. $£72\,000 \div 60 = £1200$.

15 No she hasn't. March was her best month.

4 Test ready

Sample student answers

1 The student must state their decision about which shop to buy from and why, using the final prices as evidence. They must also put the correct units (e.g. £59 not just 59) in their statement. e.g. 'Samantha should buy her trainers from Edexcel Sports as they only cost £56 instead of £59 or £60 from the other shops.'

2 $\frac{6}{15}$ is correct but is not in its simplest form, which the question demands.

 $\frac{6}{15}$ simplifies to $\frac{2}{5}$

4 Unit test

1 48%

2 No $\left(\frac{12}{25} = \frac{48}{100}\right)$

3 $\frac{18}{25}, \frac{4}{5}$, 0.82, 0.88, $\frac{9}{10}$

4 a £78.75 b £132

5 £240

6 a 8 b $\frac{5}{6}$

7 $\frac{13}{15}$

8 £432

9 $\frac{4}{5}$

10 £187 500

11 $2\frac{11}{12}$

12 $\frac{13}{24}$

13 6 rolls

14 £218.40

15 $0.75 \times 150 = 112.5$, $24.61 + 11 \times 7.99 = 112.5$

16 No

17 16

18 $\frac{2}{5} + \frac{1}{4} + \frac{3}{10} = \frac{19}{20}$; so he saves $\frac{1}{20}$

£10 500 $\div \frac{1}{4} \div 20 = £2100$

19 **a** 0.3%, $\frac{3}{10}$, 0.303, 31%, $\frac{1}{3}$

b Any sensible answer, for example, $\frac{3}{10\,000}$ and 37%.

20 Current percentage of shape shaded: 12.5%; number of larger triangles to be shaded: 15

21 Any sensible answer, for example: Antony is correct because equivalent fractions, decimals and fractions represent the same number, or, Wendy is correct because fractions, decimals and percentages are different ways of expressing the same quantity.

Mixed exercise 1

1 $\frac{16}{25}$ → **4.5**

2 $\frac{6}{25}$ → **4.5, 4.6, 4.7**

3 10 minutes   → **4.2**

4 $\frac{19}{20}$ or 0.95   → **4.5, 4.1**

5 $\frac{11}{24}$ → **4.1, 4.4**

6 Student H is plotted at (46, 40) instead of (36, 40). The line of best fit is not in the right place, it is too far below the points instead of going through the middle of the points.   → **3.7, 3.8**

7 10:50   → **1.2**

8 **a** £15.50   → **1.2, 4.2**

b £316

9 5   → **1.2**

10 **a** 544   → **4.2, 4.3, 4.5, 4.7**

b $\frac{3}{25}$

11 156   → **4.7, 3.2**

Working in two-way table:

	French	Spanish	Total
Girls	396	264	660
Boys	384	156	540
Total	780	420	1200

12 $\frac{1}{2}$ → **1.4, 2.3**

13 $Y = 4x - 3$ → **2.4, 2.7**

14 1 hour and 48 minutes   → **2.4**

15 **a** 14   → **2.3, 2.6**

b 49

c 35

16 No as for school A: $\frac{160}{360} \times 900 = 400$ and for school B: $\frac{120}{360} \times 1140 = 380$.   → **3.6**

17 **a**   → **3.3**

Height, h (metres)	Frequency
$0 < h \leqslant 4$	9
$4 < h \leqslant 8$	19
$8 < h \leqslant 12$	14
$12 < h \leqslant 16$	7
$16 < h \leqslant 20$	1

b

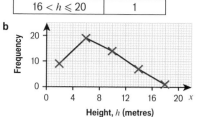

18 **a** She has not multiplied 3 by $2x$ to give $6x$ or 3 by 7 to give 21.   → **2.5, 2.6**

b Neither are correct as neither of them have used the HCF of $4x$ to give $4x(3x + 4)$.

19 **a** $\frac{8a^5b^7}{4a^2b^3}$ → **2.2**

b $3a^3b^4 \times 4a^2b^2$

20 684   → **3.2**

UNIT 5 Equations, inequalities and sequences

5.1 Solving equations 1

1 **a** $-6 \div 3$; $\times 5 + 4$ **b** 12, $-\frac{2}{5}$

2 **a** 6   **b** 10   **c** 3

    d 2   **e** 13   **f** 3

3 $y = 8 \times 7$, $y = 56$

4 **a** 50   **b** 12   **c** 20

    d 12   **e** 64   **f** 12

5 **a** 7   **b** 6   **c** 30

    d 2   **e** 1.25   **f** 2

    g 4.5   **h** 7.5

6 **a** -2, $y = 2$ **b** $+4$, $a = 7$

    c $\div 3$, $b = 4$ **d** $\times 4$, $c = 12$

    e Students' own answers, e.g. I look at whether I need to add, subtract, divide or multiply in order to get the unknown on its own.

7 **a** 2   **b** 18   **c** 5   **d** 4

8 **a** 4   **b** 3   **c** 3   **d** 4

9 **a** 13   **b** 10

10 **a** $9a = 180°$ **b** $a = 20°$

11 **a** $6c = 30$ **b** $c = 5$ **c** 10 cm

12 $9b = 360$; $b = 40$; $200°$

13 $10v = 60$ $v = 6$

Length $= 24$ cm, Width $= 6$ cm

14 **a** $4s = 36$ **b** £9

15 **a** $n - 11 = 18$ **b** 29

5.2 Solving equations 2

1 **a** $4d$ **b** $8f$ **c** $3p + 12$ **d** $b - 1$

2 **a** 6, 10, 14, 18, 22   **b** -1, 2, 5, 8, 11

3 **a** -4, $\div 3$ **b** $+2$, $\times 5$

4 **a** 2

    b i $k = 3$ **ii** $w = 5$ **iii** $w = 4.5$

5 **a** 3, 2, 20, 10   **b** 4, 18, 2, 9

6 **a** $a = 2$ **b** $a = 3$ **c** $a = 2$
    d $f = -2$ **e** $c = 2$ **f** $d = -2$

7 **a** $a = 3$ **b** $p = -\frac{1}{2}$ **c** $x = -\frac{1}{5}$
    d $y = \frac{2}{3}$ **e** $t = -\frac{5}{8}$ **f** $a = 2\frac{1}{3}$

8 $w = 5.5$

9 **a** $3a + 120° = 180°$ **b** $a = 20°$

10 $a = 30°$

11 9

12 **a** Students' drawings of a square with length and width $3y - 3$.
    b $12y - 12$ **c** $y = 4$

13 15 cm

14 6

15 **a** $c = 5$ **b** $d = 24$ **c** $e = 6$ **d** $f = -3$

16 8

17 **a** $a = 12$ **b** $b = 30$ **c** $c = 12$
    d $d = 12$ **e** $e = 16$ **f** $f = 27$
    g Students' own answers, e.g. Substitute the solution back into the equation; if the answer to the equation is correct, then the solution is correct.

18 **a** $x = 16$ **b** $x = 120$ **c** $x = 12$
    d $x = 30$ **e** $x = \frac{9}{2}$ **f** $x = \frac{10}{7}$

5.3 Solving equations with brackets

1 **a** $3a + 21$ **b** $5b - 30$

2 **a** $4a + 6$ **b** $30p - 20$
    c $18 - 21c$ **d** $-12 + 8x$

3 **a** $e = 6$ **b** $a = -1$ **c** $b = 3$
    d $y = -\frac{3}{7}$ **e** $x = 8$ **f** $x = -9$

4 **a** $d = 10$ **b** $b = 0$ **c** $d = 7$
    d $m = 7$ **e** $c = -1$ **f** $e = 3$
    g $f = 6$ **h** $m = 1.5$ **i** $n = 0.8$

5 $5(b + 5) = 45$; $b = 4$

6 Steve is 33 years old and Jenson is 3 years old.

7 $\times 2, -3, \div 5$

8 **a** $g = 5$ **b** $a = 1$ **c** $c = \frac{2}{3}$
    d $b = 3.5$ **e** $h = 3.4$ **f** $t = -11.5$

9 **a** $a = 14$ **b** $c = 5$ **c** $p = 6$
    d $a = -4$ **e** $v = 4\frac{2}{3}$ **f** $e = 4.5$
    g $h = 1\frac{1}{3}$ **h** $x = 13$ **i** $x = 1$
    j $x = 8$ **k** $x = -1$ **l** $x = -\frac{2}{3}$
    m $x = 22$ **n** $m = 5.5$ **o** $n = 0.5$
    p Students' own answers.

10 **a** $s = 28$ **b** $f = -2$ **c** $x = -8$
    d $m = 13$ **e** $y = -20$ **f** $t = 3$

11 Width $= 1\frac{2}{3}$ cm, Length $= 10\frac{1}{6}$ cm

12 **a** $x = 5$ **b** $x = \frac{5}{7}$

13 $w = \frac{11}{5}$

5.4 Introducing inequalities

1 **a** $<$ **b** $>$ **c** $>$
    d $>$ **e** $=$

2 **a**

   b

   c

   d

3 **a** $x \leqslant 6$ **b** $x > 1$ **c** $-3 < x \leqslant 1$
    d $-4 \leqslant x < 1$

4 **a** 5   **b** 0, 1, 2, 3   **c** 0, 1, 2, 3, 4
    d $-3, -2, -1, 0, 1, 2, 3$

5 **a** $2n < 6$ **b** $3n < 9$
    c $5n < 15$ **d** $3n + 1 < 10$

6 **a** $x < 4$

   b $x \geqslant -4$

   c $x \leqslant 6$

   d $x > 6$

   e $x < 2$

   f $x \geqslant -2$

   g $x \leqslant -8$

   h $x > 6$

   i $x \leqslant 10$

j $x > 5$

k $x \leqslant -1$

l $x \geqslant 4$

7 a $x > 2$ **b** $x > 3$ **c** $x > -6$
d $x \geqslant 2$ **e** $x < 2.5$ **f** $x \leqslant 1$
g $x \geqslant 2$ **h** $x \geqslant 0.5$ **i** $x < 2.2$

8 a

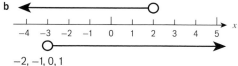

b $-1, 0, 1, 2, 3, 4$
c $x < 4$

9 $n < 9$

10 $4n + 8 > 3n + 2; n > -6$, e.g. $-5, -4, -3 \ldots$

5.5 More inequalities

1 $-5, -4, -3, -2, -1, 0, 1, 2, 3$

2 a $x < -2$ **b** $x < -5$ **c** $x < 4$ **d** $x > -12$
e $x \geqslant 1$ **f** $x > 1$ **g** $x < 9$ **h** $x \leqslant 4$

3 a $2, 3, 4, 5$
b

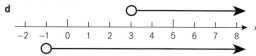

$-2, -1, 0, 1$
c

$-2, -3, -4 \ldots$
d

$4, 5, 6, \ldots$

4 5

5 a $2 \leqslant x \leqslant 3$ **b** $-3 \leqslant x \leqslant 3$
c $-\frac{7}{5} < x \leqslant 4$ **d** $-4 < x \leqslant 0$
e $-2.5 < x < 1$ **f** $-3\frac{1}{3} < x < 1$

6 a $2 < x \leqslant 4$ **b** $-1 \leqslant x \leqslant 4$
c $2 \leqslant x < 7$ **d** $-1 < x \leqslant 7$
e $-9 < x < -6$ **f** $-3 < x < 1$

7 a $2 < x < 5$ **b** $-1 < x \leqslant 4$
c $1.25 \leqslant x < 3.75$ **d** $-1 < x < 2$
e $-2 \leqslant x < 1$ **f** $0 \leqslant x \leqslant 5\frac{1}{4}$

8 a $\frac{2}{3} \leqslant x \leqslant 3$

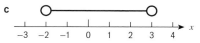

b $-\frac{7}{4} \leqslant x < \frac{3}{4}$

c $\frac{1}{3} < x \leqslant \frac{8}{3}$

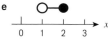

9 a No   **b** It reverses   **c** $x > 2$

10 a $x < -5$ **b** $x < -2$ **c** $x < -3$ **d** $x \leqslant -2$
e $x < -3$ **f** $x \leqslant -2$ **g** $x \leqslant 0$ **h** $x \leqslant -2.5$

11 $x < -3, -4$

12 a

$-2 < x \leqslant 5$
b

$2 < x \leqslant 5$
c

$-2 < x < 3$
d

$-2 < x \leqslant 2$
e

$1 < x \leqslant 2$
f

$1 < x < 5$
g

$-\frac{2}{3} \leqslant x < -\frac{1}{6}$
h

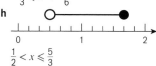

$\frac{1}{2} < x \leqslant \frac{5}{3}$

13 3.25 kg

14 a The width of Q cannot be negative, so $x - 2 > 0$ and $x > 2$.
b $2 < x < 8$

5.6 Using formulae

1 $-2, +10, \times 3, \div 8$

2 a 47   **b** -11 **c** 22

3 a 96   **b** 50   **c** -56

4 a 9   **b** 6
   c i 13   **ii** 24

5 a 2 **b** −6.5 **c** 20 **d** −0.25

6 a 5 **b** 20 **c** 10

7 a 21 **b** 56 **c** −35
d 30.1 **e** 1 **f** 5.25

8 a i 3 **ii** −3.4
b i −2.5 **ii** −14.5

9 a i 200 miles **ii** 357.5 miles
b i 4 hours **ii** 4 hours

10 Jon is 4 years old and Fiona is 7 years old.

11 a 15.9 **b** 8.5

12 64.8 g

13 30

14 5

15 a $x = y - 4$ **b** $x = y + 7$ **c** $x = y - 5$
d $t = s - 8$ **e** $x = 2 - y$ **f** $v = 6 - m$
g $t = v - a$ **h** $a = v - t$ **i** $s = u - x$
j $d = \frac{P}{5}$ **k** $x = \frac{y}{4}$ **l** $x = \frac{y}{3}$
m $x = \frac{y}{m}$ **n** $m = \frac{y}{x}$ **o** $I = \frac{P}{V}$

16 a $x = \frac{y+1}{2}$ **b** $x = \frac{y-1}{2}$ **c** $x = \frac{y-3}{5}$
d $x = \frac{y-3}{5}$ **e** $x = \frac{y+3}{4}$ **f** $x = \frac{y+2}{4}$
g $N = \frac{M+5}{7}$ **h** $N = \frac{M+7}{5}$ **i** $t = \frac{s+1}{4}$
j $W = 3V$ **k** $m = 4r$ **l** $n = 2k$
m $x = 2t$ **n** $D = TV$ **o** $m = dv$

17 a $F = 19$ **b** $t = \frac{F-7}{3}$

18 a Equation **b** Formula **c** Expression
d Equation **e** Equation **f** Formula
b i Same: Equations and formulae have an $=$ sign, and they both can both be solved to find the value of an unknown.
ii Different: A formula shows the relationship between two or more variables.

5.7 Generating sequences

1 a 8, 14 **b** 3, 5 **c** 4, 0

2 a 25, 21 **b** 5, 2

3 a 3.5, 4 **b** $\frac{2}{3}$, 1 **c** 0.3, −0.5
d −5.5, −6.5 **e** $-2\frac{3}{5}, -3\frac{2}{5}$ **f** −7.8, −7.1

4 a 3, 3.4, 3.8, 4.2, 4.6 **b** 10, 9.8, 9.6, 9.4, 9.2
c 7, 10, 13, 16, 19 **d** 7, 9, 11, 13, 15
e −3, −1, 1, 3, 5 **f** −7, −12, −17, −22, −27

5 33

6 $p = 1$

7 a i
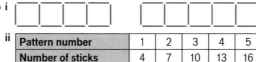

ii

Pattern number	1	2	3	4	5
Number of sticks	3	6	9	12	15

iii Yes, +3 **iv** 30

b i

ii

Pattern number	1	2	3	4	5
Number of sticks	4	7	10	13	16

iii Yes, +3 **iv** 31

c i

ii

Pattern number	1	2	3	4	5
Number of sticks	3	5	7	9	11

iii Yes, +2 **iv** 21

d i
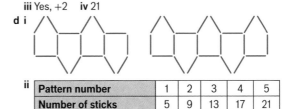

ii

Pattern number	1	2	3	4	5
Number of sticks	5	9	13	17	21

iii Yes, +4 **iv** 41

8 a Students' drawings with 10 dots and 15 dots arranged to form triangles.
b 10 dots; 15 dots
c +2, +3, +4, +5, ...
d 55 dots
e No, the difference between the terms changes.

9 a Students' drawings of 4 by 4 square and 5 by 5 square.
b 64

10 64, 125

11 a 50 000, 500 000 **b** $\frac{1}{8}, \frac{1}{16}$ **c** 0.01, 0.001
d 256, 512 **e** $\frac{16}{7}, \frac{32}{7}$ **f** 0.4, 0.2

12 a Double (or multiply by 2). **b** Halve (or divide by 2).
c Multiply by 3. **d** Multiply by −2.
e Multiply by 10. **f** Divide by 3.

13 a 9, 13, 17 **b** 25, 125, 625

14 a 8, 13, 21 **b** 24, 39, 63 **c** 40, 65, 105

15 $5x$

16 a $\frac{1}{6}, -\frac{1}{7}$ **b** $\frac{9}{10}, \frac{11}{12}$ **c** $\frac{14}{3}, \frac{17}{2}$

17 a 8 **b** 20

18 27, 81; rule: ×3 each time; or 19, 33; rule: +2, +6, +10, +14 etc. i.e. add alternate even numbers.

5.8 Using the nth term of a sequence

1 a 4 **b** 8 **c** 12 **d** 16

2 a $x = 2$ **b** $x = 11$ **c** $x = 4.5$

3

n	1	2	3	4	5
Term	3	5	7	9	11

4 a i 3, 6, 9, 12, 15 **ii** 3, 10, 17, 24, 31
iii 6, 11, 16, 21, 26 **iv** 19, 17, 15, 13, 11
v 11, 8, 5, 2, −1 **vi** 2.5, 3, 3.5, 4, 4.5
vii 26, 22, 18, 14, 10 **viii** −1, −5, −9, −13, −17
b The common difference is the coefficient of the nth term.

5 a $3n - 1$ **b** $4n - 2$
c $5n - 3$ **d** $2n + 3$
e $21 - 2n$ **f** $22 - 2n$

6 $5n + 1 = 17$
$5n = 16$
$n = \frac{16}{5}$
n is not a whole number so 17 is not a term in the sequence.

7 **a** $126 - 2n$
  b $126 - 2n = 9$
    $2n = 126 - 9$
    $2n = 117$
    $n = \frac{117}{2}$
    n is not a whole number so 9 is not a term in the sequence.

8 **a** $3n - 1$ (Yes, no)　**b** $3n + 2$ (Yes, yes)
  c $4n - 3$ (Yes, no)　**d** $5n - 1$ (No, yes)
  e $45 - 5n$ (No, no)　**f** $6n - 1$ (Yes, no)

9 **a** 40　　**b** 61　　**c** -49

10 **a** $2n - 1$ (19)　　**b** $3n$ (30)
  c $12 - 2n$ (-8)　　**d** $4n - 1$ (39)

11 **a i** 26　　**ii** $+6$
  b 56

12 **a**

  b

Pattern number	1	2	3	4	5
Number of dots	4	7	10	13	16

  c $3n + 1$　　**d** 91

13 **a** 108　　**b** 103　　**c** 104　　**d** 105

14 **a** $2n + 1$
  b No, he needs $20 \times 2 + 1 = 41$ tiles.
  c $\frac{35 - 1}{2} = 17$, so he can make pattern number 17.

15 **a**

Pattern number	1	2	3	4	5
Number of white tiles	4	5	6	7	8
Number of blue tiles	2	4	6	8	10

  b $2n$　　**c** $n + 3$　　**d** 40
  e 33　　**f** 25

16 **a** 1, 4, 9, 16, 25　　**b** 3, 12, 27, 48, 75
  c 0, 3, 8, 15, 24　　**d** 0.25, 1, 2.25, 4, 6.25
  e 5, 8, 13, 20, 29　　**f** 64, 61, 56, 49, 40

17 No, the 3rd term is $4 \times 3^2 = 4 \times 9 = 36$ or solve $3n^2 = 144$ and show that n is not an integer.

5 Check up

1 **a** $c = 1$　　**b** $f = -2$　　**c** $x = 14$
  d $a = 4$　　**e** $d = 0.75$

2 **a** $x = 3$　　**b** $x = -5$　　**c** $a = \frac{7}{3}$

3 **a** $8x - 20$　　**b** 33 cm, 33 cm, 2 cm

4 $I = \frac{1}{2}$

5 $l = 9.5$

6 **a** $M = 4P$　　**b** $x = \frac{y + 5}{3}$

7 **a** $a = 3$　　**b** $t = 3.25$

8 **a** $x = 5$　　**b** $x = 7$

9 **a**

  b

  c

10 **a** $x > -1$　　**b** $-2 \leqslant x < 1$

11 **a** $-3, -2, -1, 0, 1, 2, 3$　　**b** $-5, -4, -3, -2, -1, 0, 1$

12 **a** $x > 10$　　**b** $x \leqslant 6$　　**c** $x > 10$　　**d** $x \geqslant 10$

13 **a** $-5 < x \leqslant 2$　　**b** $3 \leqslant x < 10$

14 **a i** $6\frac{1}{2}, 8$　　**ii** 32, 64　　**iii** 6.25, -3.125
  b i Arithmetic　**ii** Geometric　**iii** Geometric

15 **a** 90　　**b** 86　　**c** $100 - 2n$
  d No. All the terms will be even numbers.

16 **a** $4n - 2$　　**b** 38

17 **a** $4n - 3$
  b Yes
    $4n - 3 = 57$
    $4n = 60$
    $n = \frac{60}{4} = 15$

19 Students' own answers.

5 Strengthen

Equations and formulae

1 **a** -7　　**b** $+10$　　**c** $\div 2$

2 **a** $m = 3$　　**b** $x = 9$

3 **a** $x = 5$　　**b** $y = 10$　　**c** $s = 6$
  d $t = 0.5$　　**e** $p = 6$　　**f** $s = 4$
  g $x = 14$　　**h** $z = 6$　　**i** $k = 10$

4 **a** $a = 4$　　**b** $m = 3$

5 **a** $t = 5$　　**b** $a = 2$

6 **a** $x = 5$　　**b** $x = 10$　　**c** $x = 6$

7 2 kg

8 **a** $a = 5$　　**b** $s = 2$　　**c** $x = -1$
  d $a = 4$　　**e** $p = 2$　　**f** $y = 5$

9 **a** $x = -3$　　**b** $a = \frac{1}{3}$　　**c** $z = \frac{1}{2}$

10 **a** $x = 3y$　　**b** $x = \frac{y + 1}{4}$

11 **a** $x = n - 5$　　**b** $x = m + 2$　　**c** $x = \frac{y}{3}$
  d $x = \frac{y + 2}{3}$　　**e** $x = 2P$

Inequalities

1 **a** F　　**b** B　　**c** C
  d D　　**e** E　　**f** A

2 **a–d** $n < -1$

3 a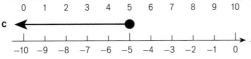

b

c

d

e

f

4 a 3, 4, 5, 6   **b** −7, −6, −5, −4
c −1, 0, 1, 2, 3   **d** 4

5 a $x > 5$ (−5)
  b $x < 3$ (÷4)
  c $x \geqslant 6$ (+10, ÷3)
  d $y < 3$ (−18, ÷6)

6 a $3 < x \leqslant 5$ **b** $2 \leqslant x < 5$ **c** $-2 \leqslant x < 2$

7 a $1 < x \leqslant 3$ **b** $4 < x \leqslant 6$ **c** $-1 \leqslant x \leqslant 5$

Sequences

1 a 19, 23   **b** −11, −15   **c** 23, 28
  d 27, 81   **e** 15, 7.5   **f** 16, −32

2 a Arithmetic   **b** Arithmetic   **c** Arithmetic
  d Geometric   **e** Geometric   **f** Geometric

3 a $2n$ **b** $4n$ **c** $5n$ **d** $-4n$

4 a +4   **b** −1 difference   **c** $4n-1$
  d i 3   **ii** 7   **e** 39

5 a $2n + 5 = 20$
$$2n = 15$$
$$n = 7.5$$
So 20 is not a term in the sequence as n is not a whole number.
  b $2n + 5 = 33$
$$2n = 28$$
$$n = 14$$
So 33 is a term in the sequence.

5 Extend

1 a 17   **b** 5

2 a 3.35   **b** 2.63   **c** 0.33
  d 20   **e** 1   **f** 16.8

3 a $d = \frac{2}{3}$ **b** $e = -\frac{5}{6}$ **c** $x = \frac{4}{5}$

4 a $127 - 7n$ **b** 19th term, −6

5 $4t - 7 = 2t + 9$
$$2t = 16$$
$$t = 8$$
The length is $4t - 7 = 4 \times 8 - 7 = 25$, so the width is $100 \div 25 = 4$.
Therefore $x = 4$.

6 210°

7 2.5 litres

8 a $2x + 18 = 4x + 6$ **b** 30 cm

9 $100 = 10^2$
$$10^x \times 100^y = 10^x \times 10^{2y} = 10^{x+2y}$$
So $10^{x+2y} = 10^z$
Therefore $z = x + 2y$

10 a $x \leqslant 6$ **b** $x < 50$ **c** $x > 5$ **d** $x > \frac{8}{3}$

11 a $x \geqslant 5$ **b** $x > 7$ **c** $x < \frac{1}{2}$

12 a 4, 5, 6, 7, 8   **b** 2
  c −15, −14, −13, −12, −11, −10, −9, −8, −7
  d 2, 3, 4, 5   **e** None   **f** None

13 a $15x + 10$ **b** 7, 49, 59

14 The nth term is $6n + 4$ and n doesn't equal an integer when 555 is a term.

15 a $x = \frac{n-M}{2}$ **b** $k = \frac{Ph}{2}$ **c** $d = 3(c - f)$

Test ready
Sample student answers

1 No. The student has found the correct value of x, but has not found the perimeter, as specified in the question.
The correct final answer is: perimeter = 60 cm.

2 No, the student has not shown how they found x.
No, they have not shown that the smallest angle is 55°.
Students' own answers, for example:
$$x + 35 + 2x + 15 + 3x - 20 = 180$$
$$6x + 30 = 180$$
$$6x = 150$$
$$x = 25$$
Angles are: $x + 35° = 60°$, $2x + 15° = 65°$, $3x - 20 = 55°$.
Therefore the smallest angle is 55°.

5 Unit test

1 a $a = 15$ **b** $b = 20$ **c** $p = 5$

2 a $b = 6$ **b** $b = -32$ **c** $b = 3$

3 a $t = \frac{d}{5}$ **b** $x = y + 6$ **c** $n = 4m$

4 $p = \frac{q - 8}{3}$

5 i $x = 12$ **ii** $x = 20$

6 Expression, formula

7 33.4 cm

8 a $x = 7.5$ **b** −4, −3, −2, −1, 0, 1, 2, 3

9 3, 6, 11, 18

10 $x = \frac{9}{2}$

11 $x = 6$

12 $x = \frac{1}{2}$

13 $x = \frac{-7}{5}$

14

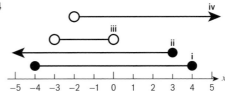

15 $2x + 10 = 5x - 80$ (angles at the base of an isosceles triangle are equal)

$$90 = 3x$$
$$x = 30$$
$$2x + 10 = 2 \times 30 + 10 = 70$$
$$y = 180 - (70 + 70) = 40$$

(angles in a triangle add to 180°)

So $y = 40°$

16 a $x = 0, 1, 2, 3, 4$ **b** $x = -3, -2, -1$

17 a $t = 2$

b

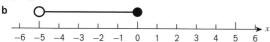

18 a $11a - 20 =$ sum, so $40 < 11a - 20 < 50$

b 6, 30, 10

19 a 33, 38   **b** $5n + 3$

20 The expression for the nth term is $7n + 1$.

$$7n + 1 = 163$$
$$7n = 162$$

162 is not a multiple of 7. Therefore, 163 isn't in the sequence.

21 16, 26

22 Triangle $= 4$, Asterisk $= 2$, Square $= 4$

23 Students' own answers.

UNIT 6 Angles

Note: for some questions in this chapter there is more than one way of reaching the correct answer.

6.1 Properties of shapes

1 Square: 4 lines, order 4
Rectangle: 2 lines, order 2
Equilateral triangle: 3 lines, order 3

2 a Accurate drawing of a 65° angle

b Accurate drawing of a 128° angle

c Accurate drawing of a rectangle and its diagonals

3 a i 360°   **ii** 360°

b 360°

4 a $a = 75°$ (angles on a straight line add up to 180°)

b $b = 105°$ (vertically opposite angles are equal)

5 A and D, C and F, E and H

6 a

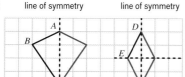

b Kite, rhombus

c Kite: 83°, 127°, 83°, 67°
Rhombus: 127°, 53°, 127°, 53°

d A kite has one pair of opposite angles that are equal. A rhombus has two pairs of opposite angles that are equal.

7 a A rhombus, B kite, C parallelogram, D rectangle, E trapezium

b i A, C, D   **ii** A, B   **iii** A, C, D

8 a Rhombus, square   **b** Square

c Kite, rhombus   **d** Isosceles trapezium

e Only parts **a** and **c** have more than one possible answer.

9 a, b, c

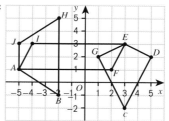

10 a 144°, 36°, 144°   **b** 65°, 115°, 65°

c 110°, 82°, 86°

11 $a = 90°$, $b = 99°$, $c = 81°$, $d = 107°$, $e = 73°$, $f = 114°$, $g = 66°$

12 a i, ii, iii

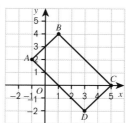

iv Rectangle

b i, ii, iii

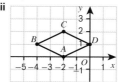

iv Rhombus

c i, ii, iii

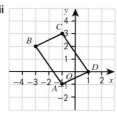

iv Parallelogram

13 $\angle ADC = 180 - 61 = 119°$ (angles on a straight line add up to 180°)
$\angle DCB = 28°$ (vertically opposite angles are equal)
$\angle ABC = 360 - 94 - 119 - 28 = 119°$ (angles in a quadrilateral add up to 360°)
So $\angle ADC = \angle ABC$ but $\angle DCB \neq \angle DAB$
One pair of opposite angles are equal so $ABCD$ is a kite.

14 $b = c = 65°$ (the parallelograms are identical)
$a = d = 115°$ (angles on a straight line add up to 180°)
$65° + 115° + 65° + 115° = 360°$

6.2 Angles in parallel lines

1 a CD and BE **b** AC (or AB) and BE (or CD)

2 a b, d, h **b** a, c, e, g, i

3 $a = 88°$ (angles on a straight line add up to 180°)
$b = 95°$ (angles on a straight line add up to 180°)
$c = 50°$ (angles on a straight line add up to 180°)
$d = 165°$ (angles around a point add up to 360°)
$e = 65°$ (angles on a straight line add up to 180°)
$f = 90°$ (angles on a straight line add up to 180°)
$g = 85°$ (angles on a straight line add up to 180°)
$h = 95°$ (vertically opposite angles are equal, or angles on a straight line add up to 180°)
$i = 85°$ (any of the three reasons)

4 q and r, p and s

5 a $a = 60°$ **b** $p = 80°$, $q = 100°$
  c $x = 120°$, $y = 60°$

6 a $a = 83°$ **b** $b = 111°$ **c** $c = 99°$, $d = 81°$

7 Angle x is not a corresponding angle to the 130° angle.
  $x = s = 180 - 130 = 50°$

8 a $t = 95°$ (corresponding angles are equal),
    $u = 85°$ (angles on a straight line add up to 180°)
  b $v = 80°$ (corresponding angles are equal, angles on a straight line add up to 180°)
  c $w = 105°$ (corresponding angles are equal, angles on a straight line add up to 180°)

9 a a (corresponding angles are equal)
    c (vertically opposite to a)
    e (corresponding angle to a)
    h (vertically opposite to the marked angle)
    g (corresponding angle to h)
  b $b = 91°$ (angles on a straight line add to 180°)
    so $a + b + c = 89° + 89° + 91° = 269°$

10 $x + y = 180°$ (angles on a straight line add up to 180°)
  $z = y$ (corresponding angles are equal)
  So $x + z = 180°$

11 a $\angle ADC = 180 - 105 = 75°$
  b $\angle BCD = 180 - 75 = 105°$
    (**ADC** and **BCD** are co-interior angles. Co-interior angles add up to 180°)
  c $\angle ABC = 180 - 105 = 75°$
    (**DAB** and **ABC** are co-interior angles. Co-interior angles add up to 180°)
  d Opposite angles in a parallelogram are equal.

12 a $x = 115°$, with reasons, e.g.
    $\angle BEF = 65°$ (corresponding angles are equal)
    $\angle ABE = 65°$ (alternate angles are equal)
    $x = 180 - 65$ (angles on a straight line add up to 180°)
  b Students' own answers.

13 $a = 66°$ (co-interior angles add up to 180°)
  $b = 27°$ (vertically opposite angles are equal, and lies on straight line with a and 87°)
  $c = 65°$ (lies on straight line with corresponding angle to 115°)
  $d = 30°$ (lies on straight line with c and vertically opposite angle to 85°)
  $e = 81°$ (corresponding angles are equal)
  $f = 27°$ (lies on straight line with 81° and corresponding angle to 72°)
  $g = 108°$ (corresponding angle with $f + 81°$)
  $h = 39°$ (alternate angle to angle on straight line with $87° + 54°$)
  $i = 116°$ (co-interior angles add up to 180°)
  $j = 108°$ (corresponding angles are equal)

14 E.g. $\angle BFH = \angle FHC = 108°$ (alternate angles are equal)
  $\angle BFD = \angle EFA = 40°$ (vertically opposite angles are equal)
  $\angle DFH = \angle BFH - \angle BFD = 108 - 40 = 68°$

6.3 Angles in triangles

1 a Equilateral triangle **b** Isosceles triangle
  c Scalene triangle **d** Right-angled triangle

2 180°

3 a 30 **b** 66 **c** 72.5

4 a $a = 30°$ (angles on a straight line add up to 180°)
    $b = 90°$ (angles on a straight line add up to 180°)
  b $c = 50°$ (angles around a point add up to 360°)
    $d = 45°$ (angles on a straight line add up to 180°)
  c $e = 53°$ (angles in a triangle add up to 180°)

5 a a and b are equal. **b** d, e and f are all equal.
  c $d = e = f = 180 \div 3 = 60°$, but the third angle in the isosceles triangle is unknown.

6 a Isosceles and right angled
  b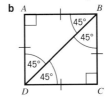

7 $60 + 45 = 105°$

8 The equal angle to BCA is ABC not BAC;
  $\angle BAC = 180 - 2 \times 47 = 86°$

9 a $a = 84°$, $b = 48°$ **b** $c = d = 54°$
  c $e = 42 + 28 = 70°$

10 a 75°
  b i No; 45° with vertical **ii** No; 30° with vertical

11 $d + b + e = \mathbf{180°}$ Angles on a straight line add up to **180°**.
  $d = a$ (angles d and a are alternate) Alternate angles are **equal**.
  $e = c$ (angles e and c are alternate) Alternate angles are **equal**.
  So $d + b + e = a + b + c = \mathbf{180°}$
  This proves that the angles in a triangle add up to **180°**.

12 a $a = 23°$ (angles in a triangle add up to 180°)
    $b = 157°$ (angles on a straight line add up to 180°)
  b $c = 44°$ (angles on a straight line add up to 180°, and base angles of an isosceles triangle are equal)
  c $d = 180 - 2 \times (180 - 148) = 116°$ (angles on a straight line add up to 180°, base angles of an isosceles triangle are equal, and angles in a triangle add up to 180°)

13 $x + y = 180° - z$ because the angles in a triangle add up to **180°**.
  $w = 180° - z$ because the angles on a **straight** line add up to 180°.
  So $x + y = w$

14 C

15 E.g. $\angle ABD = 180 - \angle CBD = 180 - 125 = 55°$ (angles on a straight line add up to 180°)
  $\angle BAD = 180 - 70 - 55 = 55°$ (angles in a triangle add up to 180°)
  So $\angle ABD = \angle BAD$, so triangle ABD is isosceles (an isosceles triangle has two equal angles)

16 $a = b = c = 180 \div 3 = 60°$ (the three angles of an equilateral triangle are equal)
  $d = e = (180 - 90) \div 2 = 45°$ (angles in a triangle add up to 180°, and the base angles in an isosceles triangle are equal)
  $f = 180 - 45 = 135°$ (angles on a straight line add up to 180°)
  $g = 180 - 12 - 135 = 33°$ (angles in a triangle add up to 180°)

17 110°

6.4 Exterior and interior angles

1 Triangle 3, pentagon 5, octagon 8, decagon 10, hexagon 6, quadrilateral 4, nonagon 9, heptagon 7

2 a 36 **b** 30 **c** 20

3 a 90° **b** 60°

4 a 103 **b** 43 **c** 87
  d 143 **e** 167

5 Students' own drawings.

6 a Irregular **b** Irregular **c** Regular
  d Regular **e** Irregular **f** Regular

7 a i 120° **ii** 72° **iii** 90°
  b i 360° **ii** 360° **iii** 360°
    Each set of exterior angles adds up to 360°.
  c 360°

8 a 15 **b** 24°

9 a 12 sides **b** 8 sides **c** 20 sides

10 a 360° **b** 360°

11 a $a = 44°$ **b** $b = 54°$

12 No; this would only be correct if the exterior angles are equal but they are not as the shape is irregular.

13 a 18 sides **b** 180° **c** 180°
  d Angles on a straight line add up to 180°.

14 $a = 112°, b = 97°, c = 85°, d = 148°$

15

Regular polygon	Exterior angle	Interior angle
pentagon	72°	108°
hexagon	60°	120°
decagon	36°	144°

16 Exterior angle = 36°, 10 sides

17 54° (exterior angle of pentagon = $\frac{360}{5} = 72°$.
  Interior angle = $180 - 72 = 108°$. x is half interior angle)

18 112.5° (exterior angle = $\frac{360}{8} = 45°$.
  Interior angle = $180 - 45 = 135°$. $a = 180 - \frac{135}{2} = 112.5°$
  (angles on a straight line add up to 180°).)

6.5 More exterior and interior angles

1 a i 4 **ii** 4
  b i 5 **ii** 5
  c i 7 **ii** 7
  d i 10 **ii** 10

2 a 720 **b** 180 **c** 1440

3 a a neither, b interior, c exterior
  b $a = 135°, b = 94°, c = 72°$

4 a Students' own drawings. **b** 360°

5 a Students' own drawings.
  b No; the interior angles are 108°, and 360 is not a multiple of 108.

6 Students' own drawings.

7 Always true; the four angles of the trapezium add up to 360°, so they can always be fitted together around a point.

8 This polygon has **5** sides.
  It is made of **3** triangles.
  The interior angles in a triangle add up to **180°**.
  So the sum of the interior angles in this polygon is
  $3 \times 180° = 540°$.

9 a Students' own drawings. **b** $4 \times 180° = 720°$
  c No

10 a $n - 2$
  b i Students' own drawings. (2 triangles = 4 − 2)
    ii Students' own drawings. (5 triangles = 7 − 2)
  c $S = 180° \times (n - 2)$

11 a i 360° **ii** $a = 156°$
  b i 540° **ii** $b = 142°$
  c i 720° **ii** $c = 146°$

12 Yes; a regular n-sided polygon can be divided into $n - 2$ triangles.

13 a i 720° **ii** 120°
  b i 540° **ii** 108°
  c i 2880° **ii** 160°
  d Student's own answer, e.g.
    Exterior angle = $360 \div n$
    Interior angle = $180 -$ exterior angle
    Sum = $n \times$ interior angle

14 a 11 sides **b** 14 sides **c** 17 sides **d** 24 sides

15 $p = 108°$ (exterior and interior angles add up to 180°)
  $q = 36°$ (an isosceles triangle has two equal sides and two equal angles)
  $r = p - q = 108 - 36 = 72°$ (the interior angles of a regular polygon are equal)

16 a 120° **b** 30°

17 a ABC, CDE
  b i 150° **ii** 15° **iii** 135°

18 $x = 54$

6.6 Geometrical problems

1 $a = 60°$ (angles in a triangle add up to 180°)
  $b = 120°$ (angles on a straight line add up to 180°)
  $c = 110°$ (angles in a pentagon add up to 540°)
  $d = 140°$ (co-interior angles add up to 180°)
  $e = 63°$ (corresponding angles are equal, and vertically opposite angles are equal)

2 a $x = 36$ **b** $x = 50$ **c** $x = 105$ **d** $x = 82.5$

3 a $x + 60$ **b** $x + x + 60 = 2x + 60$

4 a $x + 3x = 180°$ **b** $x = 45°$ **c** 45°, 135°

5 a $x + 8x = 180°$ (co-interior angles add up to 180°);
    $x = 20°$
  b $x + 2x + 60° = 180°$ (angles in a triangle add up to 180°); $x = 40°$
  c $x + 2x + 3x = 180°$ (angles in a triangle add up to 180°); $x = 30°$
  d $3x + 5x = 180°$ (corresponding angles are equal, and angles on a straight line add up to 180°)
    $x = 22.5°$
  e $4x + 5x + 90° = 360°$ (angles around a point add up to 360°); $x = 30°$

6 a Vertically opposite angles are equal.
  b $x = 50°$ **c** 90°

7 a $x + x + 20° = 90°$; $x = 35°$
    $y = x + 20° = 55°$ (alternate angles are equal)
  b $x + 90° = 2x + 15°$ (corresponding angles are equal, and vertically opposite angles are equal); $x = 75°$
    $y = 180 - (2x + 15) = 15°$ (angles on a straight line add up to 180°)
  c $160° - x = x + 40°$ (vertically opposite angles are equal); $x = 60°$
    $y = 360 - (x + 40) - 2x - x = 80°$ (angles in a quadrilateral add up to 360°, and vertically opposite angles are equal)

8 a i $4x = 140°$ **ii** $x = 35°$ **iii** 35°, 105°, 40°
  b i $100° + x = 180°$ **ii** $x = 80°$
    iii 80°, 80°, 20°

9 Equilateral; $3x + 75 = 180°$, so $x = 35$ and angles are 60°, 60°, 60°.

10 $x, 3x, 5x$: sometimes
  x, x, x: sometimes
  $x + 60°, x - 60°, 60° - 2x$: never
  $x + 60°, x - 60°, x + 180°$: never
  $60° - x, x + 90°, 30°$: always

11 a E.g. $y = x + 60° + 2x - 20°$ (exterior angle of triangle = sum of opposite interior angles)
So $y = 3x + 40°$
b $x = 37°$

12 a E.g. $x = 83 - 47 = 36$ (exterior angle of triangle = sum of opposite interior angles)
b $25°$

13 a

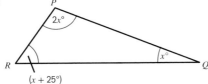

b $4x + 25° = 180°$ **c** $38.75°$

14 $x + 90 + 2x + 110 + 130 = 540$ (angles in a pentagon add up to 540°)
$3x = 210; x = 70$
$\angle ABC = 2x = 140°$

6 Check up

1 a $55°$ (alternate angles are equal)
b $88°$ (corresponding angles are equal, and vertically opposite angles are equal)

2 a $a = 77°$ (alternate angles are equal)
$b = 77°$ (vertically opposite angles are equal)
b $c = 68°$ (alternate angles are equal)
$d = 68°$ (vertically opposite angles are equal)
$e = 84°$ (corresponding angles are equal)
$f = 28°$ (angles on a straight line add up to 180°)
c $g = 90°$ (corresponding angles are equal, and angles on a straight line add up to 180°)
$h = 104°$ (co-interior angles add up to 180°)

3 a $120°$ (angles of equilateral triangle = 60°, angles on a straight line add up to 180°.)
b $108°$ (base angles of isosceles triangle are equal and angles in a triangle add up to 180°.)
c $43°$ (base angles of isosceles triangle are equal and angles in a triangle add up to 180°.)

4 $a = 108°$, reason 4
$b = 95°$, reason 6
$c = 13°$, reasons 3 and 4

5 $45°, 105°, 30°$

6 a $p = 72°$ (opposite angles in a parallelogram are equal)
$q = 108°$ (co-interior angles add up to 180°)
b $r = 91°$ (angles on a straight line add up to 180°)
$s = 104°$ (angles in a quadrilateral add up to 360°)
c $a = 38°$ (angles on a straight line add up to 180°, opposite angles of a kite are equal, and angles in a quadrilateral add up to 360°)

7 a 24 sides   **b** $20°$

8 a $x = 108°$ (angles on a straight line add up to 180°)
b $y = 60°$ (exterior angles of a regular polygon are equal and add up to 360°)
$z = 120°$ (interior and exterior angles add up to 180°)

9 $1440°$

10 $90° + a + 105° + 55° + a = 540°$ (interior angles of a pentagon add up to 540°); $a = 145°$

12 Students' own answers.

6 Strengthen

Angles between parallel lines

1 a Alternate   **b** Corresponding
   c Corresponding   **d** Vertically opposite

2 a, b
   i $p = 49°$ (alternate angles are equal)
   ii $q = 88°$ (corresponding angles are equal)
   iii $r = 37°$ (alternate angles are equal)
     $s = 37°$ (vertically opposite angles are equal)
   iv $t = 76°$ (vertically opposite angles are equal)
     $u = 76°$ (corresponding angles are equal)
     $v = 104°$ (angles on a straight line add up to 180°)

3 a $a = 63°$ (vertically opposite angles are equal)
$b = 63°$ (corresponding angles are equal)
$c = 117°$ (angles on a straight line add up to 180°)
b $d = 57°$ (angles on a straight line add up to 180°)
$e = 57°$ (alternate angles are equal)
$f = 57°$ (angles on a straight line add up to 180°)
$g = 65°$ (corresponding angles are equal)
c $h = 108°$ (corresponding angles are equal, and angles on a straight line add up to 180°)
$i = 95°$ (co-interior angles add up to 180°)

Triangles and quadrilaterals

1 a $180°$ **b** $360°$ **c** $180°$

2 a $a = 28°$ (angles in a triangle add up to 180°)
b $b = 70°$ (angles in a quadrilateral add up to 360°)
c $c = 55°$ (angles in a triangle add up to 180°)
d $a = 50°$ (opposite angles of a parallelogram are equal)
$b = 130°$ (co-interior angles add up to 180°)

3 a $80°$ **b** $120°$ **c** $40°$

4 a Side AB = side AC, angle b = angle c
b Side DE = side DF, angle e = angle f
c Side GH = side GI, angle h = angle i

5 a Isosceles   **b** $180°$
   c PQR and PRQ **d** $152°$ **e** $76°$

6 $a = 36°, b = 34°, c = 48°$

7 $\angle BAC = 30°$, reason 3
$\angle ABC + \angle ACB = 150°$, reason 1
$\angle ABC = \angle ACB = 75°$, reason 2

8 a Equilateral   **b** $60°$
   c $120°$ (angles on a straight line add up to 180°)

9 a $9x$ **b** $9x = 180°$
   c $x = 20°$ **d** $20°, 60°, 100°$

Interior and exterior angles

1 a $360°$ **b** 18 sides   **c** $12°$

2 a y **b** x **c** $60°$
   d $180°$ **e** $120°$

3 a iii $540°$ **iv** $101°$
   b iii $720°$ **iv** $155°$

4 d $1080°$ **e** $135°$

6 Extend

1 $52°$

2 $155°, 155°, 25°$

3 $\angle BFE = 114°$ (alternate angles are equal)
$\angle CFE = 48°$ (vertically opposite angles are equal)
$x = \angle BFE - \angle CFE = 66°$

4 $126°$

5 $\angle BEG = 180° - (121° + 29°) = 30°$ (angles on a straight line add to 180°)
$\angle EBC = 121°$ (corresponding angles are equal)
$\angle EBG = 121° - 62° = 59°$
$\angle BGE = 180° - (30° + 59°) = 91°$ (angles in a triangle add to 180°)

6 $\angle CBD = \angle BCD = (180 - 2(x - 30)) \div 2 = 120 - x$ (triangle BCD is isosceles)
$120 - x + x + \angle BCA = 180°$ (angles on a straight line add up to 180°)
So $\angle BCA = 60°$
$\angle EAC = 180 - 90 - x = 90 - x$ (angles in triangle EAC add up to 180°)
$\angle BAC = 180 - (x + 30) - (90 - x) = 60°$ (angles on a straight line add up to 180°)
$\angle ABC = 180 - 60 - 60 = 60°$ (angles in triangle ABC add up to 180°)
So $\angle BCA = \angle BAC = \angle ABC = 60°$; hence the triangle is equilateral.

7 a Isosceles   **b** $2x - 180$
  c i $\angle CDB = 180 - 120 - (2x - 180) = 240 - 2x$
    ii $\angle BCD = 180 - x - (240 - 2x) = x - 60$
    iii $(x - 60) + y = 180$ so $x = 240 - y$
    iv $2(240 - y) - 180 = 300 - 2y$

8 No, because 500 is not divisible by 180.

9 $x + 9x = 180°$, $x = 18°$; 20 sides

10 $AB = BC = CD = DA$ (side lengths of congruent regular octagons are equal)
Interior angle of regular octagon = 135°
At each of points A, B, C and D, the third angle is
$360 - 2 \times 135° = 90°$ (angles around a point add up to 360°)
So $ABCD$ has four equal sides and four angles of 90°; hence it is a square.

11 $x = 120°$, $y = 150°$

12 48°

6 Test ready

Sample student answers

Alternate angles are equal.
Opposite angles in a parallelogram are equal.
Angles in a triangle add up to 180°.

6 Unit test

1 a 84°   **b** 120°   **c** 165°

2 a 30 sides   **b** exterior 10°, interior 170°

3 a 160° (angles at a point add to 360°)
  b 66° (alternate angles are equal)
  c 41° (angles on a straight line add to 180° and corresponding angles are equal)

4 $\angle ADC = 180° - 110° = 70°$ (angles on a straight line add to 180°)
$\angle DAC = 70°$ (angles at the base of an isosceles triangle are equal)
$\angle ACD = 40°$ (angles in a triangle add to 180°)
$\angle CAB = 40°$ (alternate angles are equal)
$\angle ACB = 50°$ ($90° - 40°$)
$\angle ABC = 90°$ ($180° - (40° + 50°)$)
$\angle ABC$ is a right-angled triangle.

5 a $x = 38\frac{1}{3}$ **b** $136\frac{2}{3}$, 165, $76\frac{2}{3}$, $76\frac{2}{3}$, 85

6 a $a = 55°$ **b** $b = 60°$, $c = 120°$, $d = 30°$

7 a $x = 30°$ **b** Both angles are 90°

8 $\angle DEF = \angle EFA = 105°$
$(6 - 2)180° = 720°$ $720° - 2(105°) = 510°$
$\angle ABC = x$, $\angle BAF = 2x$, $\angle ABC = \angle BCD$, $\angle BAF = \angle CDE$
$6x = 510°$, $x = 510° \div 6 = 85°$, $2x = 170°$, $BAF = 170°$

9 $\angle GBC = \angle FEB = 20°$ (corresponding angles are equal)
$\angle ABE = \angle FEB = 20°$ (alternate angles are equal or vertically opposite angles are equal)
$\angle BAE = \angle BEA = 160° \div 2 = 80°$ (angles at the base of an isosceles triangle are equal, angles in a triangle add to 180°)
$\angle AED = 180° - (20° + 80°) = 80° = x$ (angles on a straight line add to 180°)

10 a $\angle BCD = 180° - y$ (angles on a straight line add to 180°)
  b $\angle BDC = 180° - (x + (180° - y)) = y - x$ (angles in a triangle add to 180°)
  c $\angle BDA = 90° + x - y$ (angles on a straight line add to 180°)
  d $\angle DBA = 180° - 2(90° + x - y)$ (angles in a triangle add to 180° and angles at the base of an isosceles triangle are equal)
    Therefore $\angle DBA = 2(y - x)$.

11 a i 162°   **ii** 168°   **iii** 171°
    iv 172.8°   **v** 176.4°   **vi** 179.64°
  b An angle of 180° is equivalent to a straight line so a polygon would never be formed. If the angle was larger than 180° then it would cease to be the interior angle and would become the exterior angle.

12 Sensible list containing the units of this module including °. Then students' own (sensible) answers.

UNIT 7 Averages and range

7.1 Mean and range

1 a Red   **b** 118

2 a 6   **b** 14   **c** 8

3 a 352   **b** 35.2 books

4 18.9 °C

5 Range = greatest − least = 5 − 2 = 3

6 9

7 a 7 minutes   **b** 4 minutes

8 a

Waiting time, w (min)	Frequency, f	$w \times f$
5	3	15
6	4	24
7	6	42
8	4	32
9	3	27
Total	20	140

  b 140 minutes   **c** 20   **d** 7 minutes
  e The data is the same and so the answers are the same but the method of finding the totals is different.

9 a 5.57 portions   **b** 3 portions

10 a

Score, s	Frequency, f	$s \times f$
1	1	1
2	2	4
3	4	12
4	3	12
5	6	30
6	5	30
7	2	14
8	3	24
9	3	27
10	1	10
Total	30	164

  b Mean = 5.5; range = 9   **c** Any suitable frequency table.
  d Mean = 5.3; range = 4   **e** close, 11W, 11Y, spread

11 Jade; Mia's range is 9 and Jade's range is 8. Since Jade's range is the smallest she has the more consistent scores.

12 a Tess's mean: 44.5, Tess's range: 28; Jo's mean:49.25, Jo's range: 96

b Tess, because she always scores consistently.

c Yes

13 1.6 children

7.2 Mode, median and range

1 a Mode = 6; median = 6; range = 8

b Mode = 22; median = 23.5; range = 12

2

```
14 | 8 9
15 | 2 5 7 8
16 | 0 2 3 4 5 7
17 | 2 3 5 6 7
18 | 0 1 6
```

Key: 14 | 9 means 149 bpm

3 a 28   **b** 79 years   **c** 63 years

4 a 9   **b** 5

c 5th value is 16, which is the middle value.

5 a 7.0 cm   **b** 7.0 cm   **c** 2.8 cm

6 No she is not correct. 5 only occurs once in the data set, 25 is the mode since this occurs 3 times.

7 Most of the values occur just once but several occur twice.

8 a 87 points   **b** 42 points   **c** 76 points

9 a i 23   **ii** 195   **iii** 2.76

b i Range = 32, range without outlier = 14

ii Range = 83, range without outlier = 21

iii Range = 6.09, range without outlier = 2.46

c Students' own answers, e.g. The range without the outlier.

10 a Outlier = 86; range without outlier = 14

b Outlier = 2.9; range without outlier = 0.6

11 9 volts

12 a 50 points

b The data is grouped; the actual values are not given.

13 a 25 cm   **b** 25 cm

14 25 years

7.3 Types of average

1 30

2 a 1, 3, 5, 7, 9, 11, 13, 15, 17, 19, 21, 23, 25, 27, 29, 31, 33, 35, 37, 39

b 10.5th   **c** 20

3 a 25   **b** 132   **c** 1.4   **d** 6100

e Set **a** Yes, Set **b** No, Set **c** No, Set **d** Yes

4 a 51.6   **b** 45.2   **c** Set **a** No and Set **b** Yes

5 a 18   **b** 1.9

c Yes because **a** is a good spread of results and **b** has outliers.

6 a i Median = 41.5; mode = 37 and 43; mean = 51.25

ii Outlier and bimodal so use median

b i Median = 1.9; mean = 2.88

ii Good spread of results so use median

c i Mean = 48; mode = 48; median = 48

ii Same results so mean or median

d i Mode = blue

ii Qualitative so use mode

7 a Mode = 26; Median = 26; Mean = 26.8

b Mode = 26; Median = 26; Mean = 30.8

c Mode = 26; Median = 26; Mean = 24

8

Average	Advantages	Disadvantages
Mean	Every value makes a difference	Affected by extreme values
Median	Not affected by extreme values	May not change if a data value changes
Mode	Easy to find; not affected by extreme values; can be used with non-numerical data	There may not be one

9 Any suitable summary.

10 2

11 40–49

12 $7750 \leqslant p < 8000$

13 3

14 a 49   **b** 2   **c** 1

d 1.8367 = 1.8 (1 d.p.)

15 151–175

16 $240 < d \leqslant 260$

7.4 Estimating the mean

1 60 minutes

2 a 18   **b** 22.5   **c** 19   **d** 4.5

3 3.92

4 a 3 hours 44 minutes   **b** 2 hours 50 minutes

c 4 hours 39 minutes

5 8.666... = 9 people as an estimate

6 a 21 minutes

b The table only gives intervals of values, not actual values.

7 a

Distance, d (miles)	Frequency, f	Midpoint of class, m	$m \times f$
$0 < d \leqslant 2$	9	1	$1 \times 9 = 9$
$2 < d \leqslant 4$	7	3	$3 \times 7 = 21$
$4 < d \leqslant 6$	8	5	$5 \times 8 = 40$
$6 < d \leqslant 8$	6	7	$7 \times 6 = 42$
Total	30	Total	112

b 30 students   **c** 112 miles

d 3.733... = 3.7 miles (1 d.p.)

8 3 hours 41 minutes

9

Temp, t (°C)	Frequency, f	Midpoint, m	$f \times m$
$16 \leqslant t < 18$	2	17	34
$18 \leqslant t < 20$	5	19	95
$20 \leqslant t < 22$	5	21	105
$22 \leqslant t < 24$	10	23	230
$24 \leqslant t < 26$	8	25	200

a 22.1 °C   **b** $22 \leqslant t < 24$

c $22 \leqslant t < 24$ **b** 10 °C

10 $175 + 360 + 825 + 390 + 300 + 170 = 2220$

$2220 \div 40 = 55.5$ kg

11 a 41.4 cm

b Yes, there is one outlier in the interval $10 < l \leqslant 20$, so the median would be a better average to use.

7.5 Sampling

1 a 20   **b** 8   **c** 16

2 C Putting all the names in a hat and choosing one without looking.

3 **a** People in the Midlands
  b Too many people to survey in total.
  c 1000
  d No, taking data from one town would not be representative of the population.
4 **a** His local area
  b His office is in the local area with working age people, so this is not representative of everyone.
5 **a** All in one street so not representative of the town.
  b Ask people in different streets too.
6 No. The people in the shopping centre could live near the centre and do not drive very far. The people in the motorway service area could be travelling long distances. The survey would be more accurate if it is taken in a wide variety of places.
7 **a** Biased
  b Ask students in different classes too.
8 **a** Yes — all students have the same chance of selection.
  b No — all students do not have the same chance of being selected.
9 **a** 03, 83, 41, 33, 81, 39, 22, 09
  b Could pick the 3rd, 83rd, 41st, 33rd, 81st and 39th person.
  c Random numbers could give the 3rd, 33rd, 39th, 22nd and 9th person.
10 Method 1 — put all room numbers into a hat and pick out 20 without looking.
  Method 2 — use a random number list to select 20 numbers between 1 and 200.
11 55
12 **a** 38
  b Students' own answers, e.g. the sample was representative of the population; students answered truthfully; did not change their minds.
13 **a** $\frac{3}{5}$ **b** 9
14 50
15 **a** 120
  b 5–16: 3; 17–21: 3; 22–40: 4; over-40: 2
  c 12
  d $(32 + 28 + 43 + 17) \div 10 = 12$

7 Check up

1 4.3 people
2 **a** 40.35714 = 40.4 minutes (1 d.p.)
  b 50 minutes
3 5.7 − 4.3 = 1.4
4 **a** 1.96   **b** 4   **c** 2
5 **a** Jade had the highest mean score. (Ria's mean = 149 ÷ 5 = 29.8).
  b Jade is more consistent as her range is smaller, showing that her scores are less spread out. Jade's range = 28; Ria's range = 53 − 11 = 42.
6 **a** 3.4 kg   **b** 4.3 kg   **c** 2.1 kg
7 **a** Median
  b Mean would be affected by the large break value and the mode is the two lowest values in the data set so it doesn't give a good representation of the data.
8 **a** 27.1129 = 27 emails   **b** 39 emails
  c 31–40   **d** 31–40
9 There is no mode. The median works best because the mean is affected by the large data value which is not a typical salary for this data set.

10 Random sample, e.g. put all the names in a hat and pick out 3 without looking.
11 Monday morning — people are at school or work so not representative.
  Shopping centre — needs to include a more varied area so all people can be included in sample.
12 120 students
14 **a**

Age (yr)	Frequency, f	Class midpoint, m	$f \times m$
18–24	32	21	672
25–29	41	27	1107
30–40	70	35	2450
41–59	53	50	2650
60–70	14	65	910

  b 37.1   **c** 30–40
  d 30–40, unequal class sizes

7 Strengthen

Averages and range

1 **a** 20   **b** 4
2 **a** Students' own answers
  b 8   **c** 192   **d** 24 seconds
3 14.2 °C
4 88.83333... = £88.83
5 **a** 33.2 runs
  b Kyle (Kyle's total = 166; George's total = 158)
  c 48 runs
  d George as he has the smaller range and is therefore more consistent.
6 **a** Sam   **b** Ellie = 5, Sam = 62
  c Ellie   **d** Ellie (Ellie's mean = 37; Sam's mean = 32)
  e Ellie as she scores more consistently and she has the higher mean score.
7 **a** **i** 8   **ii** 8
  b **i** 5   **ii** 10
  c

Number of emails	Frequency	Number of emails × frequency
1	8	8
2	5	10
3	3	9
4	2	8
Total	18	35

  d 18   **e** 35   **f** 1.9 (1 d.p.)
8 **a**

Number of people	Frequency	Number of people × frequency
1	7	7
2	10	20
3	8	24
4	6	24
5	1	5
Total	32	80

  b 32   **c** 80   **d** 2.5   **e** 4
  f **i** 32   **ii** 16.5th value   **iii** 2
9 **a** 2.8 kg   **b** 2.8 kg   **c** 3.9 kg
  d 21, 11   **e** 2.8 kg

Averages and range for grouped data

1 a

Puzzles completed	Frequency, f	Midpoint of class, m	$m \times f$
1–5	11	3	33
6–10	15	8	120
11–15	23	13	299
16–20	16	18	288
Total	**65**	**Total**	**740**

b 65   **c** 740

d Raw data is lost once it has been grouped.

e 11.3846... = 11.4 (1 d.p.)

f 11–15   **g** 33rd data value   **h** 11–15

2 a

Score	Frequency, f	Midpoint of class, m	$m \times f$
0–20	9	10	$10 \times 9 = 90$
21–40	6	30.5	183
41–60	2	50.5	101
61–80	2	70.5	141
Total	**19**	**Total**	**515**

b 19   **c** 515

d Raw data is lost once it has been grouped.

e 27.10526... = 27   **f** 80

Sampling

1 A, C and E

2 It would be better to take the survey on several days of the week. Saturday morning may not provide a representative sample of people. It should also be considered that people who have gone to the town centre may be less likely to do their weekly shopping online than the people who have stayed at home.

3 a $\frac{3}{10}$ **b** 60

7 Extend

1 a 6   **b** white

2 11

3 £30.10

4 8

5 $\frac{5x+7}{3}$

6 a Liz = 28.345 m, Katie = 27.578 m

b Liz = 28.4 m, Katie = 29.8 m

c 20.47 m

d Median is insensitive to extreme values

e Median

f Students' own answers, e.g. Liz because her distances were more consistent or Katie because 3 of her throws were further than Liz's best throw.

7 Boys' range = 26 cm, boys' median = 180.5 cm
Girls' range = 26 cm, girls' median = 168.5 cm
Boys are taller on average but same spread in data.

8 a 3.3   **b** 8

9 a £10 222 (nearest £1)   **b** $9000 < p \leqslant 12\,000$
c $9000 < p \leqslant 12\,000$ **d** £13 000

10 a

Time (min:sec)	Frequency	Midpoint	Frequency × midpoint
$2{:}02 < t \leqslant 2{:}04$	2	2:03	4:06
$2{:}04 < t \leqslant 2{:}06$	24	2:05	50:00
$2{:}06 < t \leqslant 2{:}08$	19	2:07	40:13
$2{:}08 < t \leqslant 2{:}10$	4	2:09	8:36
	49		102:55

b 2 minutes 6 seconds   **c** 8 seconds

11 a $40 < l \leqslant 50$

b

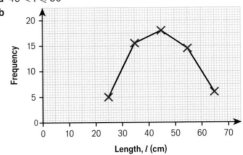

12 a 180

b Use a random sample, e.g. use random numbers to select the number on 180 of the tickets at random.

Test ready

Sample student answers

1 They have correctly identified that the boys' range is smaller than the girls' and thus their scores are more consistent. To get the second mark they need to also compare an average.

2 a No.

b Missing out the zeros does not affect the total which is still 16. He needs to divide by the total number of values, which is 8 (including the zeros). So the correct mean is $\frac{16}{8} = 2$.

7 Unit test

1 a Kim did not put the numbers into ascending order.

b 11

c 7.14

2 Sample size is too small.
Only people who don't work/study are at home at this time, so not representative.

3 4

4 a £0.60 (or 60p)   **b** £0.58 (or 58p)

5 Median for males = 87 cm
Median for females = 100.5 cm
Range for males = 25 cm
Range for females = 25 cm
Overall the males are shorter than the females (using medians). The range is the same for both males and females.

6 a Mean = 3.1 (1 d.p.)   **b** Median = 3
c Modal score = 3   **d** Range = 5

7 $\frac{3x+4y}{7}$

8 a Mean = 396.7 (1 d.p.)
  b Range = 399
  c Modal class = $400 \leqslant n < 500$
  d Median is in class $400 \leqslant n < 500$
  e

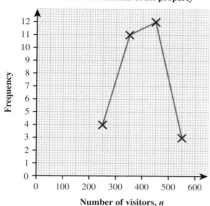

Visitors to National Trust property

Frequency vs Number of visitors, n

9 20 KS4 students

10 a Mode. There is a clear modal value and it is not biased to either end of the data set.
  b Mean. It considers all data.
  c Mode. It can be used on non-numerical data.
  d Median. It is not affected by extreme values.

11 Students' own answers in the form of:
  Lesson ____ made me think the hardest because____.

Mixed exercise 2

1 Year 7 mean = 10.8, Year 8 mean = 11.6   → **3.3, 7.1**

2 $x = 98°$, for example, $180 - 153 = 27$ as angles on a straight line total $180°$ and $360 - 75 - 40 - 120 - 27 = 98$ as exterior angles total $360°$.   → **6.5**

3 12, 23, 34, rule: +11   → **5.7**

4 5 and 7   → **7.1**

5 A length cannot be negative.   → **Unit 5**

6 No, as Ewan should have multiplied by 4 first to give $x + 5 = 28$ and then subtracted 5 to give $x = 23$.   → **5.3**

7 $x = 22°$, for example:
  angle $SQP = 180° - 68° - 68° = 44°$ as the angles in triangle total $180°$ and triangle PQS is isosceles
  angle $SQR = 180° - 44° = 136°$ as angles on a straight line total $180°$
  $x = (180° - 136°) \div 2 = 22°$ as the angles in triangle total $180°$ and triangle SQR is isosceles.   → **6.3**

8 a 3, 23, 123   → **5.7**
  b Subtract 8 and then divide by 5
  c Divide by 4 and then add 3

9 $x = 34°$ → **6.2, 6.3**

10 a $5x + 2x - 1 + 2x - 1 + 5x + 2x - 1 + 2x - 1 = 95$
  $18x - 4 = 95$ → **5.2**
  b $x = 5.5$

11 14   → **7.1**

12 No, as the smallest term in sequence A is 7 and the largest term in sequence B is 1.   → **5.7, 5.8**

13 a $b = 115°$ → **6.2, 5.2, 5.3**
  b $b = 124°$, as $5a - 10 = 50$, so $a = 12$, substitute $a = 12$ into $3a + 20$ to give $56°$ and $b = 180 - 56$ as the angles on a straight line total $180°$.

14 Yes, as 175 is halfway between 100 and 250 and more of the tomatoes have mass less than 175 g, so the mean will be less than 175 g.
  Or: Total mass of 10 tomatoes $= 6 \times 100 + 3 \times 250 + 1 \times 25$
  $= 1375$ g,
  so the mean mass $= 1375 \div 10 = 137.5$ g.   → **7.1**

15 25, 49, 81   → **5.8**

16 $90 < 8x + 10 < 180$
  $80 < 8x < 170$
  $10 < x < 21.25$
  The greatest whole number value of x is 21.   → **5.4**

17 a 10.375   → **7.4, 7.3**
  b The estimate should be lower as most of the teachers are in the lower half of each of the first intervals, so lower than the midpoints used for the first estimate.
  c $0 < y \leqslant 5$
  d $5 < y \leqslant 10$

18 $a = 107°$ as $180 - 152 = 28$ because the adjacent angles in a rhombus total $180°$ and $a = 180 \times 6 \div 8 - 28 = 107$, the other angle in the rhombus subtracted from the interior angle of a regular polygon.   → **6.1, 6.4, 6.6**

19 No, as Inequality 1: $2 < x \leqslant 7$ has these integer solutions: 3, 4, 5, 6, 7.
  Inequality 2: $2 \leqslant x < 7$ has these integer solutions: 2, 3, 4, 5, 6.   → **5.4**

UNIT 8 Perimeter, area and volume 1

8.1 Rectangles, parallelograms and triangles

1 a At 90°   **b** i and iv

2 a 16 cm   **b** 26 m   **c** 17 m   **d** 250 mm

3 a Area 25 cm^2, perimeter 20 cm
  b Area 370 mm^2, perimeter 94 mm
  c Area 8 m^2, perimeter 12 m

4 a i $l \times w$ **ii** $2l + 2w$
  b $A = l \times w$ **c** $P = 2l + 2w$

5 A rectangle with dimensions 12 cm × 4 cm

6 $32 \div 5 \approx 6$, $19 \div 3 \approx 6$, $6 \times 6 = 36$

7 $12 \div 4 = 3$, $3 \times 5 = 15$ cm

8 a, b

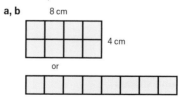

9 a 72 cm^2 **b** 63 cm^2 **c** 2432 mm^2 **d** 14.64 cm^2

10 a Perimeter = 40 cm, area = 45 cm^2
  b Perimeter = 23.6 cm, area = 27.6 cm^2
  c The two measurements for base and height have to be perpendicular.

11 a 40 cm^2 **b** Students' drawings.
  c Half   **d** 20 cm^2

12 a 13.5 cm^2 **b** 20 cm^2 **c** 17.15 cm^2
  d 294 mm^2 **e** No

13 a Perimeter 26 cm, area 30 cm^2
  b Perimeter 19.8 cm, area 13.86 cm^2
  c Perimeter 36 cm, area 54 cm^2

14 16 cm^2

15 a 4800 cm^2 **b** 1440 cm

16 3 triangles sketched and labelled, with area 18 cm^2

17 a 20 cm^2 **b** 9 cm^2 **c** 6 cm^2

18 a $a = 10$ cm **b** b $= 4$ cm **c** c $= 5$ cm

19 Area of triangle $= 20$ cm^2. Area of rectangle $= 100$ cm^2.
Length of rectangle $= 100 \div 8 = 12.5$ cm

20 a Sketch of two rectangles 3.7 m $\times$ 2.6 m,
two rectangles 2.9 m $\times$ 2.6 m

  b Estimate $2 \times 4 \times 3 + 2 \times 3 \times 3 = 42$ m^2

  c $7 \times £12 = £84$

  d Students' own answers.

8.2 Trapezia and changing units

1 a 26 mm **b** 610 cm **c** 540 mm **d** 1.27 m

2 a 8 **b** 4 **c** 30 **d** 15

3 a $h = 4$ **b** $x = 3$ **c** $y = 3$ **d** $p = 5.5$

4 32 cm^2

5 a 56 cm^2 **b** 57.8 cm^2 **c** 31 cm^2 **d** 46.4 cm^2

6 Area of trapezium $= \frac{1}{2}(3 + 5) \times 4 = 16$ cm^2

Any triangle of area 16 cm^2, for example height 8 cm
base 4 cm

7 Area 18 cm^2, perimeter 17.6 cm

8 a $40 = \frac{1}{2} \times (6 + 10) \times h$ **b** $40 = 8h$ **c** $h = 5$ cm

9 7 mm

10 a 27.5 cm^2 **b** 2.75 cm^2 **c** 140 mm^2
  d 1400 mm^2 **e** 14 000 mm^2 **f** 80 mm^2
  g 0.9 cm^2 **h** 0.09 cm^2 **i** 900 mm^2

11 a 4.41 cm^2 **b** 441 mm^2

12 a 1 m^2 = 10 000 cm^2

  a

13 a 70 000 cm^2 **b** 5 m^2 **c** 0.5 m^2 **d** 32 000 cm^2

14 a, b Students' own answers.

15 a 1.05 m^2 **b** 10 500 cm^2 **c** 10 500 cm^2
  d Students' own answers

16 a 2.8 cm^2 **b** 20 250 cm^2 **c** 11.2 cm^2

17 a 13.2 cm^2 **b** 1320 mm^2

18 Slide A has 0.5 bacteria per mm^2.
Slide B has 0.6 bacteria per mm^2.
So slide B has more bacteria per square millimetre.

8.3 Area of compound shapes

1 1000

2 a Area 24 cm^2, perimeter 20 cm
  b Area 30 cm^2, perimeter 30 cm
  c Area 64 cm^2, perimeter 32.4 cm

3 a 50 000 cm^2 **b** 22 000 cm^2 **c** 2 m^2 **d** 0.72 m^2

4 a 80 000 m^2 **b** 400 000 m^2
  c 35 000 m^2 **d** 5000 m^2
  e 4 ha **f** 12 ha
  g 22.5 ha **h** 0.3 ha

5 a 1 km^2 = 1 000 000 m^2

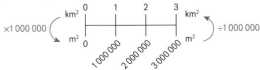

6 a 1 ha or 10 000 m^2 **b** 100

  c

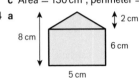

7 a Windermere **b** 983 ha or 9.83 km^2

8 a Area C 60 cm^2; Area D 40 cm^2
  b 100 cm^2 **c** 44 cm^2
  d It is the same. It doesn't matter how the shape is
divided up.

9 a i 4 cm, 7 cm **ii** 9.6 m, 9.7 m
  b i 40 cm **ii** 38.6 m
  c i 71 cm^2 **ii** 57 m^2

10 24 cm

11 Area 64.5 cm^2, perimeter 39 cm

12 a $12 \times 8 - 8 = 88$ cm^2, $4 \times 8 + 6 \times 4 + 4 \times 8 = 88$ cm^2
  b Students' own answers.

13 a Area $= 295$ cm^2, perimeter $= 82$ cm
  b Area $= 106.62$ m^2, perimeter $= 63.6$ m
  c Area $= 130$ cm^2, perimeter $= 58$ cm

14 a

8 cm 2 cm 6 cm 5 cm

  b base $= 5$ cm, height $= 2$ cm
  c 5 cm^2 **d** 35 cm^2

15 a 216.5 cm^2 **b** 30 cm^2

16 a Students' own answers. **b** Students' own answers.
  c 153 cm^2
  d Split it into rectangle and triangles and find areas
separately.

17 36 cm^2

18 49.5 cm^2

19 360 cm^2

8.4 Surface area of 3D solids

1 a Cuboid, triangular prism, square-base pyramid,
hexagonal prism, cube

  b i Cuboid: 6 faces, 12 edges, 8 vertices;
triangular prism: 5 faces, 9 edges, 6 vertices;
square-based pyramid: 5 faces, 8 edges, 5 vertices;
hexagonal prism: 8 faces, 18 edges, 12 vertices;
cube: 6 faces, 12 edges, 8 vertices

  ii cuboid: rectangles; triangular prism: triangles and
rectangles;
square-based pyramid: square and triangles;
hexagonal prism: hexagons and rectangles;
cube: squares

2 a 12 cm^2 **b** 24 cm^2 **c** 18 cm^2

3 Net of cuboid, e.g.

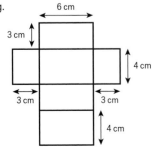

4 Net of cuboid, e.g.

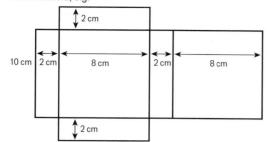

5 164 cm^2

6 **a i** 36 cm^2 **ii** bottom
    b i 18 cm^2 **ii** back
    c i 8 cm^2 **ii** left hand side (or opposite side)
    d $2 \times 36 + 2 \times 18 + 2 \times 8 = 124$ cm^2
    e Students' own answers, e.g. You can work out the area of individual faces, using the fact that there are three pairs of identical faces.

7 **a** 160 cm^2 **b** 15.5 m^2 **c** 1350 mm^2
    d $6 \times$ area of one face

8 **a**

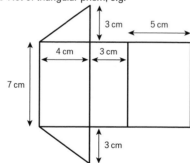

    b 198 cm^2 or 234 cm^2

9 **a** Net of triangular prism, e.g.

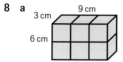
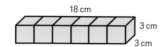

    b i 35 cm^2, 21 cm^2, 28 cm^2, 6 cm^2, 6 cm^2
      ii 96 cm^2

10 **a** 660 cm^2 **b** 2864 cm^2

11 **a** 7.5 m^2 **b** 2

12 **a** Cuboid 7.46 m^2, prism 3.51 m^2, total 10.97 m^2
     b Estimate $11 \times 13 = £143$

13 8 cm

14 **a** 96 cm^2 **b** 520

8.5 Volume of prisms

1 Student sketches of:
    a a rectangle   **b** a square   **c** a triangle   **d** a pentagon

2 **a** 12 cm^2 **b** 18 mm^2 **c** 100 cm^2

3 **a** 64   **b** 210   **c** 72   **d** 30

4 **a** 2.4 m   **b** 0.52 m   **c** 430 cm   **d** 70 cm

5 9 cm^3

6 **a** 4 cm^3 **b** 8 cm^3 **c** 12 cm^3 **d** 20 cm^3

7 Volume of 1 cube $= 3 \times 3 \times 3 = 27$ cm^3
    Volume of 4 cubes $= 27 \times 4 = 108$ cm^3

8 **a** 48 cm^3 **b** 3.6 m^3 **c** 7500 mm^3
    d 125 cm^3 **e** 4^3 or $4 \times 4 \times 4 = 64$ cm^3

9 **a** 12 cm^3 **b** 4 cm^3
    c $4 \times 3 = 12$ cm^3, students' own answers, e.g. It is equal to the volume.
    d $24 \times 2 = 48$ cm^3, $2.4 \times 1.5 = 3.6$ m^3, $500 \times 15 = 7500$ mm^3, $25 \times 5 = 125$ cm^3

10 **a** 40 cm^3 **b** 72 cm^3 **c** 300 cm^3

11 **a**

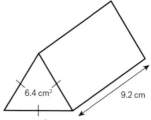

    b 58.88 cm^3

12 7 cm

13 Students' sketches of 3 cuboids with volume 24 cm^3.

14 60

15 DVDs on bottom of box: $3 \times 2 = 6$
    Number of layers: $450 \div 15 = 30$
    $6 \times 30 = 180$, $180 \div 12 = 15$, 15 minutes

16 **a** 1.3 m^3
     b £90.20 (or £89.55 if using unrounded answer in part **a**)

8.6 More volume and surface area

1 **a** 1 000 000   **b** 6000   **c** 5 000 000
    d 7200   **e** 3 710 000   **f** 35

2 1000

3 Teaspoon 5 ml, drink can 330 ml, bucket 5 litres, juice carton 1 litre

4 **a** $h = 9$ **b** $h = 3$ **c** $h = 20$

5 **a** 498.5 cm^2 **b** 421.5 cm^3

6 **a** 24 000 mm^3 **b** 2400 mm^3 **c** 5700 cm^3
    d 570 mm^3 **e** 30 cm^3 **f** 3 cm^3
    g 0.48 cm^3 **h** 0.45 cm^3 **i** 4500 mm^3

7 **a** 10.416 cm^3 and 10 416 mm^3
    b 10 416 mm^3 **c** Students' own answers

8 **a** 1 m^3 **b** 1 m^3 = 1 000 000 cm^3
    c

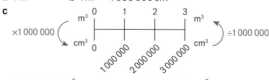

9 **a** 5 000 000 cm^3 **b** 3 500 000 cm^3
    c 6 870 000 cm^3 **d** 300 000 cm^3
    e 9 m^3 **f** 8.2 m^3
    g 0.6 m^3 **h** 3.159 m^3
    i 0.94 m^3 **j** 6.452 m^3

10 **a** 35 cm^3 = 35 ml   **b** 300 cm^3 = 300 ml
    c 3000 cm^3 = 3000 ml = 3 litres
    d 3500 cm^3 = 3500 ml = 3.5 litres
    e 40 ml = 40 cm^3 **f** 400 ml = 400 cm^3
    g 4000 ml = 4000 cm^3 **h** 3000 ml = 3000 cm^3

11 **a** 2 litres = 2000 cm^3 **b** 4 litres = 4000 cm^3
    c 4.5 litres = 4500 cm^3 **d** 0.4 litres = 400 cm^3
    e 1 m^3 = 1 000 000 cm^3 = 1 000 000 ml = 1000 litres
    f 2 m^3 = 2 000 000 cm^3 = 2 000 000 ml = 2000 litres
    g 2.4 m^3 = 2 400 000 cm^3 = 2 400 000 ml = 2400 litres
    h 0.5 m^3 = 500 000 cm^3 = 500 000 ml = 500 litres

12 200 cm^3

13 a $80\,000\,\text{cm}^3$ **b** $80\,000\,\text{ml}$, 80 litres

14 $1280\,\text{ml}$

15 a $2\,\text{cm}$ **b** $8\,\text{cm}$

16 $500\,\text{cm}^2$

17 a $2.55\,\text{cm}^3$
   b Volume of block $= 300\,\text{cm}^3$, $300 \div 2.55 = 117.6$, which rounds down to 117.
   c Down

18 682 (682.666... round down)

19 a $350\,\text{m}^3 = 350\,000$ litres **b** $105\,\text{m}^2$

20 9

8 Check up

1 a $17.5\,\text{cm}$ **b** Trapezium

2 a i $30\,\text{cm}$ **ii** $48\,\text{cm}^2$
   b i $22\,\text{cm}$ **ii** $15\,\text{cm}^2$

3 $9\,\text{cm}$

4 $5\,\text{cm}$

5 a $34\,\text{cm}$ **b** $57\,\text{cm}^2$

6 $260\,\text{mm}^2$

7 $13.75\,\text{m}^2$

8 $402\,\text{cm}^2$

9 a $376\,\text{cm}^2$ **b** $480\,\text{cm}^3$

10 $48\,\text{cm}^3$

11 a $1500\,\text{mm}^2$ **b** $5.47\,\text{m}^2$
   c $12\,000\,\text{cm}^2$ **d** $9.8\,\text{m}^2$
   e $0.0186\,\text{m}^2$ **f** $300\,000\,\text{mm}^2$

12 a 6000 litres **b** $450\,\text{ml}$
   c $200\,\text{cm}^3$ **d** $8.4\,\text{cm}^3$
   e $0.549\,\text{m}^3$ **f** $1\,600\,000\,\text{cm}^3$

14 a $8\,\text{cm}$, $6\,\text{cm}$, $48\,\text{cm}^3$ **b** $24\,\text{cm}^3$
   c No, because shorter side will have length 0.

8 Strengthen

2D shapes

1 a $18\,\text{cm}$ **b** $16\,\text{cm}$ **c** $22\,\text{cm}$ **d** $10.5\,\text{cm}$

2 a $6\,\text{cm}$, $4\,\text{cm}$ **b** $2\,\text{cm}$, $3\,\text{cm}$

3 a i $1\,\text{cm}$ **ii** $18\,\text{cm}$
   b i $4\,\text{cm}$, $4\,\text{cm}$ **ii** $30\,\text{cm}$

4 a Sketch of L-shape divided into two rectangles
   b $15\,\text{cm}^2$, $4\,\text{cm}^2$ **c** $19\,\text{cm}^2$

5 a i $3\,\text{cm}$, $4\,\text{cm}$ **ii** $6\,\text{cm}^2$, $12\,\text{cm}^2$ **iii** $18\,\text{cm}^2$
   b i $4\,\text{cm}$, $6\,\text{cm}$ **ii** $20\,\text{cm}^2$, $44\,\text{cm}^2$ **iii** $64\,\text{cm}^2$

6 Students' drawings showing lines perpendicular to each other, e.g.

7 a $9\,\text{cm}^2$ **b** $9\,\text{cm}^2$ **c** No

8 a $35\,\text{cm}^2$ **b** $8\,\text{cm}^2$ **c** $70\,\text{cm}^2$

9 a $12\,\text{cm}^2$, $4\,\text{cm}^2$; $16\,\text{cm}^2$ **b** $28\,\text{cm}^2$, $7\,\text{cm}^2$, $35\,\text{cm}^2$

10 a $300\,\text{cm}^2$ **b** $150\,\text{cm}^2$ **c** $150\,\text{cm}^2$

11 a $40\,\text{cm}^2$ **b** $36\,\text{cm}^2$

12 $h = 3\,\text{cm}$

13 $b = 5\,\text{cm}$

3D solids

1 a

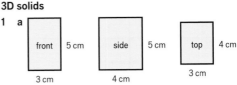

   b Labelled sketches showing measurements: back ($3\,\text{cm}$ by $5\,\text{cm}$), other side ($4\,\text{cm}$ by $5\,\text{cm}$), bottom ($3\,\text{cm}$ by $4\,\text{cm}$)
   c Top and bottom, two sides, front and back
   d $15\,\text{cm}^2$, $20\,\text{cm}^2$, $12\,\text{cm}^2$ **e** $94\,\text{cm}^2$

2 a

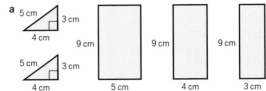

   b Triangles $6\,\text{cm}^2$, rectangles $45\,\text{cm}^2$, $36\,\text{cm}^2$, $27\,\text{cm}^2$
   c $120\,\text{cm}^2$

3 a $36\,\text{cm}^3$ **b** $140\,\text{cm}^3$ **c** $2250\,\text{cm}^3$

Measures

1 a i $1\,\text{cm}^2$ **ii** $2\,\text{cm}^2$ **iii** $3\,\text{cm}^2$
   b i $100\,\text{mm}^2$ **ii** $200\,\text{mm}^2$ **iii** $300\,\text{mm}^2$
   c

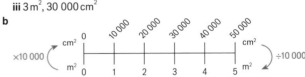

2 a i $1\,\text{m}^2$, $10\,000\,\text{cm}^2$ **ii** $2\,\text{m}^2$, $20\,000\,\text{cm}^2$
   iii $3\,\text{m}^2$, $30\,000\,\text{cm}^2$
   b

3 a $500\,\text{mm}^2$ **b** $3\,\text{cm}^2$ **c** $60\,000\,\text{cm}^2$ **d** $8\,\text{m}^2$

4 a $3000\,\text{mm}^3$ **b** $9\,\text{cm}^3$
   c $4720\,\text{mm}^3$ **d** $2\,000\,000\,\text{cm}^3$
   e $1\,200\,000\,\text{cm}^3$ **f** $9\,\text{m}^3$
   g $3.2\,\text{m}^3$ **h** $5000\,\text{ml} = 5$ litres
   i $875\,\text{cm}^3 = 875\,\text{ml} = 0.875$ litres
   j $2300\,\text{ml} = 2300\,\text{cm}^3$ **k** $1345\,\text{ml} = 1345\,\text{cm}^3$

8 Extend

1 70% of $50 \times 80 = 2800\,\text{m}^2$

2 $WX = 90 \div 12 = 7.5\,\text{cm}$, $2x + 2 \times 7.5 = 28$, $2x = 13$,
   $AB = x = 6.5\,\text{cm}$

3 $3.9\,\text{m}^2$

4 240

5 Yes. Area of wall $= 150\,\text{m}^2$

6 Yes. Area $= \frac{1}{2}(20 + 12) \times 15 = 240\,\text{m}^2$
   $240 \div 3 = 80$ litres, 80 litres $= 16$ tins,
   $16 \times 11.99 = £191.84$ or $16 \times 12 = £192$

7 a $14x$ **b** $x = 2\,\text{cm}$ **c** $36\,\text{cm}^2$

8 Rectangle dimensions are x, $x + 5$
   Perimeter $= x + x + 5 + x + x + 5 + 5 + x + x + 5 + x + 5 + x + 5$
   $= 8x + 30 = 54\,\text{cm}$, $x = 3\,\text{cm}$
   Area of rectangle $= x \times (x + 5) = 3 \times 8 = 24\,\text{cm}^2$
   Area of 8-sided shape $= 3 \times 24 = 72\,\text{cm}^2$

9 a $2\,\text{m}$ **b** $6\,\text{m}$ **c** $\sqrt{50} = 7.1\,\text{m}$

10 a $50 = \frac{1}{2}(a+6) \times 10$

b $50 = \frac{1}{2} \times 10 \times (a+6)$
$50 = 5 \times (a+6)$
$10 = a+6$
$a = 4\,cm$

11 $192\,cm^3$

12 Students' own answers, e.g. a cuboid with dimensions $2\,cm \times 2\,cm \times 9\,cm$ or a right-angled triangular prism of height $6\,cm$, base $3\,cm$ and length $4\,cm$.

8 Test ready

Sample student answers

a Students' own answers, e.g. She has not shown her working.

b Students' own answers, e.g. Label the diagram with length, height and width measurements. Some may note that she could have drawn a cuboid measuring $5\,cm \times 10\,cm \times 15\,cm$.

8 Unit test

1 $85.5\,cm$

2 a $28\,cm$ **b** $6\,cm$

3 a $5\,cm$ **b** $30\,cm^2$

4 a $0.37\,m^2$ **b** $1\,500\,000\,cm^3$

5 a $1500\,cm^3$ **b** $1010\,cm^2$

6 a $2 \times 2 \times 2 = 8\,cm^3$
$8\,cm^3 \times 8 = 64\,cm^3$
Jody is correct.

b Drawing of cuboid with accurate dimensions labelled: $8\,cm$ by $4\,cm$ by $2\,cm$.

c $112\,cm^2$

7 $4000\,cm^3$

8 $x = 10\,cm$

9 $14.7\,ha$

10 a 300 litres   **b** $2.3\,m^2$ **c** Yes, this is enough paint.

11 a $88\,m^2$

b Rectangle sketch with area 88, e.g. $11\,m \times 8\,m$. Perimeter correctly calculated, e.g. for $11\,m \times 8\,m$ rectangle, perimeter is $38\,m$.

c Students' own answers

d $37.5\,m$

e Smallest possible perimeter is for a square.

12 Students' own answers

Area of parallelogram and area of rectangle are both essentially base × height. The formula for area of triangle is similarly base × height, but this then needs to be halved.

UNIT 9 Graphs

9.1 Coordinates

1 a 4   **b** 7   **c** 4.5   **d** 8.5

2 a 5   **b** 3.5   **c** 1   **d** 3

3 a −2   **b** −5   **c** 1   **d** 4

4 a B   **b** A

5 a, c, f

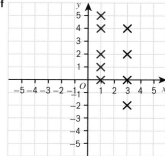

b (1, 5), (1, 1), (1, 0), (1, 4), (1, 2)

d All points are on a vertical line through $x = 1$.

e Vertical line through $x = 3$

6 a (4, −1), (4, 0), (4, 1), (4, 2), (4, 3), (4, 4)

b All have x-coordinate 4.

c i $x = 4$ **ii** $x = -2$ **iii** $y = 3$

7 Line P is $x = -4$; Line Q is $x = 5$; Line R is $y = -3$; Line S is $y = 1$

8

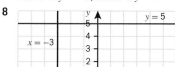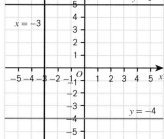

9 a $y = 0$ **b** $x = 0$

10 a (−2, −2), (−1, −1), (0, 0), (1, 1), (2, 2), (3, 3), (4, 4), (5, 5)

b x and y coordinates are the same

c (5, 5), (−2, −2)

d Each y-coordinate is equal to the x-coordinate.

e $y = x$

11

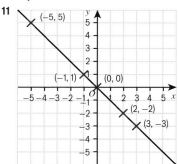

c Students' own answers, e.g. (4, −4), (1, −1), (5, −5).

12 Graph A: $y = -3$ Graph B: $y = x$ Graph C: $x = 3$

13 a $x = 2$ **b** $y = x$ **c** $x = -2, y = 2, y = -2$

14

Line segment	Start point	End point	Midpoint
AB	(4, 1)	(4, 3)	(4, 2)
CD	(1, 4)	(5, 4)	(3, 4)
EF	(2, –5)	(5, 1)	(3.5, –2)
GH	(1, –4)	(3, –2)	(2, –3)
JK	(–3, –3)	(1, –3)	(–1, –3)
LM	(–4, 1)	(–4, 5)	(–4, 3)
PQ	(–2, 4)	(1, –2)	(–0.5, 1)
RS	(–1, 5)	(2, 0)	(0.5, 2.5)

15 Answers should match.

16 (4.5, 6)

17 a (5, 7)   **b** (3.5, 8.5)   **c** (–1, 5)   **d** (–1, –1)

18 (13, 9)

9.2 Linear graphs

1 Students' own answers, for example: $1 + 3$; $8 \div 2$; $5 - 1$.

2 a 1   **b** 0.2   **c** 20   **d** 1.6

   e –18   **f** –2   **g** –8   **h** 2

3 a i 8   **ii** –1

   b i –3   **ii** 3

   c i 0   **ii** $-1\frac{1}{2}$

   d i $3\frac{1}{2}$ **ii** $2\frac{3}{4}$

4 a 4.2   **b** 2.5   **c** 0.6   **d** 1.4

5 a (–3, 2), (0, 5), (1, 6)

   b, c, d, h

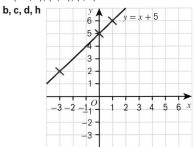

   e Students' own answers, e.g. (–1, 4), (–2, 3).

   f Add 5

   g $y = x + 5$

6 a (–1, –1), (0, 1), (2, 5)

   b, c, d, h

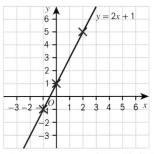

   e Students' own answers, e.g. (–2, –3), (1, 3).

   f Multiply by 2 then add 1

   g $y = 2x + 1$

7 a i

x	–3	–2	–1	0	1	2	3
$y = x + 1$	–2	–1	0	1	2	3	4

   ii

x	–3	–2	–1	0	1	2	3
$y = 2x - 3$	–9	–7	–5	–3	–1	1	3

   b

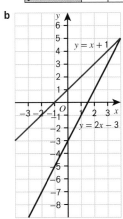

8 a

x	–3	–2	–1	0	1	2	3
$y = 3x - 2$	–11	–8	–5	–2	1	4	7

   b

x	–3	–2	–1	0	1	2	3
$y = 2x + 4$	–2	0	2	4	6	8	10

   c

x	–3	–2	–1	0	1	2	3
$y = 4x - 6$	–18	–14	–10	–6	–2	2	6

   d

x	–3	–2	–1	0	1	2	3
$y = 0.5x + 1$	–0.5	0	0.5	1	1.5	2	2.5

9 a, b

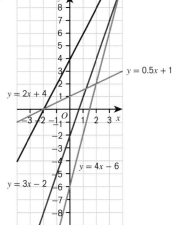

   c Students' own answers, for example checking the working in their table.

10 a i

x	-3	-2	-1	0	1	2	3
$y=-x+1$	4	3	2	1	0	-1	-2

ii

x	-3	-2	-1	0	1	2	3
$y=-2x-3$	3	1	-1	-3	-5	-7	-9

iii

x	-3	-2	-1	0	1	2	3
$y=-4x+2$	14	10	6	2	-2	-6	-10

b

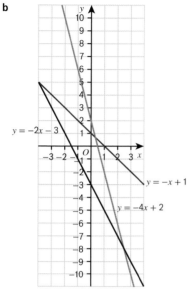

11 a, b, c, d $y=-3x+2$ $y=-4x+6$

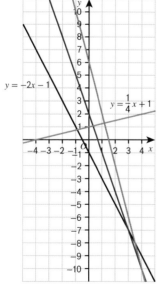

e i Upwards, because the gradient is positive (5).
  ii Downwards, because the gradient is negative (-4).

12 a

x	-2	-1	0	1	2	3	4
y	0	0.5	1	1.5	2	2.5	3

b

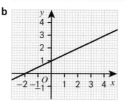

c 1.3
d i 1.6   **ii** 0.8   **iii** 2.4
e 3.6
f Students' own answers, e.g. For **d iii**, I read up from the x-axis and across to the y-axis, whereas for **e** I read across from the y-axis and down to the x-axis.

13 a i 0.2   **ii** -0.1 **iii** 1.4
  b i -1.6 **ii** 0.8   **iii** 2.8
  c The values are read from the graph.
  d i 4   **ii** -2 **iii** -0.4

14 a

x	-2	-1	0	1	2	3
y	-0.6	0.2	1	1.8	2.6	3.4

b

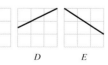

c i 1.4   **ii** Estimates in the range -0.38 to -0.35

9.3 Gradient

1 By definition, lines that are parallel never intersect.

2 a 5   **b** -2 **c** $1\frac{1}{3}$ **d** $\frac{1}{2}$
  e 0.06   **f** 0.006   **g** 0.6   **h** 0.06

3 a i 3   **ii** 6
  b i 1   **ii** 2   **iii** 3
  c 3
  d -1
  e Students' own answers, e.g. It is always the same for each line.

4 a 3   **b** -1

5 a Gradient line $A=1$ Gradient line $B=-2$
  Gradient line $C=-3$ Gradient line $D=-\frac{1}{2}$
  Gradient line $E=2$ Gradient line $F=\frac{1}{2}$
  Gradient line $G=0$
  b i E,A,F **ii** C,B,D
  c Students' own answers, e.g. Ignoring any minus signs, the larger the number, the steeper the gradient.

6 a i B **ii** C
  b

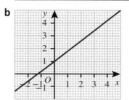

      A B C D E

7 a Parallel   **b** 2   **c** Gradient

8 Students' own answers, e.g. He has not used the scales on the axes properly. The correct gradient is 0.6.

9 0.12

10 a i

x	−3	−2	−1	0	1	2	3
$y = x + 1$	−2	−1	0	1	2	3	4

    ii

x	−3	−2	−1	0	1	2	3
$y = -2x + 3$	9	7	5	3	1	−1	−3

  b i, ii

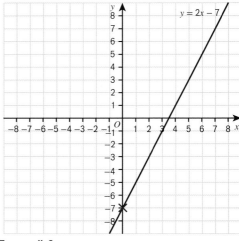

  c Gradients are 1, −2
  d i 1   **ii** −2
  e Students' own answers, e.g. The coefficient is equal to the gradient.

11 $y = -4x - 1$

12 a $y = 5x + 2$
  b $y = 3x + 2$ and $y = 3x$
  c $y = -3x + 4$

13 a −1   **b** $\frac{1}{2}$ **c** −1

9.4 $y = mx + c$

1 a 3   **b** −2

2 a i −10   **ii** 5
  b i 10   **ii** 20
  c i 10   **ii** 10
  d i 10   **ii** 5
  e i 2   **ii** 5
  f i 5   **ii** 2

3 c $y = 10 - x$ **d** $y = 10 - 2x$ **e** $y = \frac{10 - 2x}{5}$

4 a 2   **b** −6   **c** 3

5 1

6 a y-intercept graph $A = 2$
    y-intercept graph $B = 0$
    y-intercept graph $C = -1$
    y-intercept graph $D = -2$
  b Students' own answers, e.g. In equations with form $y = mx + c$, c is equal to the y-intercept.
  c (0, 4)

7 a i 3   **ii** −2   **iii** $y = 3x - 2$
  b $y = -2x - 3$

8 $A\; y = x + 1$, $B\; y = 2x - 1$, $C\; y = -5x + 1$, $D\; y = -\frac{1}{2}x + 3$,
  $E\; y = \frac{1}{2}x - 3$

9 a $y = 5x - 6$ **b** $y = 6x + 6$ or $y = -5x + 6$ **c** $y = 5x - 6$

10 a $y = 3x$, $y = -3x$ **b** $y = 3x - 3$, $y = x - 3$

11 Students' own answers, e.g. If $x = -2$,
  $y = 3x + 11 = (3 \times -2) + 11 = -6 + 11 = 5$.
  No, the line passes through
  (−2, 5), not (−2, −5).

12 a $1 = (3 \times 2) + c$ **b** $c = -5$
  c $y = 3x - 5$

13 a $y = 2x - 2$ **b** $y = 4x + 6$
  c $y = \frac{1}{2}x + 11$ **d** $y = 3x - 27$

14 a $y = 2x - 1$ **b** $y = \frac{1}{2}x + 3$
  c $y = 2x - 6$ **d** $y = -x + 2$

15 a, c, d

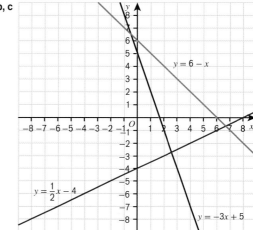

  b i −7   **ii** 2

16 a, b, c

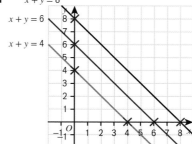

17 a i 8 (0, 8)   **ii** 8 (8, 0)
  b i 6 (0, 6)   **ii** 6 (6, 0)
  c i 4 (0, 4)   **ii** 4 (4, 0)

18 a

  b Students' own answers, e.g. The graphs are all parallel with gradient −1.

19 a, b, c, d

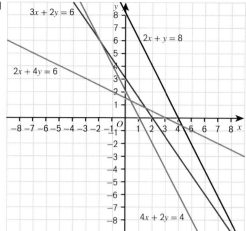

$3x + 2y = 6$
$2x + y = 8$
$2x + 4y = 6$
$4x + 2y = 4$

20 a, b $y = -x + 8$; gradient $= -1$, y-intercept $= 8$
$y = -x + 6$; gradient $= -1$, y-intercept $= 6$
$y = -x + 4$; gradient $= -1$, y-intercept $= 4$
$y = -2x + 8$; gradient $= -2$, y-intercept $= 8$
$y = -\frac{3}{2}x + 3$; gradient $= -\frac{3}{2}$, y-intercept $= 3$
$y = -2x + 2$; gradient $= -2$, y-intercept $= 2$
$y = -\frac{1}{2}x + \frac{3}{2}$; gradient $= -\frac{1}{2}$, y-intercept $= \frac{3}{2}$

9.5 Real-life graphs

1 a £3.50 **b** £120 **c** £45
2 a £6 **b** £1.50 **c** £90
3 a 5 **b** 2.5 **c** $y = 2.5x + 5$
4

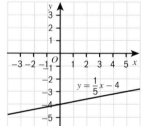

$y = \frac{1}{5}x - 4$

5 a

Fuel used (litres)	0	1	2	3	4	5
Distance travelled (km)	0	10	20	30	40	50

b

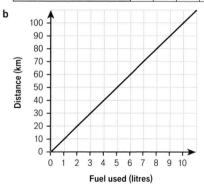

c 6 litres **d** 17 litres

6 a

Water used (m³)	0	1	2	3	4	5	6	7	8	9	10
Cost (£)	30	32	34	36	38	40	42	44	46	48	50

b

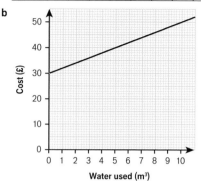

c $5\frac{1}{2}$ cubic metres

d 5500 litres

7 a i £25 **ii** £2.50
b Cost (£)
c 2.5
d quantity in kg

8 a Line 1 = B Line 2 = A Line 3 = C
b Higher or larger

9 a

Hours worked	0	1	2	3	4	5
Total cost (£)	60	90	120	150	180	210

b

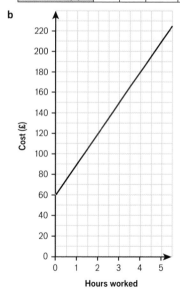

c $y = 30x + 60$
d i Gradient is hourly cost.
ii y-intercept is callout fee.

10 a Line 1 = A Line 2 = B
b £100
c £26
d Plumber A £240, Plumber B £180, difference £60
e Plumber A is cheaper at £560; plumber B would cost £580.

11 a Option A – one-off joining fee of £600 then no cost per game
Option B – flat rate of £20 per game but no joining fee
b Option A is cheaper for more than 30 games.

12 a £40   **b** £45

13 a

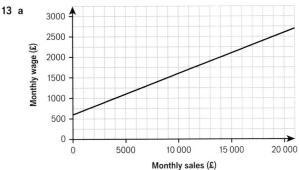

b £600

c In his current job, with average sales of £12 000, his average monthly pay will be £1800. On average he would be £200 better off in the new job.

14 a 1 C, 2 A, 3 B
b £110
c C is cheapest. It charges £240, A charges £260, B charges £320.

9.6 Distance–time graphs

1 400 m/s, 45 km/h, 40 m/min

2 a i 40 km   **ii** 120 km   **iii** 20 km
b i 5 hours   **ii** 15 minutes   **iii** 7.5 minutes

3 a, b

Satbir's journey

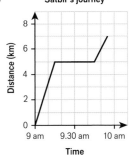

c i 20 km/hour   **ii** 12 km/hour
d Last part is uphill as speed is slower.

4 a 40 km   **b** 2 hours   **c** 20 km/h
d 30 minutes   **e** 40 km/h
f Students' own answers, e.g. He was travelling faster when the graph is steeper.

5 a 60 km/h   **b** 45 minutes
c 30 km/h   **d** 30 km

6 a Being stationary
b Travelling towards the starting position

7 a 1.6 km   **b** 16 minutes   **c** 12 minutes
d Walking to surgery at 6 km/h – other parts 4.5 km/h and 5 km/h

8 a

Bob's journey

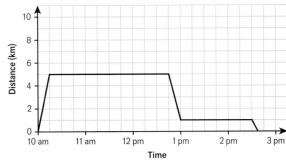

b First leg is fastest at 20 km/h, second leg 16 km/h, last leg 12 km/h

9 a B
b Southampton to Winchester arrives at 11:20 and Winchester to Southampton arrives at 11:40.
c Southampton to Winchester is faster.

10 a 3000 feet   **b** 3 hours 15 minutes
c 1 hour 30 minutes   **d** 9.30 am

11 Jon's average speed was $32 \div 20 = 1.8$ km/min
$= 1.8 \times 60 = 108$ km/h
So Jon had the faster average speed.

12 a i B   **ii** A
b i C   **ii** A

13 a i 5 m/s   **ii** 5 m/s^2 **b** $y = 5x + 5$

9.7 More real-life graphs

1 Between B and C

2 a Positive correlation   **b** $y = 5x + 12$

3 a B
b The speed at which each container fills
c A is Line 2, B is Line 1, C is Line 3.

4 a B   **b** A is Line 1, B is Line 2

5 a Students' own answers, e.g.   **b**

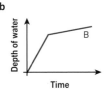

6 A is Line 1, B is Line 2.

7 The container **a** in Q5 looks like container A in Q6 but upside-down. Their graphs of depth of water against time are reflections of each other in the line $y = x$.

8 a 20°C   **b** 54 cm
c i 29 to 2 s.f.   **ii** 0.29 to 2 d.p.
d 0.29   **e** Yes
f No. A straight line has a constant gradient.

9 a −4
b The depth of oil is decreasing at a rate of 4 cm/h.

10 a Students' own answers, e.g. There should not be a line connecting the percentage in January to the origin.
b 83% or 84%
c Students' own answers, e.g. It is increasing.

11 a Positive correlation: the more hours of study the better the test score.
b 74 or 75
c $y = 9x + 34$ (answer from the line of best fit may vary)
d 74 or 75
e It is not easy to read off accurately from a graph.
f Students' own answers

12 a Between 2.1 and 2.2

b Results outside the range of the data are always less reliable. Assume that the decrease continues at the same rate.

9 Check up

1 a

x	-2	-1	0	1	2
$y = 2x - 1$	-5	-3	-1	1	3

b

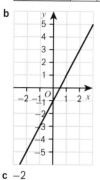

c -2

2 2

3 $(4, 4)$

4 a $x = -5$ **b** $y = 2$

  c $y = x$ **d** $y = -\frac{1}{2}x + 3$

5 a $y = 3x + 4$ and $y = 3x - 7$ **b** $y = x + 3$ and $y = 4x + 3$

  c $y = 4x + 3$ **d** $y = 3x + 4$ and $y = 4x + 3$

6 a 5 minutes

  b Between 10 minutes and 16 minutes; the gradient is steepest.

  c 2.4 km/h

7 Graph A

8 a $y = x + 10$ **b** ≈ 58

9 a £80   **b** £40   **c** £60

11 Students' own answers e.g.
Triangle: e.g. $F(-4, 3)$ $M(-1, 3)$ $I(-4, 0)$
Square: e.g. $F(-4, 3)$ $M(-1, 3)$ $J(-1, 0)$ $I(-4, 0)$
Trapezium: $I(-4, 0)$ $J(-1, 0)$ $G(-1, -2)$ $L(-2.5, -2)$
Kite: $K(-2.5, 4)$ $M(-1, 3)$ $L(-2.5, -2)$ $F(-4, 3)$

9 Strengthen

Algebraic straight-line graphs

1 a

x	0	1	2	3	4
$y = 3x$	0	3	6	9	12

b $(0, 0)$ $(1, 3)$ $(2, 6)$ $(3, 9)$ $(4, 12)$

c, d, e

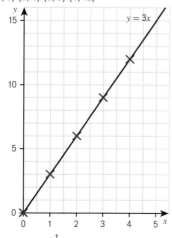

f 1.5   **g** $\approx 1\frac{1}{3}$

2 a

x	0	1	2	3	4
$y = 4x - 3$	-3	1	5	9	13

b, c, d

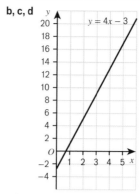

e $(0, -3)$

3 a Students' own answers, for example $(1, 0)$.

b, c

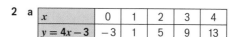

4 a Positive   **b** Negative

5 a Positive A, B, D, E; Negative C

  b Line A 3; Line B 2; Line C -3; Line D 2; Line E 1

  c Lines B and D are parallel.

  d E, B, A

  e E, B, A

  f The bigger the gradient the steeper the line.

6 a $\frac{1}{2}$ **b** $\frac{1}{3}$ **c** $-\frac{1}{4}$

7 a Gradient always $= 2$

  b

Line	y-intercept
$y = 2x + 3$	3
$y = 2x + 1$	1
$y = 2x$	0
$y = 2x - 2$	-2
$y = 2x - 4$	-4

  c $(0, -5)$

8 a $y = 2y + 4$; it has the same gradient, 2

  b $y = 2y + 4$; it has the same y-intercept, 4

9 a Line A gradient $= 4$; Line B gradient -2

  b Line A intercept 2; Line B intercept 1

  c Line A $y = 4x + 2$; Line B $y = -2x + 1$

10 $y = 5x + 10$

11 a, b

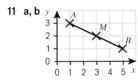

c

	x-coordinate	*y*-coordinate
A	1	3
B	5	1
M	$\frac{1+5}{2} = 3$	$\frac{3+1}{2} = 2$

12 a $(5, 6)$ **b** $(2, 5)$ **c** $(1, 2)$ **d** $(2\frac{1}{2}, 6\frac{1}{2})$

Distance–time graphs and scatter graphs

1 a i Yes   **ii** No
   b Horizontal line
   c 20 minutes
   d 18 minutes
   e 10 miles
   f 12 minutes
   g Return home; 50 miles per hour
   h 8.20 pm
   i

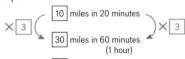

2 a i 4   **ii** 4   **iii** $y = 4x + 4$
   b 112 ice creams

Real-life graphs

1 a

	Translator A	Translator B
Fixed fee	£50	£150
Cost for 1000 words	£140	£190
Rate per word	9p	4p
Cost for 1500 words	£185	£210
Cost for 2500 words	£275	£250

   b i Translator A   **ii** Translator B

2 Diagram A; description 1; graph W
   Diagram B; description 5; graph V
   Diagram C; description 2; graph Y
   Diagram D; description 4; graph Z
   Diagram E; description 3; graph X

9 Extend

1 Gradient of A is $\frac{50}{750} = \frac{1}{15} \approx 0.067 = 6.7\%$
   Gradient of B is $\frac{5}{80} = \frac{1}{16} = 0.0625 = 6.25\%$
   Gradient of C is 6%
   6.7% is greater than 6.25% and 6%, so hill A is the steepest.

2 Speed of particle A = 1200 m ÷ 120 s = 10 m/s, so particle B moved faster.

3 a £2   **b** £2   **c** £3 ÷ 5hr = £0.6/hr

4 a Current rate = £25.60; offer rate = £25.30 – he is 30p better off if he changes.
   b Draw a graph. You will see that the cost is equal at 75 minutes. The original rate is better below 75 minutes; the offer rate is better above 75 minutes.

5 Approx. 2030 to 2040

6 8 cm^2

---

7 a

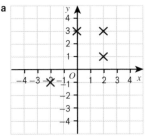

   b Kite
   c $y = x + 1$

8 a $y = -2x + 7$ **b** Gradient = -2, intercept = 7
   c Students' own answers, e.g. The equations can be rearranged to $y = -2x + 3$ and $y = -2x + 7$; both have a gradient of -2, so they are parallel.

9 a $y = 4x$ (or: $v = 4t$)   **b** 40 m/s   **c** 200 m

10 a 1991
   b Yes; 14.28... % increase, or 14% increase is only 3192 thousand

11 Students' own answers, e.g. $y = 3$, $y = -x + 4$, $x = 3$, $y = x - 4$, $x = -3$, $y = -x - 4$, $y = -3$, $y = x + 4$.

9 Test ready

Sample student answers

Students' own answers, e.g. Student A forgot to join the points with a straight line. Student B drew the graph line for values of x from -1 to 2 rather than from -2 to 3.

9 Unit test

1 a $x = 4$
   b, c

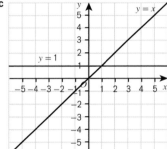

2 $(2, 3)$

3 a

x	-1	0	1	2	3
y	-5	-1	3	7	11

b

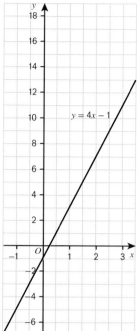

$y = 4x - 1$

4 a $y = 8.6$ **b** $x = 0.65$

5 a $y = 5x + 3$ **b** $y = 2x - 5$, $y = 2x + 3$

   c $y = -\frac{1}{2}x + 6$ **d** $y = -\frac{1}{2}x + 6$

   e $y = 0.5x + 2$ **f** None of them

6 $A: y = -2x - 3$
   $B: y = x + 2$

7 $y = 2x - 1$

8

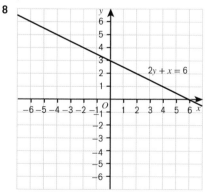

$2y + x = 6$

9 a 13:30   **b** 3 km/h

   c 2 hours

   d Left at 16:00, travelled 0.5 km in 12 minutes, stopped for 36 minutes, travelled 1 km in 42 minutes, arrived home at 17:30.

   e 13:30 to 14:00, identified with reason (steepest section or equivalent)

10 a 320 m/s   **b** decreases

   c Two points identified, e.g. (0, 340) and (35 000, 200) The speed of sound decreases by 1 m/s for every additional 250 m above sea level (or equivalent).

11

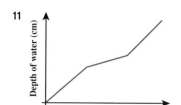

12 a, c

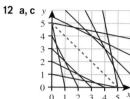

   b $y = -5x + 5$
     $y = -2x + 4$
     $x + y = 3$
     $y = -0.5x + 2$
     $y = -0.2x + 1$

   d $y = -\frac{1}{5}x + 5$
     $y = -\frac{1}{2}x + \frac{11}{2}$
     $y = -x + 7$
     $y = -2x + 11$
     $y = -5x + 25$

   e The gradients of the reflected lines are the reciprocals of the original lines.

13 Students' own answers

UNIT 10 Transformations

10.1 Translation

1 5 right, 2 up

2 a C **b** C **c** A

3 a 2 right, 4 up

   b 2 right, 2 down

   c 7 right, 5 up

4 a 2 right, 4 up   **b** 4 right, 4 down   **c** 4 up

5 a $\begin{pmatrix} 2 \\ 4 \end{pmatrix}$ **b** $\begin{pmatrix} 4 \\ -4 \end{pmatrix}$ **c** $\begin{pmatrix} 0 \\ 4 \end{pmatrix}$

6

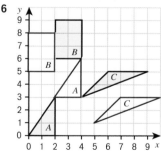

7 Students' own answers, e.g. He has translated shape A by $\begin{pmatrix} 4 \\ 6 \end{pmatrix}$ instead of $\begin{pmatrix} 2 \\ 3 \end{pmatrix}$.

8

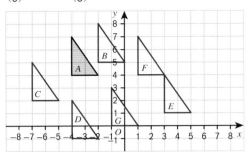

9 a $\begin{pmatrix} 6 \\ -1 \end{pmatrix}$ **b** $\begin{pmatrix} 0 \\ -7 \end{pmatrix}$ **c** $\begin{pmatrix} 6 \\ -9 \end{pmatrix}$ **d** $\begin{pmatrix} 10 \\ -7 \end{pmatrix}$

e $\begin{pmatrix} 0 \\ -8 \end{pmatrix}$ **f** $\begin{pmatrix} 6 \\ -2 \end{pmatrix}$ **g** $\begin{pmatrix} -6 \\ 2 \end{pmatrix}$ **h** $\begin{pmatrix} -4 \\ 6 \end{pmatrix}$

   i Students' own answers, e.g. They are the same but with each number multiplied by -1.

10 A translation by vector $\begin{pmatrix} 6 \\ -3 \end{pmatrix}$

11

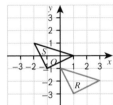

12 a, b Students' own answers.

10.2 Reflection

1 AB and CD, IJ and KL, MN and OP

2 a D **b** B **c** E

   d A **e** C **f** F

3 A; students' own answers, e.g. Corresponding vertices are the same distance from the mirror line but on opposite sides.

4 a

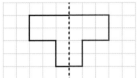

b

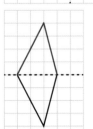

c

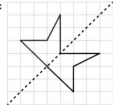

5 a, b, c

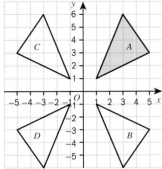

d Yes; students' own answers.

6 a, b, c, d

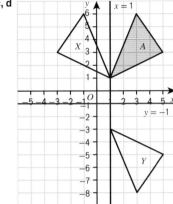

7 a, b

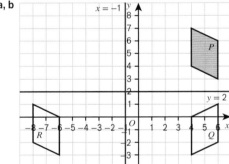

c Yes

8 Students' own answers, e.g. The line $y = 1$ runs through Shape S.

9 a, b

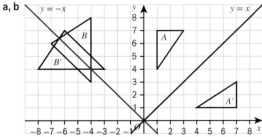

10 Students' own answers, e.g. Draw the mirror line. Reflect each vertex in the mirror line. Join the vertices with a ruler.

11 a $y = 2$ **b** Reflection in the line $y = 2$.

12 a Reflection in the line $y = 0$ or x-axis

   b Reflection in the line $y = -\frac{1}{2}$

   c Reflection in the line $y = x$

   d Reflection in the line $y = -x$

13 a Students' own answers, e.g. She should write 'in the line $x = 0$ or in the y-axis.'
  b Reflection in the line $x = 0$

14 Reflection in the line $y = x$

10.3 Rotation

1 a $360°$ **b** $180°$ **c** $90°$

2 a $90°$ anticlockwise
  b $180°$
  c $90°$ clockwise

3 a

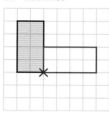

  b

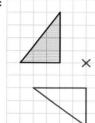

  c

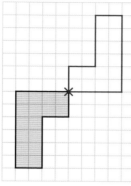

  d

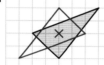

4 a Students' own answers, e.g. In Q3a and 3b, the object and rotated image share a common vertex, but this does not happen with Q3c .
  b Students' own answers, e.g. In Q3d the centre of rotation is inside the shape.

5

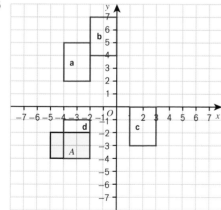

6 a, b, c

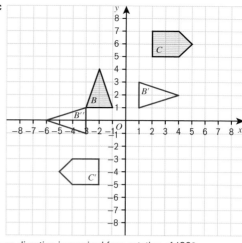

  d Yes; no direction is required for a rotation of $180°$.

7 a Clockwise $90°$ **b** $(-1, 2)$
  c Rotation of $90°$ clockwise about $(-1, 2)$

8 a $90°$ anticlockwise about $(0, 0)$
  b $180°$ about $(0, 2)$
  c $90°$ anticlockwise about $(-2, 2)$

9 Rotation of $180°$ (clockwise or anticlockwise) around $(0, 0)$

10 Students' own answers.

10.4 Enlargement

1 3, 3

2 a 6   **b** 3   **c** 2
  d 25   **e** $\frac{1}{2}$ **f** 24

3 a

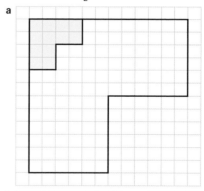

  b   **c**

4 a No
  b i The shape gets bigger.
     ii The shape gets smaller.
     iii The shape stays the same size.

5 a

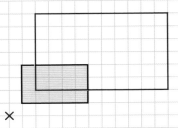

b

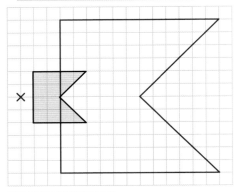

c

d

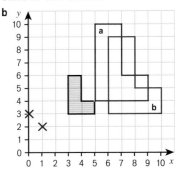

6 a Lines drawn correctly through pairs of vertices.
c Students' own answers

7 a, b

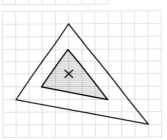

c Students' own answers, e.g. The enlarged shape is the same size but its position is different.

8 a i, ii

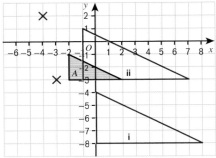

b i, ii

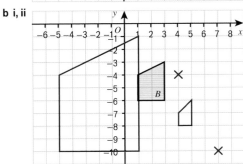

9 a

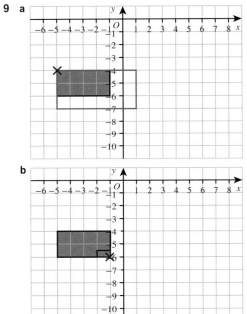

b

10 a Length = 24 cm and width = 18 cm
b Length = 16 cm and width = 12 cm

10.5 Describing enlargements

1 B

2 a 2   **b** 3   **c** $\frac{1}{2}$ **d** $2\frac{1}{2}$

3 a 2   **b** $\frac{1}{2}$ **c** $1\frac{1}{2}$ **d** 3
   e $\frac{1}{4}$

4 a i 4   **ii** 12   **b** 4 × scale factor = 12
   c 3

5 a 2   **b** $\frac{1}{2}$

6 a 2   **b** 10 cm

7 Yes; students' own answers, e.g. In Q4, 3 = 12 ÷ 4.

8 **a** 3   **b** (2, 2)

  c Shape A is enlarged by scale factor 3 using centre of enlargement (2, 2).

9 **a** Enlarged by a scale factor of 3 from centre $(-6, -3)$

  b Enlarged by a scale factor of 2 from centre $(-6, 6)$

  c Enlarged by a scale factor of 2 from centre (1, 1)

10 Enlarged by a scale factor of 2.5 from centre (1, 0)

11 **a** **i** Enlarged by a scale factor of 1.5 from centre $(6, -6)$

  ii Enlarged by a scale factor of $\frac{2}{3}$ from centre $(6, -6)$

  b **i** Enlarged by a scale factor of 3 from centre $(2, -3)$

  ii Enlarged by a scale factor of $\frac{1}{3}$ from centre $(2, -3)$

  c $\frac{1}{4}$

10.6 Combining transformations

1 Translation, column vector; reflection, equation of a line; rotation, angle, direction and centre; enlargement, scale factor and centre

2 **a** Reflection   **b** Enlargement
  c Translation   **d** Rotation

3 **a** Direction of rotation   **b** Centre of enlargement
  c Translation vector   **d** Equation of the mirror line

4 **a, b**

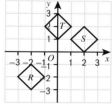

  c A translation by $\begin{pmatrix} 2 \\ 4 \end{pmatrix}$

  d Students' own answers, e.g. The column vector of the single translation in part **c** is equal to the sum of the column vectors in parts **a** and **b**.

5 **a, b**

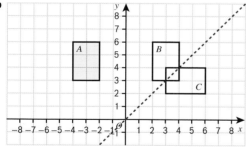

  c Rotation of 90° clockwise about (0, 0)

6

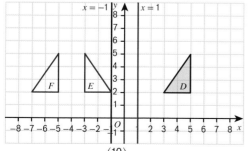

Translation with vector $\begin{pmatrix} 10 \\ 0 \end{pmatrix}$

7

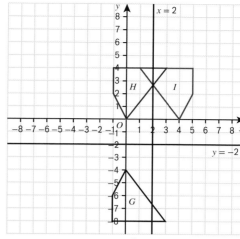

Rotation through 180° about $(2, -2)$

8 **a** Translation   **b** Rotation

9

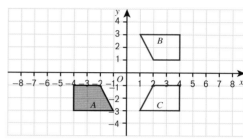

Reflection in the y-axis

10 **a, b**

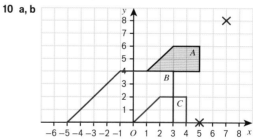

  c Translation with vector $\begin{pmatrix} -1 \\ -4 \end{pmatrix}$

11 **a** Reflection in the line $y = -1$

  b Reflection in the line $x = 1.5$

12 $a = -6, b = 2$

10 Check up

1 a, b, c, d

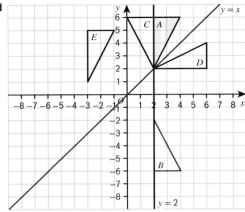

2 a $\begin{pmatrix} -9 \\ -2 \end{pmatrix}$ **b** $\begin{pmatrix} 1 \\ -6 \end{pmatrix}$ **c** $\begin{pmatrix} 10 \\ -4 \end{pmatrix}$

3 a $x = 0$ or y-axis   **b** $y = -1$ **c** $y = -x$

4 a, b

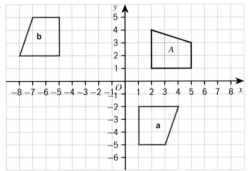

5 a Rotation of 90° anticlockwise around (0, 0)
   b Rotation of 90° clockwise around (−0.5, −0.5)

6

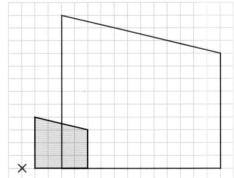

7

8 a Enlargement by scale factor of 3 from centre (−6, 7)
   b Enlargement by scale factor of 0.5 from centre (6, −6)

10 a

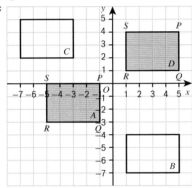

   b Students' own answers

10 Strengthen

Translations, reflections and rotations

1 a Left   **b** Down   **c** 3 left, 4 down

2 c $\begin{pmatrix} 3 \\ 2 \end{pmatrix}$ **d** $\begin{pmatrix} -3 \\ 2 \end{pmatrix}$ **e** $\begin{pmatrix} -2 \\ -1 \end{pmatrix}$

   f $\begin{pmatrix} 2 \\ -1 \end{pmatrix}$ **g** $\begin{pmatrix} 4 \\ -3 \end{pmatrix}$ **h** $\begin{pmatrix} -3 \\ 4 \end{pmatrix}$

3 6 right, 4 down

4 a 6 right, 4 down

   b, c

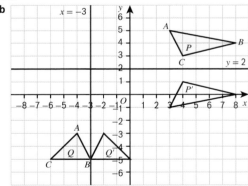

5 a

	Coordinates of vertex			
	P	*Q*	*R*	*S*
Shape *A*	(−1, 0)	(−1, −3)	(−5, −3)	(−5, 0)
Shape *D*	(5, 4)	(5, 1)	(1, 1)	(1, 4)

   b $\begin{pmatrix} 6 \\ 4 \end{pmatrix}$

6 a, b

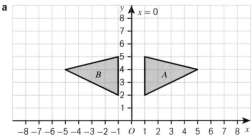

7 a

   b Reflection in the y-axis or $x = 0$

8

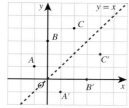

9 A, 180° clockwise   B, 90° clockwise
    C, 90° anticlockwise   D, 180° anticlockwise

10 c   **i**   **ii**

     iii   **iv**

d Students' own answers, e.g. A direction is necessary for **i**
and **iv**, but not for **ii** and **iii**, as 180° in either direction gives
the same image.

11 a, b, c, d

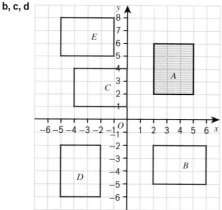

e Students' own answers, e.g. It changes the centre of
rotation and changes the position of the image.

12 a, b See above

13 b (0, 0)

    c Rotation 90° anticlockwise about (0, 0)

    d i Rotation of 90° clockwise about (1.5, −1.5)
       ii Rotation of 180° about (0, −2)

14 a i Students' own answers, e.g. It is not in the same
orientation.
       ii Students' own answers, e.g. It is not in the same
orientation.

    b Students' own answers, e.g. By finding the centre of
rotation and rotating rectangle A onto rectangle B.

Enlargements

1   B and E

2 a 3   **b** B: 2, E: $\frac{1}{2}$

    c

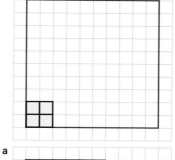

3 a

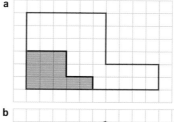

    b

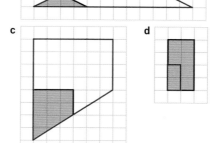

    c   **d**

4 a, b, c

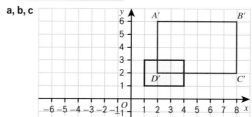

5 a, b

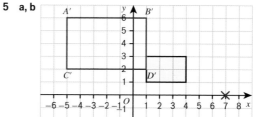

    c The centre of enlargement determines the position of
the image.

6 a 3   **b** 3   **c** (0, 5)   **d** (0, 5)

    e Enlargement by a scale factor of 3 from a centre of
enlargement at (0, 5)

10 Extend

1. Reflection in the line $y = -x$

   Translation $\begin{pmatrix} -4 \\ -4 \end{pmatrix}$

   Rotation of 180° about (0, 0)

2. Enlargement by a scale factor of $\frac{1}{4}$ from a centre of enlargement at (4, 3)

3. Rotation of 60° clockwise around O

4. **a** Irregular pentagon   **b** 60.5

   c
   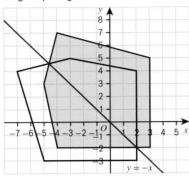

   d Yes, reflection does not change the area of a shape.

5. **a** $\begin{pmatrix} -x \\ -y \end{pmatrix}$ **b** $\begin{pmatrix} -6 \\ 1 \end{pmatrix}$

6. $\begin{pmatrix} a+c \\ b+d \end{pmatrix}$

7. **a** 8 and 6

   b i 10 cm   **ii** 6 cm^2

   iii Length = 6 cm and width = 4 cm

   iv 20 cm   **v** 24 cm^2

8. It is multiplied by a factor of x^2

9. **a** 36 cm^2 **b** Students' own answers.

10. 1.5 cm^2

10 Test ready

Sample student answers

1. **a** Student B is correct. You must count the squares from one corner to the corresponding corner of the second shape, not just the nearest corner.

   b e.g. Draw a dot onto two corresponding corners so they make sure they start and finish in the right place.

2. The student correctly identified the transformation as a reflection but did not correctly describe the reflection by giving the equation of the mirror line. Correct answer would be 'Reflection in the x-axis (or in the line with equation $y = 0$).'

10 Unit test

1. Scale factor = 3

2.
   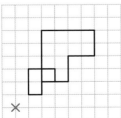

3. **a, b, c**
   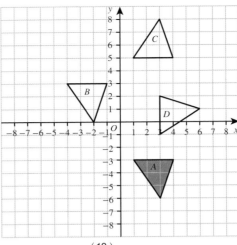

4. **a** Translation by the vector $\begin{pmatrix} 10 \\ -2 \end{pmatrix}$

   b Enlargement by a scale factor of 2.5 with a centre of enlargement (−8, 7)

   c Reflection in the line $y = -2$

5.
   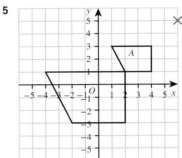

6. She is missing the coordinates of the point the shape is being rotated about (the centre of rotation).

7. Reflection in the line $y = x$

8. **a, b**
   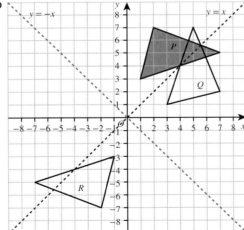

9. Area = 30

10 a Coordinate grid correctly drawn with axes labelled from
     −8 to 8.
  b–f Students' own answers.
11 a Translation, enlargement, reflection, rotation
  b Translation: translation, column vector
     Enlargement: enlargement, scale factor, centre of
     enlargement
     Reflection: reflection, equation of line of reflection
     Rotation: rotation, coordinates of point of rotation, direction
     of rotation, angle of rotation

Mixed exercise 3

1 **B** down   → **10.1**
2 a It is not a straight line.   → **9.2**
  b   → **9.2**

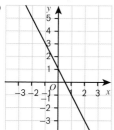

3 200   → **8.1, 8.2**
4 a i C   **ii** A   **iii** D
    iv B   **v** E   → **9.1–9.4**
  b $y = -4$ → **9.1**
5 a $(24x + 8)$ cm   → **2.1, 8.3**
  b $(6x + 2)$ cm   → **2.2, 8.1**
6 $p = -7, q = 8$ → **9.1**
7 $x = 24$ → **8.5**
8 $x = 7$ → **5.2, 8.3**
9 a Toby has drawn the enlarged triangle in the wrong position.
     B' should be at (5, 4).   → **10.4**
  b   → **10.4**

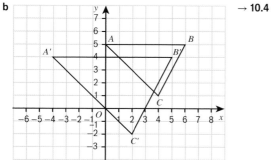

10   → **9.6**

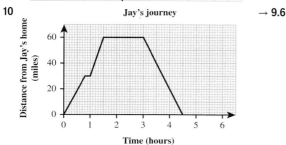

11 4600 cm^2 → **8.4**
12 81 cm^2 → **8.1**

13 Yes, as area $= 9.9\,\text{m}^2$.
    Number of packs of decking $= 9.9 \div 1.7 = 5.82...$
    Six packs of decking costs £216.
    25% off £216 = £162, which is less than £175.   → **8.2, 8.3**
14 a   → **10.6**

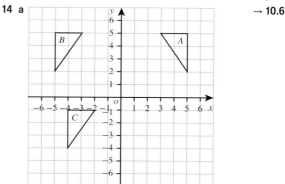

  b No. This would give the shaded triangle, which is not in the
     same position as triangle C.   → **10.6**

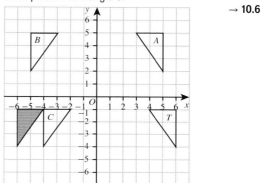

Index